Python 工程应用系列丛书
东北大学秦皇岛分校教材建设基金资助

Python 高维数据分析

Python Multidimensional Data Analysis

主编　赵煜辉

西安电子科技大学出版社

内 容 简 介

本书从矩阵计算如特征值分解和奇异值分解出发，讨论了正规方程的最小二乘法模型引出欠秩线性方程组的求解方法问题；然后介绍了两种有损的降维方法，即主成分分析(主成分回归)和偏最小二乘回归，包括模型、算法和多个实例，并扩展到线性回归的正则化方法，给出了岭回归和Lasso的原理算法和实例；最后通过红外光谱的标定迁移实例将线性模型扩展到迁移学习领域。

本书每章都有基于Python语言和Sklearn机器学习库的红外光谱数据集分析的实例。红外光谱集是关于物质吸光率的纯数据，可以与其标签标示的数据物质浓度直接进行回归分析，读者在阅读中可以把精力最大限度地集中在高维数据的建模、算法实现和分析过程上。

本书既可作为信息管理和信息系统专业、计算机相关专业和大数据专业的教学用书，也可作为从事光谱分析、化学分析的工程人员及化学计量学研究人员的参考书，还适合对数据分析和研究感兴趣的其他Python工程师学习阅读。本书引用的原始文献和数据对上述人员是非常有帮助的。

图书在版编目(CIP)数据

Python高维数据分析：英文/赵煜辉主编. —西安：西安电子科技大学出版社，2020.8
ISBN 978-7-5606-5577-2

Ⅰ.①P…　Ⅱ.①赵…　Ⅲ.①软件工具—程序设计—高等学校—教材—英文　Ⅳ.①TP311.561

中国版本图书馆CIP数据核字(2019)第281480号

策划编辑　刘小莉
责任编辑　张　梅　雷鸿俊
出版发行　西安电子科技大学出版社(西安市太白南路2号)
电　　话　(029)88242885　88201467　　邮　　编　710071
网　　址　www.xduph.com　　电子邮箱　xdupfxb001@163.com
经　　销　新华书店
印刷单位　陕西日报社
版　　次　2020年8月第1版　2020年8月第1次印刷
开　　本　787毫米×1092毫米　1/16　印　张　17.25
字　　数　408千字
印　　数　1～2000册
定　　价　43.00元
ISBN 978-7-5606-5577-2/TP
XDUP 5879001-1

前　言

在人工智能时代，各个领域都需要统计处理像语音、图像、视频、蛋白质、DNA、物质的光谱等这样的维度在几十、几百甚至成千上万的高维数据。如果直接处理不仅十分耗时且大部分情况下无法直接计算。如何分析高维数据是人工智能应用的一个基本问题，涉及矩阵计算和数据降维方法。降维方法有许多，如典型相关分析、Fisher's 线性判别分析、独立成分分析等，本书着重介绍了主成分分析、偏最小二乘法和正则化方法这三种最基础也是最有代表性的数据分析方法。

矩阵计算是学习高维数据分析的基础，因此本书第 1 章主要介绍了矩阵计算的基本概念、常见的矩阵分解方法和二次型等，本章的特征值分解和奇异值分解是后面章节的基础，也是机器学习的基础知识。通过第 1 章的学习，读者就拥有了打开高维数据分析大门的钥匙。第 2 章给出了正定矩阵的最小二乘解的证明和相关的性质，进而引出欠秩矩阵的一般求解方法问题。如果已经对矩阵计算基础知识比较了解，那么读者可以有选择地学习这两章内容。

第 3 章从分析葡萄酒成分数据引入主成分分析问题，不但介绍了主成分分析的基础知识，如理论推导、算法以及 Python 程序实例和结果展示，而且在主成分回归、在线过程分析中主成分的应用实例等方面进行了扩展，让读者从几个简单的 Python 程序中就能够学习到解决实际问题的基本方法。

第 4 章详细介绍的偏最小二乘算法是从自变量和因变量的关系来分析数据的方法，其数据降维方法是求解与目标数据(因变量)相关性最大的潜变量。本章先给出 SIMPLS 和 NIPALS 两种常见偏最小二乘算法，然后介绍偏最小二乘的集成学习方法——Stack PLS 算法，目的是教会读者扩展这些基本数据的方法。本章详细介绍了通过交叉验证选择潜变量数量的方法，也介绍了编写数据分析程序的许多细节知识，读者可以通过模仿和实践这些程序开发过程，得到用 Python 语言进行数据分析的训练。

第 5 章通过介绍岭回归和 Lasso 这两个理论模型，说明了对数据进行正则化约束的基本方法，给出了算法的数学模型，并调用 Sklearn 包的 Lasso 和岭回归的 Python 程序。

第 6 章介绍了迁移方法，以扩展前面介绍的已有域建立的线性回归模型到其他场景。通过分析一个不同红外光谱仪测定的光谱信号存在差异的现象，说明了建立的红外光谱偏最小二乘模型不适用其他仪器，接着介绍了这个标定迁移学习问题的研究背景以及相关方法，然后给出了两个基于特征的标定迁移方法及其 Python 程序、实验结果和数据分析过程。

本书是机器学习和人工智能教学的一门以数学基础开始的课程，其中介绍了大量化学计量学数据和光谱数据 Python 程序实例，能够让读者从理解最小二乘法的原理和其在处理实际数据中面临的问题开始，循序渐进地处理高维数据的降维和回归问题。本书多数章节都采用了近红外光谱数据的 Python 程序实例。红外光谱数据的物理和化学意义简单，以纯数值型高维小样本数据为处理对象，可以使读者精力集中在算法的设计、模型的推导和问题的解决上，不需要考虑不同应用中复杂的数据预处理需要。

本书初稿已经在东北大学秦皇岛分校2016级信息管理与信息系统专业的本科教学中试用，并根据学生反馈意见进行了修改。在使用本书后，同学们普遍认为该书提高了他们的英文文献的阅读能力，不再惧怕阅读机器学习领域的科研文献，并且书中给出的代码实现过程，很好地解释了文献中的思想，使同学们掌握了英文文献阅读的同时，实现了文献所述方法的学习，Python代码能力也得到了有效锻炼，学到了用Python解决真实数据的处理和分析问题的方法。

赵煜辉博士对全书内容做了整体安排，并且编写了第2章、第5章、第6章内容；高书礼、于金龙、赵子恒分别编写整理了第1章、第3章、第4章内容。全书代码由赵子恒、于金龙编写校对。

本书是东北大学秦皇岛分校教材建设基金资助项目。

由于我们经验不足、水平有限，错漏之处在所难免，希望广大教师和读者提出宝贵意见和建议。

注：由于本书为黑白印刷，书中有些彩色插图无法呈现应有效果，可用手机扫图旁二维码查看彩图。

编　者
2020年2月

Contents

Chapter 1 Basis of Matrix Calculation

1.1 Fundamental Concepts

The purpose of this chapter is to review important fundamental concepts in linear algebra, as a foundation for the rest of the course. We first discuss the fundamental building blocks, such as an overview of matrix multiplication from a "big block" perspective, linear independence, subspaces and related ideas, rank, etc., upon which the rigor of linear algebra rests. We then discuss vector norms, and various interpretations of the matrix multiplication operation. We close the chapter with a discussion on determinants.

1.1.1 Notation

Throughout this course, we shall indicate that a matrix $\boldsymbol{A}$ is of dimension $m\times n$, and whose elements are taken from the set of real numbers, by the notation $\boldsymbol{A}\in \boldsymbol{C}^{m\times n}$. This means that the matrix $\boldsymbol{A}$ belongs to the Cartesian product of the real numbers, taken $m\times n$ times, one for each element of $\boldsymbol{A}$. In a similar way, the notation $\boldsymbol{A}\in \boldsymbol{C}^{m\times n}$ means the matrix is of dimension $m\times n$, and the elements are taken from the set of complex numbers. By the matrix dimension $m\times n$, we mean $\boldsymbol{A}$ consists of m rows and n columns.

Similarly, the notation $\boldsymbol{a}\in \mathbb{R}^m(\boldsymbol{C}^m)$ implies a vector of dimension m whose elements are taken from the set of real (complex) numbers. By "dimension of a vector", we mean its length, i. e., that it consists of m elements.

Also, we shall indicate that a scalar a is from the set of real (complex) numbers by the notation $\boldsymbol{a}\in\mathbb{R}(\boldsymbol{C})$. Thus, an upper case bold character denotes a matrix, a lower case bold character denotes a vector, and a lower case non-bold character denotes a scalar.

By convention, a vector by default is taken to be a column vector. Further, for a matrix $\boldsymbol{A}$, we denote its i-th column as $\boldsymbol{a}_i$. We also imply that its j-th row is $\boldsymbol{a}_j^{\mathrm{T}}$, even though this notation may be ambiguous, since it may also be taken to mean the transpose of the j-th column. The context of the discussion will help to resolve the ambiguity.

1.1.2 "Bigger-Block" Interpretations of Matrix Multiplication

Let us define the matrix product $\boldsymbol{C}$ as

$$\underset{m\times n}{\boldsymbol{C}} = \underset{m\times k}{\boldsymbol{A}}\ \underset{k\times n}{\boldsymbol{B}} \tag{1.1.1}$$

The three interpretations of this operation now follow:

1.1.2.1 Inner-Product Representation

If $\boldsymbol{a}$ and $\boldsymbol{b}$ are column vectors of the same length, then the scalar quantity $\boldsymbol{a}^{\mathrm{T}}\boldsymbol{b}$ is referred to as the inner product of $\boldsymbol{a}$ and $\boldsymbol{b}$. If we define $\boldsymbol{a}_i^{\mathrm{T}} \in \mathbb{R}^k$ as the i-th row of $\boldsymbol{A}$ and $\boldsymbol{b}_j \in \mathbb{R}^k$ as the j-th column of $\boldsymbol{B}$, then the element $\boldsymbol{c}_{ij}$ of $\boldsymbol{C}$ is defined as the inner product $\boldsymbol{a}_i^{\mathrm{T}}\boldsymbol{b}_j$. This is the conventional small-block representation of matrix multiplication.

1.1.2.2 Column Representation

This is the next bigger-block view of matrix multiplication. Here we look at forming the product one column at a time. The j-th column $\boldsymbol{c}_j$ of $\boldsymbol{C}$ may be expressed as a linear combination of columns $\boldsymbol{a}_i$ of $\boldsymbol{A}$ with coefficients which are the elements of the j-th column of $\boldsymbol{B}$. Thus,

$$\boldsymbol{c}_j = \sum_{i=1}^{k} \boldsymbol{a}_i b_{ij} \quad j = 1, 2, \cdots, n \tag{1.1.2}$$

This operation is identical to the inner-product representation above, except we form the product one column at a time. For example, if we evaluate only the p-th element of the j-th column $\boldsymbol{c}_j$, we see that equation (1.1.2) degenerates into $\sum_{i=1}^{k} a_{pi} b_{ij}$. This is the inner product of the p-th row and j-th column of $\boldsymbol{A}$ and $\boldsymbol{B}$ respectively, which is the required expression for the the (p, j) element of $\boldsymbol{C}$.

1.1.2.3 Outer-Product Representation

This is the largest-block representation. Let us define a column vector $\boldsymbol{a} \in \mathbb{R}^m$ and a row vector $\boldsymbol{b}^{\mathrm{T}} \in \mathbb{R}^n$. Then the outer product of $\boldsymbol{a}$ and $\boldsymbol{b}$ is an $m \times n$ matrix of rank one and is defined as $\boldsymbol{a}\boldsymbol{b}^{\mathrm{T}}$. Now let $\boldsymbol{a}_i$ and $\boldsymbol{b}_i^{\mathrm{T}}$ be the i-th column and row of $\boldsymbol{A}$ and $\boldsymbol{B}$ respectively. Then the product $\boldsymbol{C}$ may also be expressed as

$$\boldsymbol{C} = \sum_{i=1}^{k} \boldsymbol{a}_i \boldsymbol{b}_i^{\mathrm{T}} \tag{1.1.3}$$

By looking at this operation one column at a time, we see this form of matrix multiplication performs exactly the same operations as the column representation above. For example, the j-th column $\mathbf{c}_j$ of the product is determined from equation 1.1.3 to be $\mathbf{c}_j = \sum_{i=1}^{k} \boldsymbol{a}_i b_{ij}$, which is identical to equation (1.1.2) above.

1.1.2.4 Matrix Pre- and Post-Multiplication

Let us now look at some fundamental ideas distinguishing matrix pre- and post-multiplication. In this respect, consider a matrix $\boldsymbol{A}$ pre-multiplied by $\boldsymbol{B}$ to give $\boldsymbol{Y} = \boldsymbol{B}\boldsymbol{A}$. (All matrices are assumed to have conformable dimensions). Then we can interpret this multiplication as $\boldsymbol{B}$ operating on the column of $\boldsymbol{A}$ to give the columns of the product. This follows because each column $\boldsymbol{y}_i$ of the product is a transformed version of the corresponding column of $\boldsymbol{A}$; i.e., $\boldsymbol{y}_i = \boldsymbol{B}\boldsymbol{a}_i$, $i = 1, 2, \cdots, n$. Likewise, let's consider $\boldsymbol{A}$ post-multiplied by a matrix $\boldsymbol{C}$ to give $\boldsymbol{X}=\boldsymbol{AC}$. Then, we interpret this multiplication as $\boldsymbol{C}$ operating on the rows of $\boldsymbol{A}$, because each row $\boldsymbol{X}_i^{\mathrm{T}}$ of the product is a transformed version of the corresponding row of $\boldsymbol{A}$; i. e., $\boldsymbol{X}_j^{\mathrm{T}}=\boldsymbol{a}_j^{\mathrm{T}}\boldsymbol{C}$, $j=1, 2, \cdots, m$, where we define $\boldsymbol{a}_j^{\mathrm{T}}$ as the j-th row of $\boldsymbol{A}$.

Example 1:

Consider an orthonormal matrix $\boldsymbol{Q}$ of appropriate dimension. We know that multiplication by an orthonormal matrix results in a rotation operation. The operation $\boldsymbol{QA}$ rotates each column of $\boldsymbol{A}$. The operation $\boldsymbol{AQ}$ rotates each row.

There is another way to interpret pre-multiplication and post-multiplication. Again consider the matrix $\boldsymbol{A}$ pre-multiplied by $\boldsymbol{B}$ to give $\boldsymbol{Y}=\boldsymbol{BA}$. Then according to equation (1.1.2), the j-th column $\boldsymbol{y}_i$ of $\boldsymbol{Y}$ is a linear combination of the columns of $\boldsymbol{B}$, whose coefficients are the j-th column of $\boldsymbol{A}$. Likewise, for $\boldsymbol{X}=\boldsymbol{AB}$, we can say that the i-th row $\boldsymbol{X}_i^{\mathrm{T}}$ of $\boldsymbol{X}$ is a linear combination of the rows of $\boldsymbol{B}$, whose coefficients are the i-th row of $\boldsymbol{A}$.

Either of these interpretations is equally valid. Being comfortable with the representations of this section is a big step in mastering the field of linear algebra.

1.1.3 Fundamental Linear Algebra

1.1.3.1 Linear Independence

Suppose we have a set of n m-dimensional vectors $\{\boldsymbol{a}_1, \boldsymbol{a}_2, \cdots, \boldsymbol{a}_n\}$, where $\boldsymbol{a}_i \in \mathbb{R}^m$, $i=1, 2, \cdots, n$. This set is linearly independent under the conditions①

$$\sum_{j=1}^{n} c_j \boldsymbol{a}_j = 0 \text{ if and only if } c_1, c_2, \cdots, c_n = 0 \tag{1.1.4}$$

In words, equation (1.1.4) means that a set of vectors is linearly independent if and only if the only zero linear combination of the vectors has coefficients which are all zero.

A set of n vectors is linearly independent if an n-dimensional space may be formed by taking all possible linear combinations of the vectors. If the dimension of the space is less than n, then the vectors are linearly dependent. The concept of a vector space and the dimension of a vector space is made more precise later.

Note that a set of vectors $\{\boldsymbol{a}_1, \boldsymbol{a}_2, \cdots, \boldsymbol{a}_n\}$, where $n>m$ cannot be linearly independent.

Example 2:

$$\boldsymbol{A}=[\boldsymbol{a}_1 \quad \boldsymbol{a}_2 \quad \boldsymbol{a}_3]=\begin{bmatrix} 1 & 2 & 1 \\ 0 & 3 & -1 \\ 0 & 0 & 1 \end{bmatrix} \tag{1.1.5}$$

This set is linearly independent. On the other hand, the set

① Equation (1.1.4) is called a linear combination of the vectors $\boldsymbol{a}_j$. Each vector is multiplied by a weight (or coefficient) c_j, and the result summed.

$$B = [b_1 \quad b_2 \quad b_3] = \begin{bmatrix} 1 & 2 & -3 \\ 0 & 3 & -3 \\ 0 & 0 & 0 \end{bmatrix} \tag{1.1.6}$$

is not. This follows because the third column is a linear combination of the first two. -1 times the first column plus -1 times the second equals the third column. Thus, the coefficients c_j in equation (1.1.4) resulting in zero are any scalar multiple of equation (1.1.1).

1.1.3.2 Span, Range and Subspaces

In this section, we explore these three closely-related ideas. In fact, their mathematical definitions are almost the same, but the interpretation is different for each case.

Span

The span of a vector set $[a_1, a_2, \cdots, a_n]$, written as span$[a_1, a_2, \cdots, a_n]$, where $a_i \in \mathbb{R}^m$, is the set of points mapped by

$$\text{span}[a_1, a_2, \cdots, a_n] = \{y \in \mathbb{R}^m \mid y = \sum_{j=1}^{n} c_i a_j, \ c_j \in \mathbb{R}\} \tag{1.1.7}$$

In other word, span$[a_1, a_2, \cdots, a_n]$ is the set of all possible linear combinations of the vectors a. If the vectors are linearly independent, then the dimension of this set of linear combinations is n. If the vectors are linearly dependent, then the dimension is less.

The set of vectors in a span is referred to as a vector space. The dimension of a vector space is the number of linearly independent vectors in the linear combination which forms the space. Note that the vector space dimension is not the dimension (length) of the vectors forming the linear combinations.

Example 3:

Consider the following 2 vectors in Fig. 1.1.

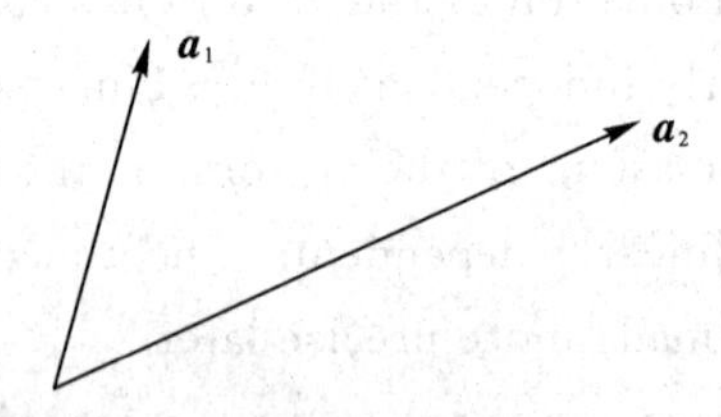

Fig. 1.1　The span of these vectors is the (infinite extension of the) plane of the paper

Subspaces

Given a set (space) of vectors $[a_1, a_2, \cdots, a_n] \in \mathbb{R}^m$, $m \geqslant n$, a subspace S is a vector subset that satisfies two requirements:

1. If x and y are in the subspace, then $x+y$ is still in the subspace.

2. If we multiply any vector x in the subspace by a scalar c, then cx is still in the subspace.

These two requirements imply that for a subspace, any linear combination of vectors which are in the subspace is itself in the subspace. Comparing this idea with that of span, we see a subspace defined by the vectors $[\boldsymbol{a}_1, \boldsymbol{a}_2, \cdots, \boldsymbol{a}_n]$ is identical to span$[\boldsymbol{a}_1, \cdots, \boldsymbol{a}_n]$. However, a subspace has the interpretation that the set of vectors comprising the subspace must be a subset of a larger space. For example, the vectors $[\boldsymbol{a}_1, \boldsymbol{a}_2]$ in Fig. 1.1 define a subspace (the plane of the paper) which is a subset of the three-dimensional universe.

Hence formally, a k-dimensional subspace $\boldsymbol{S}$ of span$[\boldsymbol{a}_1, \cdots, \boldsymbol{a}_n]$ is determined by span$[\boldsymbol{a}_{i1}, \boldsymbol{a}_{i2}, \cdots, \boldsymbol{a}_{ik}]$, where the distinct indices satisfy $\{i_1, i_2, \cdots, i_k\} \subset \{1, 2, \cdots, n\}$; that is, the vector space $\boldsymbol{S}=$span$[\boldsymbol{a}_{i1}, \boldsymbol{a}_{i2} \cdots, \boldsymbol{a}_{ik}]$ is a subset of span$[\boldsymbol{a}_1, \boldsymbol{a}_2, \cdots, \boldsymbol{a}_n]$.

Note $[\boldsymbol{a}_{i1}, \boldsymbol{a}_{i2}, \cdots, \boldsymbol{a}_{ik}]$ that is not necessarily a basis for the subspace $\boldsymbol{S}$. This set is a basis only if it is a maximally independent set. This idea is discussed shortly. The set $\boldsymbol{a}_i$ need not be linearly independent to define the span or subset.

Range

The range of a matrix $\boldsymbol{A} \in \mathbb{R}^{m\times n}$, denoted $R(\boldsymbol{A})$, is a subspace (set of vectors) satisfying

$$R(\boldsymbol{A}) = \{\boldsymbol{y} \in \mathbb{R}^m \mid \boldsymbol{y} = \boldsymbol{A}\boldsymbol{x}, \text{ for } \boldsymbol{x} \in \mathbb{R}^n\} \tag{1.1.8}$$

We can interpret the matrix-vector multiplication $\boldsymbol{y} = \boldsymbol{A}\boldsymbol{x}$ above according to the column representation for matrix multiplication (1.1.2), where the product $\boldsymbol{C}$ has only one column. Thus, we see that $\boldsymbol{y}$ is a linear combination of the columns $\boldsymbol{a}_i$ of $\boldsymbol{A}$, whose coefficients are the elements $\boldsymbol{x}_i$ of $\boldsymbol{x}$. Therefore, equation (1.1.8) is equivalent to equation (1.1.7), and $R(\boldsymbol{A})$ is thus the span of the columns of $\boldsymbol{A}$. The distinction between range and span is that the argument of range is a matrix, while for span it is a set of vectors. If the columns of $\boldsymbol{A}$ are (not) linearly independent, then R$(\boldsymbol{A})$ will (not) span n dimensions. Thus, the dimension of the vector space $R(\boldsymbol{A})$ is less than or equal to n. Any vector $\boldsymbol{y} \in$ R$(\boldsymbol{A})$ is of dimension (length) m.

Example 4:

$$\boldsymbol{A} = \begin{bmatrix} 1 & 5 & 3 \\ 2 & 4 & 3 \\ 3 & 3 & 3 \end{bmatrix} \text{(the last column is the average of the first two)} \tag{1.1.9}$$

$R(\boldsymbol{A})$ is the set of all linear combinations of any two columns of $\boldsymbol{A}$. In the case when $n<m$ (i. e., $\boldsymbol{A}$ is a tall matrix), it is important to note that $R(\boldsymbol{A})$ is indeed a subspace of the m-dimensional "universe" $\mathbb{R}^m$. In this case, the dimension of $R(\boldsymbol{A})$ is less than or equal to n. Thus, $R(\boldsymbol{A})$ does not span the whole universe, and therefore is a subspace of it.

1.1.3.3 Maximally Independent Set

This is a vector set which cannot be made larger without losing independence, and smaller without remaining maximal; i. e. it is a set containing the maximum number of independent vectors spanning the space.

1.1.3.4 A Basis

A basis for a subspace is any maximally independent set within the subspace. It is not unique.

Example 5:

A basis for the subspace **S** spanning the first 2 columns of

$$\boldsymbol{A} = \begin{bmatrix} 1 & 2 & 3 \\ & 3 & -3 \\ & & 3 \end{bmatrix}, \text{ i. e., } \boldsymbol{S} = \left\{ \begin{bmatrix} 1 \\ 0 \\ 0 \end{bmatrix}, \begin{bmatrix} 2 \\ 3 \\ 0 \end{bmatrix} \right\}$$

is

$$\boldsymbol{e}_1 = (1, 0, 0)^{\mathrm{T}}$$

$$\boldsymbol{e}_2 = (0, 1, 0)^{\mathrm{T}}$$

or any other linearly independent set is span$[\boldsymbol{e}_1, \boldsymbol{e}_2]$. Any vector in **S** is uniquely represented as a linear combination of the basis vectors.

1.1.3.5 Orthogonal Complement Subspace

If we have a subspace **S** of dimension n consisting of vectors $[\boldsymbol{a}_1, \boldsymbol{a}_2, \cdots, \boldsymbol{a}_n]$, $\boldsymbol{a}_i \in \mathbb{R}^m$, $i=1, 2, \cdots, n$, for $n \leqslant m$, the orthogonal complement subspace $\boldsymbol{S}_\perp$ of **S** of dimension $m-n$ is defined as

$$\boldsymbol{S}_\perp = \{\boldsymbol{y} \in \mathbb{R}^m \mid \boldsymbol{y}^{\mathrm{T}}\boldsymbol{x} = 0 \text{ for all } \boldsymbol{x} \in \boldsymbol{S}\} \tag{1.1.10}$$

i. e., any vector in $\boldsymbol{S}_\perp$ is orthogonal to any vector in **S**. The quantity $\boldsymbol{S}_\perp$ is pronounced "S-prep".

Example 6:

Take the vector set defining **S** from Example 5:

$$\boldsymbol{S} \equiv \begin{bmatrix} 1 & 2 \\ 0 & 3 \\ 0 & 0 \end{bmatrix} \tag{1.1.11}$$

then, a basis for $\boldsymbol{S}_\perp$ is

$$\begin{bmatrix} 0 \\ 0 \\ 1 \end{bmatrix} \tag{1.1.12}$$

1.1.3.6 Rank

Rank is an important concept which we will use frequently throughout this course. We briefly describe only a few basic features of rank here. The idea is expanded more fully in the following sections.

1. The rank of a matrix is the maximum number of linearly independent rows or columns. Thus, it is the dimension of a basis for the columns (rows) of a matrix.

2. Rank of **A** (denoted rank(**A**)), is the dimension of $R(\boldsymbol{A})$.

3. If $\boldsymbol{A}=\boldsymbol{BC}$, and $r_1=\text{rank}(\boldsymbol{B})$, $r_2=\text{rank}(\boldsymbol{C})$, then, $\text{rank}(\boldsymbol{A}) \leqslant \min(r_1, r_2)$.

4. A matrix $\boldsymbol{A} \in \mathbb{R}^{m\times n}$ is said to be rank deficient if its rank is less than $\min(m, n)$. Otherwise, it is said to be full rank.

5. If **A** is square and rank deficient, then $\det(\boldsymbol{A})=0$.

6. It can be shown that $\text{rank}(\boldsymbol{A})=\text{rank}(\boldsymbol{A}^{\mathrm{T}})$. More is said on this point later.

A matrix is said to be full column (row) rank if its rank is equal to the number of columns (rows).

Example 7:

The rank of **A** in Example 6 is 3, whereas the rank of **A** in Example 4 is 2.

1.1.3.7 Null Space of A

The null space $N(\boldsymbol{A})$ of **A** is defined as

$$N(\boldsymbol{A}) = \{\boldsymbol{x} \in \mathbb{R}^n \neq 0 \mid \boldsymbol{A}\boldsymbol{x} = 0\} \tag{1.1.13}$$

From previous discussions, the product $\boldsymbol{A}\boldsymbol{x}$ is a linear combination of the columns $\boldsymbol{a}_i$ of **A**, where the elements x_i of $\boldsymbol{x}$ are the corresponding coefficients. Thus, from equation (1.1.13), $N(\boldsymbol{A})$ is the set of non-zero coefficients of all zero linear combinations of the columns of **A**. If the columns of **A** are linearly independent, then $N(\boldsymbol{A})=\varnothing$ by definition, because there can be no coefficients except zero which result in a zero linear combination. In this case, the dimension of the null space is zero, and **A** is full column rank. The null space is empty if and only if **A** is full column rank, and is non-empty, when **A** is column rank deficient②. Note that any vector in $N(\boldsymbol{A})$ is of dimension n. Any vector in $N(\boldsymbol{A})$ is orthogonal to the rows of **A**, and is thus in the orthogonal complement of the span of the rows of **A**.

Example 8:

Let **A** be as before in Example 4. Then $N(\boldsymbol{A})=c\ (1, 1, -2)^{\mathrm{T}}$, where c is a real constant.

A further example is as follows. Take 3 vectors $[\boldsymbol{a}_1, \boldsymbol{a}_2, \boldsymbol{a}_3]$, where $\boldsymbol{a}_i \in \mathbb{R}^3$, $i=1, 2, 3$, that are constrained to lie in a 2-dimensional plane. Then there exists a zero linear combination of these vectors. The coefficients of this linear combination define a vector $\boldsymbol{x}$ which is in the null space of $\boldsymbol{A}=[\boldsymbol{a}_1, \boldsymbol{a}_2, \boldsymbol{a}_3]$. In this case, we see that **A** is rank deficient.

Another important characterization of a matrix is its nullity. The nullity of **A** is the dimension of the null space of **A**. In Example 8 above, the nullity of **A** is one. We then have the following interesting property:

$$\mathrm{rank}(\boldsymbol{A}) + \mathrm{nullity}(\boldsymbol{A}) = n \tag{1.1.14}$$

1.1.4 Four Fundamental Subspaces of a Matrix

The four matrix subspaces of concern are: the column space, the row space, and their respective orthogonal complements. The development of these four subspaces is closely linked to $N(\boldsymbol{A})$ and $R(\boldsymbol{A})$. We assume for this section that $\boldsymbol{A} \in \mathbb{R}^{m\times n}$, $r \leqslant \min(m, n)$, where $r=\mathrm{rank}(\boldsymbol{A})$.

1.1.4.1 The Column Space

This is simply $R(\boldsymbol{A})$. Its dimension is r. It is the set of all linear combinations of the

② Column rank deficient is when the rank of the matrix is less than the number of columns.

columns of $\boldsymbol{A}$.

1.1.4.2 The Orthogonal Complement of the Column Space

This may be expressed as $R(\boldsymbol{A})_{\perp}$, with dimension $m-r$. It may be shown to be equivalent to $N(\boldsymbol{A}^{\mathrm{T}})$, as follows: By definition, $N(\boldsymbol{A}^{\mathrm{T}})$ is the set x satisfying:

$$\underbrace{\begin{bmatrix} \text{———} \\ \text{———} \\ \text{———} \end{bmatrix}}_{\boldsymbol{A}^{\mathrm{T}}} \begin{bmatrix} x_1 \\ x_2 \\ \vdots \\ x_n \end{bmatrix} = 0 \tag{1.1.15}$$

where columns of $\boldsymbol{A}$ are the rows of $\boldsymbol{A}^{\mathrm{T}}$. From equation (1.1.15), we see that $N(\boldsymbol{A}^{\mathrm{T}})$ is the set of $\boldsymbol{x} \in \mathbb{R}^m$ which is orthogonal to all columns of $\boldsymbol{A}$ (rows of $\boldsymbol{A}^{\mathrm{T}}$). This by definition is the orthogonal complement of $\boldsymbol{R}(\boldsymbol{A})$.

1.1.4.3 The Row Space

The row space is defined simply as $\boldsymbol{R}(\boldsymbol{A}^{\mathrm{T}})$, with dimension r. The row space is the range of the rows of $\boldsymbol{A}$, or the subspace spanned by the rows, or the set of all possible linear combinations of the rows of $\boldsymbol{A}$.

1.1.4.4 The Orthogonal Complement of the Row Space

This may be denoted as $\boldsymbol{R}(\boldsymbol{A}^{\mathrm{T}})_{\perp}$. Its dimension is $n-r$. This set must be that which is orthogonal to all rows of $\boldsymbol{A}$: i.e., for $\boldsymbol{x}$ to be in this space, $\boldsymbol{x}$ must satisfy

$$\begin{matrix}\text{rows of} \\ \boldsymbol{A}\end{matrix} \rightarrow \begin{bmatrix} \text{———} \\ \text{———} \\ \text{———} \end{bmatrix} \begin{bmatrix} x_1 \\ \vdots \\ x_n \end{bmatrix} = 0 \tag{1.1.16}$$

Thus, the set $\boldsymbol{x}$, which is the orthogonal complement of the row space satisfying equation (1.1.16), is simply $N(\boldsymbol{A})$.

We have noted before thatrank $(\boldsymbol{A})$ = rank $(\boldsymbol{A}^{\mathrm{T}})$. Thus, the dimension of the row and column subspaces are equal. This is surprising, because it implies the number of linearly independent rows of a matrix is the same as the number of linearly independent columns. This holds regardless of the size or rank of the matrix. It is not an intuitively obvious fact and there is no immediately obvious reason why this should be so. Nevertheless, the rank of a matrix is the number of independent rows or columns.

1.1.5 Vector Norms

A vector norm is a means of expressing the length or distance associated with a vector. A norm on a vector space $\mathbb{R}^n$ is a function f, which maps a point in $\mathbb{R}^n$ into a point in $\mathbb{R}$. Formally, this is stated mathematically as $f: \mathbb{R}^n \rightarrow \mathbb{R}$.

The norm has the following properties:

1. $f(\boldsymbol{x}) \geqslant 0$ for all $\boldsymbol{x} \in \mathbb{R}^n$.

2. $f(\boldsymbol{x})=0$ if and only if $\boldsymbol{x}=0$.
3. $f(\boldsymbol{x}+\boldsymbol{y})\leqslant f(\boldsymbol{x})+f(\boldsymbol{y})$ for $\boldsymbol{x}, \boldsymbol{y}\in\mathbb{R}^n$.
4. $f(a\boldsymbol{x})=|a|f(\boldsymbol{x})+f(\boldsymbol{y})$ for $a\in\mathbb{R}$, $\boldsymbol{x}\in\mathbb{R}^n$.

We denote the function $f(\boldsymbol{x})$ as $\| x \|$. The p-norms: This is a useful class of norms, generalizing on the idea of the Euclidean norm. They are defined by

$$\| x \|_p = (|x_1|^p + |x_2|^p + \cdots + |x_n|^p)^{1/p} \tag{1.1.17}$$

If $p=1$:

$$\| x \|_1 = \sum_i |\boldsymbol{x}_i|$$

which is simply the sum of absolute values of the elements.

If $p=2$:

$$\| x \|_2 = \left(\sum_i \boldsymbol{x}_i^2\right)^{\frac{1}{2}} = (\boldsymbol{x}^{\mathrm{T}}\boldsymbol{x})^{\frac{1}{2}}$$

which is the familiar Euclidean norm.

If $p=\infty$:

$$\| \boldsymbol{x} \|_\infty = \max_i |\boldsymbol{x}_i|$$

which is the largest element of $\boldsymbol{x}$. This may be shown in the following way. As $p\to\infty$, the largest term within the round brackets in equation (1.1.17) dominates all the others. Therefore equation (1.1.17) may be written as

$$\| x \|_\infty = \lim_{p\to\infty}\left[\sum_{i=1}^{n} \boldsymbol{x}_i^p\right]^{\frac{1}{p}} = \lim_{p\to\infty}[\boldsymbol{x}_k^p]^{\frac{1}{p}} = \boldsymbol{x}_k \tag{1.1.18}$$

Where k is the index corresponding to the largest element $\boldsymbol{x}_i$. Note that the $p=2$ norm has many useful properties, but is expensive to compute. Obviously, the 1-and ∞-norms are easier to compute, but are more difficult to deal with algebraically. All the p-norms obey all the properties of a vector norm.

1.1.6 Determinants

Consider a square matrix $\boldsymbol{A}\in\mathbb{R}^{m\times m}$. We can define the matrix $\boldsymbol{A}_{ij}$ as the submatrix obtained from $\boldsymbol{A}$ by deleting the i-th row and j-th column of $\boldsymbol{A}$. The scalar number $\det(\boldsymbol{A}_{ij})$ (where $\det(\cdot)$ denotes determinant) is called the minor associated with the element a_{ij} of $\boldsymbol{A}$. The signed minor $\boldsymbol{c}_{ij}\triangleq(-1)^{j+i}\det(\boldsymbol{A}_{ij})$ is called the cofactor of a_{ij}.

The determinant of $\boldsymbol{A}$ is the m-dimensional volume contained within the columns (rows) of $\boldsymbol{A}$. This interpretation of determinant is very useful as we see shortly. The determinant of a matrix may be evaluated by the expression

$$\det(\boldsymbol{A}) = \sum_{j=1}^{m} a_{ij}c_{ij} \quad i\in 1, 2, \cdots, m \tag{1.1.19}$$

or

$$\det(\boldsymbol{A}) = \sum_{i=1}^{m} a_{ij}c_{ij} \quad j\in 1, \cdots, 2, m \tag{1.1.20}$$

Both the above are referred to as the cofactor expansion of the determinant. Equation

(1.1.19) is along the ith row of $\boldsymbol{A}$, whereas equation (1.1.20) is along the j-th column. It is indeed interesting to note that both versions above give exactly the same number, regardless of the value of i or j.

Equations (1.1.19) and (1.1.20) express the $m\times m$ determinant $\det(\boldsymbol{A})$ in terms of the cofactors $\boldsymbol{c}_{ij}$ of $\boldsymbol{A}$, which are themselves $(m-1)\times(m-1)$ determinants. Thus, $m-1$ recursions of equation (1.1.19) or equation (1.1.20) will finally yield the determinant of the $m\times m$ matrix $\boldsymbol{A}$.

From equation (1.1.19) it is evident that if $\boldsymbol{A}$ is triangular, then $\det(\boldsymbol{A})$ is the product of the main diagonal elements. Since diagonal matrices are in the upper triangular set, then the determinant of a diagonal matrix is also the product of its diagonal elements.

1.1.7 Properties of Determinants

Before we begin this discussion, let us define the volume of a parallelopiped defined by the set of column vectors comprising a matrix as the principal volume of that matrix. We have the following properties of determinants, which are stated without proof:

1. $\det(\boldsymbol{AB})=\det(\boldsymbol{A})\det(\boldsymbol{B})\ \boldsymbol{A},\ \boldsymbol{B}\in\mathbb{R}^{m\times m}$. The principal volume of the product of matrices is the product of principal volumes of each matrix.

2. $\det(\boldsymbol{A})=\det(\boldsymbol{A}^{\mathrm{T}})$. This property shows that the characteristic polynomials③ of $\boldsymbol{A}$ and $\boldsymbol{A}^{\mathrm{T}}$ are identical. Consequently, as we see later, eigenvalues of $\boldsymbol{A}^{\mathrm{T}}$ and $\boldsymbol{A}$ are identical.

3. $\det(c\boldsymbol{A})=c^m\det(\boldsymbol{A})$, $c\in\mathbb{R}$, $\boldsymbol{A}\in\mathbb{R}^{m\times m}$. This is a reflection of the fact that if each vector defining the principal volume is multiplied by c, then the resulting volume is multiplied by c^m.

4. $\det(\boldsymbol{A})=0\Leftrightarrow\boldsymbol{A}$ is singular.

This implies that at least one dimension of the principal volume of the corresponding matrix has collapsed to zero length.

5. $\det(\boldsymbol{A})=\prod_{i=1}^{m}\lambda_i$, where λ_i are the eigen(singular) values of $\boldsymbol{A}$. This means the parallelopiped defined by the column or row vectors of a matrix may be transformed into a regular rectangular solid of the same m-dimensional volume whose edges have lengths corresponding to the eigen(singular) values of the matrix.

6. The determinant of an orthonormal④ matrix is ±1. This is easy to see, because the vectors of an orthonormal matrix are all unit length and mutually orthogonal. Therefore the corresponding principal volume is ∓1.

7. If $\boldsymbol{A}$ is nonsingular, then $\det(\boldsymbol{A}^{-1})=[\det(\boldsymbol{A})]^{-1}$.

8. If $\boldsymbol{B}$ is nonsingular, then $\det(\boldsymbol{B}^{-1}\boldsymbol{AB})=\det(\boldsymbol{A})$.

---

③ The characteristic polynomial of a matrix is defined in Section 1.2.7.2.

④ An orthonormal matrix is defined in Section 1.2.7.2.

9. If $\boldsymbol{B}$ is obtained from $\boldsymbol{A}$ by interchanging any two rows (or columns), thendet$(\boldsymbol{B})=-\det(\boldsymbol{A})$.

10. If B is obtained from A by by adding a scalar multiple of one row to another (or a scalar multiple of one column to another), then $\det(\boldsymbol{B})=\det(\boldsymbol{A})$.

A further property of determinants allows us to compute the inverse of **A**. Define the matrix $\tilde{\boldsymbol{A}}$ as the adjoint of $\boldsymbol{A}$:

$$\tilde{\boldsymbol{A}}=\begin{bmatrix} c_{11} & c_{12} & \cdots & c_{1m} \\ c_{21} & c_{22} & \cdots & c_{2m} \\ \vdots & & & \vdots \\ c_{m1} & c_{m2} & \cdots & c_{mm} \end{bmatrix} \tag{1.1.21}$$

where the c_{ij} are the cofactors of **A**. According to equation (1.1.19) or (1.1.20), the i-th row $\tilde{\boldsymbol{a}}_i^{\mathrm{T}}$ of $\tilde{\boldsymbol{A}}$ times the i-th column a_i is $\det(\boldsymbol{A})$; i.e.,

$$\tilde{\boldsymbol{a}}_i^{\mathrm{T}}\boldsymbol{a}_i=\det(\boldsymbol{A}) \quad i=1, 2, \cdots, m \tag{1.1.22}$$

It can also be shown that

$$\tilde{\boldsymbol{a}}_i^{T}\boldsymbol{a}_j=0 \quad i\neq j \tag{1.1.23}$$

Then, combining equation (1.1.22) and (1.1.23) for $i, j\in\{1, 2, \cdots, m\}$ we have the following interesting property:

$$\tilde{\boldsymbol{A}}\boldsymbol{A}=\det(\boldsymbol{A})\boldsymbol{I} \tag{1.1.24}$$

Where $\boldsymbol{I}$ is the $m\times m$ identity matrix. It then follows from equation (1.1.24) that the inverse $\boldsymbol{A}^{-1}$ of $\boldsymbol{A}$ is given as

$$\boldsymbol{A}^{-1}=[\det(\boldsymbol{A})]^{-1}\tilde{\boldsymbol{A}} \tag{1.1.25}$$

Neither equation (1.1.19) nor (1.1.25) are computationally efficient ways of calculating a determinant or an inverse respectively. Better methods which exploit the properties of various matrix decompositions are made evident later in the course.

1.2 The Most Basic Matrix Decomposition

1.2.1 Gaussian Elimination

In this section we discuss the concept of Gaussian elimination in some detail. But we present a very quick review by example of the elementary approach to Gaussian elimination. Given the system of equations.

$$\boldsymbol{A}\boldsymbol{x}=\boldsymbol{b} \tag{1.2.1}$$

Where $\boldsymbol{A}\in\mathbb{R}^{3\times 3}$ is nonsingular. The above system can be expanded into the form.

$$\begin{bmatrix} a_{11} & a_{12} & a_{13} \\ a_{21} & a_{22} & a_{23} \\ a_{31} & a_{32} & a_{33} \end{bmatrix}\begin{bmatrix} x_1 \\ x_2 \\ x_3 \end{bmatrix}=\begin{bmatrix} b_1 \\ b_2 \\ b_3 \end{bmatrix} \tag{1.2.2}$$

To solve the system we transform this system into the following upper triangular system by Gaussian elimination:

$$\begin{bmatrix} a_{11} & a_{12} & a_{13} \\ & a'_{22} & a'_{23} \\ & & a''_{33} \end{bmatrix} \begin{bmatrix} x_1 \\ x_2 \\ x_3 \end{bmatrix} = \begin{bmatrix} b_1 \\ b'_2 \\ b''_3 \end{bmatrix} \rightarrow \boldsymbol{Ux} = \boldsymbol{b}' \tag{1.2.3}$$

using a sequence of elementary row operations as follows

$$\begin{aligned} &\text{row } 2': = \text{row}2 - \frac{a_{21}}{a_{11}}\text{row}1 \\ &\text{row } 3': = \text{row}3 - \frac{a_{31}}{a_{11}}\text{row}1 \\ &\qquad and \\ &\text{row } 3'': = \text{row}3' - \frac{a'_{32}}{a'_{22}}\text{row}2' \end{aligned} \tag{1.2.4}$$

The prime indicates the respective quantity has been changed. Each elementary operation preserves the original system of equations. Each operation is designed to place a zero in the appropriate place below the main diagonal of **A**.

Once **A** has been triangularized, the solution $\boldsymbol{x}$ is obtained by applying backward substitution to the system $\boldsymbol{Ux}=\boldsymbol{b}$. With this procedure $\boldsymbol{x}_n$ is first determined from the last equation of (1.2.3). Then $\boldsymbol{x}_{n-1}$ may be determined from the second last row, etc. The algorithm may be summarized by the following schema:

$$\begin{aligned} &\text{for } i = n, \cdots, 1 \\ &\quad \boldsymbol{x}_i := \boldsymbol{b}_i \\ &\quad \text{for } j = i+1, \cdots, n \\ &\qquad \boldsymbol{x}_i := \boldsymbol{x}_i - \boldsymbol{u}_{ij}\boldsymbol{x}_{\mathrm{i}} \\ &\quad \boldsymbol{x}_i := \frac{\boldsymbol{x}_i}{\boldsymbol{u}_{ii}} \end{aligned} \tag{1.2.5}$$

What about the Accuracy of Back Substitution?

With operations on floating point numbers we must be concerned about the accuracy of the result since the floating point numbers themselves contain error. We want to know if it is possible that the small errors in the floating point representation of real numbers can lead to large errors in the computed result. In this vien, we can show that the computed solution $\bar{\boldsymbol{x}}$ obtained by back subtitution satisfies the expression.

$$(\boldsymbol{U}+\boldsymbol{F})\hat{\boldsymbol{x}} = \boldsymbol{b}' \tag{1.2.6}$$

Where $|\boldsymbol{F}| \leqslant nu|\boldsymbol{U}| + O(u^2)$, and u is machine epsilon. (Note the use of absolute value notation as discussed in the last lecture). The above equation says that $\hat{\boldsymbol{x}}$ is the exact solution to a slightly perturbed system. We see that all elements of $\boldsymbol{F}$ are of $O(u)$; hence we conclude that the error in the solution induced by the backward substitution process is of the same order as that due to floating point representation alone; Hence back substitution is stable. By a numerically stable algorithm we mean one that produces relatively small errors in its output values for small errors in the input values. This error performance is in contrast to the ordinary form of Gaussian elimination as we see later.

The total number of flops required for Gaussian elimination of a matrix $\boldsymbol{A}\in\mathbb{R}^{n\times n}$ may be shown to be $O\left(\frac{2n^3}{3}\right)$, (one "flop" is one floating point arithmetic operation; i. e., a floating point add, subtract, multiply or divide). It is easily shown that backward substitution requires $O(n^2)$ flops. Thus the number of operations required to solve $\boldsymbol{Ax}=\boldsymbol{b}$ is dominated by the Gaussian elimination process.

1.2.2 The LU Decomposition

Suppose we can find a lower triangular matrix $\boldsymbol{L}\in\mathbb{R}^{n\times n}$ with ones along the main diagonal and an upper triangular matrix $\boldsymbol{U}\in\mathbb{R}^{n\times n}$ such that:

$$\boldsymbol{A}=\boldsymbol{LU} \tag{1.2.7}$$

This decomposition of $\boldsymbol{A}$ is referred to as the LU decomposition. To solve the system $\boldsymbol{Ax}=\boldsymbol{b}$, or $\boldsymbol{LUx}=\boldsymbol{b}$ we define the variable $\boldsymbol{z}$ as $\boldsymbol{z}=\boldsymbol{Ux}$ and then

$$\begin{aligned}&\text{Solve } \boldsymbol{L}_z=\boldsymbol{b} \text{ for } z\\&\text{And } \boldsymbol{U}_x=\boldsymbol{z} \text{ for } x\end{aligned} \tag{1.2.8}$$

Since both systems are triangular, they are easy to solve. The first system requires only forward elimination; and the second only back-substitution. Forward elimination is the analogous process to backward substitution, but performed on a lower triangular system instead of an upper triangular one. Forward substitution requires an equal number of flops as back substitution and is just as stable. Thus once the LU factorization is complete, the solution of the system is easy: the total number of flops required to solve $\boldsymbol{Ax}=\boldsymbol{b}$ is $2n^2$, once the LU factorization of $\boldsymbol{A}$ is complete.

1.2.3 The LDM Factorization

If no zero pivots are encountered during the Gaussian elimination process, then there exist unit lower triangular matrices $\boldsymbol{L}$ and $\boldsymbol{M}$ and a diagonal matrix $\boldsymbol{D}$ such that

$$\boldsymbol{A}=\boldsymbol{LDM}^{\mathrm{T}} \tag{1.2.9}$$

Justification

Since $\boldsymbol{A}=\boldsymbol{LU}$ exists, let $\boldsymbol{U}=\boldsymbol{DM}^{\mathrm{T}}$ be upper triangular, where $d_i=u_{ii}$; hence, $\boldsymbol{A}=\boldsymbol{LDM}^{\mathrm{T}}$ which was to be shown. Each row of $\boldsymbol{M}^{\mathrm{T}}$ is the corresponding row of $\boldsymbol{U}$ divided by its diagonal element.

We then solve the system $\boldsymbol{Ax}=\boldsymbol{b}$ which is equivalent to $\boldsymbol{LDM}^{\mathrm{T}}\boldsymbol{x}=\boldsymbol{b}$ in three steps:

1. let $y=\boldsymbol{DM}^{\mathrm{T}}\boldsymbol{x}$ and solve $\boldsymbol{Ly}=\mathrm{Pr}$ (n^2 flops)
2. let $\boldsymbol{y}=\boldsymbol{M}^{\mathrm{T}}\boldsymbol{x}$ and solve $\boldsymbol{Dz}=\boldsymbol{y}$ (n flops)
3. solve $\boldsymbol{M}^{\mathrm{T}}\boldsymbol{x}=\boldsymbol{z}$ (n^2 flops)

1.2.4 The LDL Decomposition for Symmetric Matrices

For a symmetric non-singular matrix $\boldsymbol{A}\in\mathbb{R}^{n\times n}$, the factors $\boldsymbol{L}$ and $\boldsymbol{M}$ are identical.

Proof

Let $\boldsymbol{A}=\boldsymbol{LDM}^{\mathrm{T}}$. The matrix $\boldsymbol{M}^{-1}\boldsymbol{AM}^{-\mathrm{T}}=\boldsymbol{M}^{-1}\boldsymbol{LD}$ is symmetric (from left hand side) and lower triangular (from right hand side). Hence, they are both diagonal. But $\boldsymbol{D}$ is nonsingular, so $\boldsymbol{M}^{-1}\boldsymbol{L}$ is also diagonal. The matrices $\boldsymbol{M}$ and $\boldsymbol{L}$ are both unit lower triangular (ULT). It can be easily shown that the inverse of a ULT's matrix is also ULT, and furthermore, the product of ULT's is ULT. Therefore $\boldsymbol{M}^{-1}$ is ULT, and so is $\boldsymbol{M}^{-1}\boldsymbol{L}$. Thus $\boldsymbol{M}^{-1}\boldsymbol{L}=\mathrm{I}$; $\boldsymbol{M}=\boldsymbol{L}$.

This means that for a symmetric matrix $\boldsymbol{A}$ the LU factorization requires only $\frac{n^3}{3}$ flops, instead of $\frac{2}{3}n^3$ as for the general case. This is because only the lower factor need be computed.

1.2.5 Cholesky Decomposition

We now consider several modifications to the LU decomposition, which ultimately lead up to the Cholesky decomposition. These modifications are 1) the LDM decomposition, 2) the LDL decomposition on symmetric matrices, and 3) the LDL decomposition on positive definite symmetric matrices. The Cholesky decomposition is relevant only for square symmetric positive definite matrices and is an important concent in signal processing. Several examples of the use of the Cholesky decomposition are provided at the end of the section.

For $\boldsymbol{A}\in\mathbb{R}^{n\times n}$ symmetric and positive definite, there exists a lower triangular matrix $\boldsymbol{G}\in\mathbb{R}^{\mathrm{n}\times\mathrm{n}}$ with positive diagonal entries, such that $\boldsymbol{A}=\boldsymbol{GG}^{\mathrm{T}}$.

Proof

Consider $\boldsymbol{A}$ which is positive definite and symmetric. Note that covariance matrices fall into this class. Therefore $\boldsymbol{x}^{\mathrm{T}}\boldsymbol{Ax}>0$, $0\neq\boldsymbol{x}\in\mathbb{R}^{n\times n}$, and hence $\boldsymbol{x}^{\mathrm{T}}\boldsymbol{LDL}^{\mathrm{T}}\boldsymbol{x}>0$. If $\boldsymbol{A}$ is positive definite, then $\boldsymbol{L}$ is full rank; Let $\boldsymbol{y}\triangleq\boldsymbol{L}^{\mathrm{T}}\boldsymbol{x}$. Then, $\boldsymbol{y}^{\mathrm{T}}\boldsymbol{Dy}>0$, if and only if all elements of $\boldsymbol{D}$ are positive. Therefore if $\boldsymbol{A}$ is positive definite, then $d_{ii}>0$, $i=1, 2, \cdots, n$.

Because $\boldsymbol{A}$ is symmetric, the $\boldsymbol{A}=\boldsymbol{LDL}^{\mathrm{T}}$. Because the d_{ii} are positive, then $\boldsymbol{G}=\boldsymbol{L}\cdot\mathrm{diag}(\sqrt{d_{11}}, \cdots, \sqrt{d_{nn}})$. Then $\boldsymbol{GG}^{\mathrm{T}}=\boldsymbol{A}$ as desired.

As discussed earlier, this decomposition is stable without pivoting. Therefore, in solving the system $\boldsymbol{Ax}=\boldsymbol{b}$, where $\boldsymbol{A}$ is symmetric and positive definite (e. g., for the case where $\boldsymbol{A}$ is a sample covariance matrix), the Cholesky decomposition requires only half the flops for the LU decomposition phase of the computation, and does not require pivoting. Both these factors significantly reduce the execution time of the algorithm.

1.2.6 Applications and Examples of the Cholesky Decomposition

1.2.6.1 Generating Vector Processes with Desired Covariance

We may use the Cholesky decomposition to generate a random vector process with a desired covariance matrix $\boldsymbol{\Sigma} \in \mathbb{R}^{n \times n}$. Since must be symmetric and positive definite, let

$$\boldsymbol{\Sigma} = \boldsymbol{G}\boldsymbol{G}^{\mathrm{T}} \tag{1.2.10}$$

be the Cholesky factorization of $\boldsymbol{\Sigma}$. Let $\boldsymbol{w} \in \mathbb{R}^{n \times n}$ be a random vector with uncorrelated elements such that $\boldsymbol{E}(\boldsymbol{w}\boldsymbol{w}^{\mathrm{T}}) = \boldsymbol{I}$. Such $\boldsymbol{w}$ are easily generated by random number generators on the computer. Then, define $\boldsymbol{x}$ as:

$$\boldsymbol{x} = \boldsymbol{G}\boldsymbol{w} \tag{1.2.11}$$

The vector process $\boldsymbol{x}$ has the desired covariance matrix because

$$\boldsymbol{E}(\boldsymbol{w}\boldsymbol{w}^{\mathrm{T}}) = \boldsymbol{E}(\boldsymbol{G}\boldsymbol{w}\boldsymbol{w}^{\mathrm{T}}\boldsymbol{G}^{\mathrm{T}}) = \boldsymbol{G}\boldsymbol{E}(\boldsymbol{w}\boldsymbol{w}^{\mathrm{T}})\boldsymbol{G}^{\mathrm{T}} = \boldsymbol{G}\boldsymbol{G}^{\mathrm{T}} = \boldsymbol{\Sigma} \tag{1.2.12}$$

This procedure is particularly useful for computer simulations when it is desired to create a random vector process with a specified covariance matrix.

1.2.6.2 Whitening a Process

This example is essentially the inverse of the one just discussed. Suppose we have a stationary vector process $\boldsymbol{x}_i \in \mathbb{R}^n$, $i = 1, 2, \cdots, n$. This process could be the signals received from the elements of an array of n sensors, it could be sets of n sequential samples of any time-varying signal, or sets of data in a tapped-delay line equalizer of length n, at time instants t_1, t_2, $\cdots$, etc. Let the process $\boldsymbol{x}$ consist of a signal part $\boldsymbol{s}_i$ and a noise part $\boldsymbol{v}_i$:

$$\boldsymbol{x}_i = \boldsymbol{s}_i + \boldsymbol{v}_i, \ \mathrm{gig} = 1, 2, 3, \cdots \tag{1.2.13}$$

Where we assume the covariance of the noise $\boldsymbol{E}(\boldsymbol{v}\boldsymbol{v}^{\mathrm{T}}) \triangleq \boldsymbol{\Sigma}$ is not diagonal. The noise is thus correlated or coloured. This discussion requires $\boldsymbol{\Sigma}$ that is known or can be estimated.

Let $\boldsymbol{G}$ be the Cholesky factorization of $\boldsymbol{\Sigma}$ such that $\boldsymbol{G}\boldsymbol{G}^{\mathrm{T}} = \boldsymbol{\Sigma}$. Premultiply both sides of equation (1.2.13) with $\boldsymbol{G}^{-1}$:

$$\boldsymbol{G}^{-1}\boldsymbol{x}_i = \boldsymbol{G}^{-1}\boldsymbol{s}_i + \boldsymbol{G}^{-1}\boldsymbol{v}_i \tag{1.2.14}$$

The noise component is now $\boldsymbol{G}^{-1}\boldsymbol{v}_i$. The corresponding noise covariance matrix is

$$\boldsymbol{E}(\boldsymbol{G}^{-1}\boldsymbol{v}\boldsymbol{v}^{\mathrm{T}}\boldsymbol{G}^{-\mathrm{T}}) = \boldsymbol{G}^{-1}\boldsymbol{E}(\boldsymbol{v}\boldsymbol{v}^{\mathrm{T}})\boldsymbol{G}^{-\mathrm{T}} = \boldsymbol{G}^{-1}\boldsymbol{\Sigma}\boldsymbol{G}^{-\mathrm{T}} = \boldsymbol{G}^{-1}\boldsymbol{G}\boldsymbol{G}^{\mathrm{T}}\boldsymbol{G}^{-\mathrm{T}} = \boldsymbol{I} \tag{1.2.15}$$

Thus, by premultiplying the original signal $\boldsymbol{x}$ by the inverse Cholesky factor of the noise, the resulting noise is white. Whitening sequences by premultiplying by an inverse square-root of the covariance matrix (as opposed to some other factor of $\boldsymbol{\Sigma}^{-1}$) is an important concept in signal processing. The inverse Cholesky factor is most commonly applied in these situations because it is stable and easy to compute.

Since the received signal $\boldsymbol{x} = \boldsymbol{s} + \boldsymbol{v}$, the joint probability density function $\boldsymbol{p}(\boldsymbol{x} \mid \boldsymbol{s})$ of the received signal vector $\boldsymbol{x}$, given the noiseless signal $\boldsymbol{s}$, in the presence of Gaussian noise samples $\boldsymbol{v}$ with covariance matrix $\boldsymbol{\Sigma}$ is simply the pdf of the noise itself, and is given by the multi-dimensional Gaussian probability density function discussed in Section 1.2.1:

$$p(\boldsymbol{x} \mid \boldsymbol{s}) = (2\pi)^{-\frac{n}{2}} \, |\boldsymbol{\Sigma}|^{-\frac{1}{2}} \exp\left[-\frac{1}{2}(\boldsymbol{x} - \boldsymbol{s})^{\mathrm{T}}\boldsymbol{\Sigma}^{-1}(\boldsymbol{x} - \boldsymbol{s})\right] \tag{1.2.16}$$

In contrast, suppose we transform the vector $\boldsymbol{x} - \boldsymbol{s}$ by pre-multiplication with the inverse Cholesky factor $\boldsymbol{G}^{-1}$ of $\boldsymbol{\Sigma}^{-1}$, to form $\boldsymbol{y} = \boldsymbol{G}^{-1}(\boldsymbol{x} - \boldsymbol{s})$. This transformation whitens the noise. From the discussion above, we see that the covariance matrix of the variable $\boldsymbol{y}$ is the identity.

Therefore,

$$\begin{aligned} p(\boldsymbol{y}) &= \frac{1}{(2\pi)^{\frac{n}{2}} \ |\boldsymbol{\Sigma}|^{\frac{1}{2}}} \exp\left[-\frac{1}{2}\boldsymbol{y}^{\mathrm{T}}\boldsymbol{y}\right] \\ &= \frac{1}{(2\pi)^{\frac{n}{2}} \ |\boldsymbol{\Sigma}|^{\frac{1}{2}}} \exp\left[-\frac{\boldsymbol{y}_1^2}{2}\right]\cdots\exp\left[-\frac{\boldsymbol{y}_n^2}{2}\right] \\ &= p(y_1)p(y_2)\cdots p(y_n) \end{aligned} \tag{1.2.17}$$

Thus, in the equation (1.2.17) where we have whitened the noise, any processing on the signal involving detection or estimation may be done by processing each whitened variable y_i independently of the rest, since these variables are shown to be independent. In contrast, if we wish to process the original signal $\boldsymbol{x}$ in coloured noise, we must jointly process the entire vector $\boldsymbol{x}$, due to the fact there are dependencies amongst the elements introduced through the quadratic form in the exponent of equation (1.2.16). The whitening process thus significantly simplifies processing on the signal when the noise is coloured.

As a further example of the use of the Cholesky decomposition, we consider the multi-variate Gaussian pdf of a zero-mean random vector with covariance $\boldsymbol{\Sigma}$. The exponent of the distribution is then $\boldsymbol{x}^{\mathrm{T}}\boldsymbol{\Sigma}^{-1}\boldsymbol{x}=\boldsymbol{x}^{\mathrm{T}}\boldsymbol{G}^{-\mathrm{T}}\boldsymbol{G}^{-1}\boldsymbol{x}$ where $\boldsymbol{G}$ is the Cholesky factor of $\boldsymbol{\Sigma}$. If we let $\boldsymbol{w}=\boldsymbol{G}^{-1}\boldsymbol{x}$, then the exponent of the distribution becomes $\boldsymbol{w}^{\mathrm{T}}\boldsymbol{w}$. But from we see that the covariance of $\boldsymbol{w}$ is $\boldsymbol{I}$; i. e., $\boldsymbol{w}$ is white. Thus, we see that the matrix $\boldsymbol{\Sigma}^{-1}$ in the exponent of the Gaussian pdf has the role of transforming the original random variable $\boldsymbol{x}$ into another random variable $\boldsymbol{w}$ whose elements are uncorrelated with unit variance. Thus $\boldsymbol{\Sigma}^{-1}$ whitens and normalizes the original variables $\boldsymbol{x}$.

These variables are shown to be independent. In contrast if we wish to process the original signal $\boldsymbol{x}$ in coloured noise, we must jointly process the entire vector $\boldsymbol{x}$, due to the fact there are dependencies amongst the elements introduced through the quadratic form in the exponent of equation (1.2.16). The whitening process thus significantly simplifies processing on the signal when the noise is coloured.

As a further example of the use of the Cholesky decomposition, we consider the multi variate Gaussian pdf of a zero mean random vector with covariance $\boldsymbol{\Sigma}$. The exponent of the distribution is then $\boldsymbol{x}^{\mathrm{T}}\boldsymbol{\Sigma}^{-1}\boldsymbol{x}=\boldsymbol{x}^{\mathrm{T}}\boldsymbol{G}^{-\mathrm{T}}\boldsymbol{G}^{-1}\boldsymbol{x}$ where $\boldsymbol{G}$ is the Cholesky factor of $\boldsymbol{\Sigma}$. If we let $\boldsymbol{w}=\boldsymbol{G}^{-1}\boldsymbol{x}$, then the exponent of the distribution becomes $\boldsymbol{w}^{\mathrm{T}}\boldsymbol{w}$. But from equation (1.2.15), we see that the covariance of $\boldsymbol{w}$ is $\boldsymbol{I}$; i. e., $\boldsymbol{w}$ is white. Thus we see that the matrix $\boldsymbol{\Sigma}^{-1}$ in the exponent of the Gaussian pdf has the role of transforming the original random variable $\boldsymbol{x}$ into another random variable $\boldsymbol{w}$ whose elements are uncorrelated with unit variance. Thus $\boldsymbol{\Sigma}^{-1}$ whitens and normalizes the original variables $\boldsymbol{x}$.

1.2.7 Eigendecomposition

Eigenvalue decomposition is to decompose a matrix into the following form:

$$\boldsymbol{A} = \boldsymbol{Q\Sigma Q}^{\mathrm{T}} \tag{1.2.18}$$

Where $\boldsymbol{Q}$ is the eigenvector of this matrix $\boldsymbol{A}$, and the orthogonal matrix is invertible. $\boldsymbol{\Sigma}=\mathrm{diag}(\lambda_1, \lambda_2, \cdots, \lambda_n)$ is a diagonal array with each diagonal element being an eigenvalue.

First of all, just to be clear, a matrix is really just a linear transformation, because when you multiply a matrix times a vector you get a vector, you're essentially taking a linear transformation of that vector.

When the matrix is under the condition of high dimension, then the matrix is higher dimensional space under a linear transformation. The linear change may not be said by pictures, but as you can imagine, this transformation also have a lot of changes of direction. The first N eigenvectors obtained from eigenvalue decomposition, correspond to the major N change directions of this matrix. We can approximate this matrix by using the first N directions. That's what I said before: extract the most important features of this matrix. So to summarize, eigenvalue decomposition gives you eigenvalues and eigenvectors, and eigenvalues represent how important a feature is, and eigenvectors represent what a feature is. You can think of each eigenvector as a linear subspace, and we can do a lot of things with these linear subspaces.

However, eigenvalue decomposition also has many limitations, as the transformation of the matrix must be a square matrix.

In image processing, one method is eigenvalue decomposition. We all know that an image is actually a matrix of pixel values, so let's say I have a 100×100 image, and I do eigenvalue decomposition on this image matrix, and I'm actually extracting the features of this image, and the extracted features are the vectors that correspond to the eigenvectors. However, how important these features are in the image is represented by the eigenvalue. Such as the 100×100 image matrix decomposition, $\boldsymbol{A}$ get $\boldsymbol{A}$ 100×100 of eigenvector matrix $\boldsymbol{Q}$, and $\boldsymbol{A}$ 100×100 only on the diagonal elements of the matrix $\boldsymbol{E}$ is not zero, the characteristic values of elements on the diagonal matrix $\boldsymbol{E}$ is but also from large to small arrangement (modulus, for a single number, is take the absolute value). That is to say the images are extracted from $\boldsymbol{A}$ 100 features, and the importance of the characteristics of 100 by 100 digital, said the 100 Numbers in diagonal matrix $\boldsymbol{E}$. In practice, we found that there are 100 features extracted from the point of their characteristic value size, only the top 20 most (the 20 May not, have a plenty of 10, have a plenty of 30 or more) features the corresponding eigenvalue is very big, is close to 0, behind that is behind the contribution of the characteristics of image is almost negligible. We know that after eigenvalue decomposition of image matrix $\boldsymbol{A}$, matrix $\boldsymbol{Q}$ and matrix $\boldsymbol{E}$ can be obtained.

So if you take the inverse, you multiply the three matrices on the right hand side and you get $\boldsymbol{A}$. Now that we know that only the first 20 eigenvalues in the matrix $\boldsymbol{E}$ are more important, we may try to set all of them except the first 20 in the matrix $\boldsymbol{E}$ to 0, that is, only the first 20 main features of the image are taken to restore the image, and all the rest are discarded to see what happens at this time.

1.2.7.1　Eigenvalues and Eigenvectors

Suppose we have a matrix $\boldsymbol{A}$:

$$\boldsymbol{A} = \begin{bmatrix} 4 & 1 \\ 1 & 4 \end{bmatrix} \tag{1.2.19}$$

We investigate its eigenvalues and eigenvectors.

Suppose we take the product $\boldsymbol{A}\boldsymbol{x}_1$, where $\boldsymbol{x}_1 = [0, 1]^{\mathrm{T}}$, as shown in Fig. 1.2. then,

$$\boldsymbol{A}\boldsymbol{x}_1 = \begin{bmatrix} 4 \\ 1 \end{bmatrix} \tag{1.2.20}$$

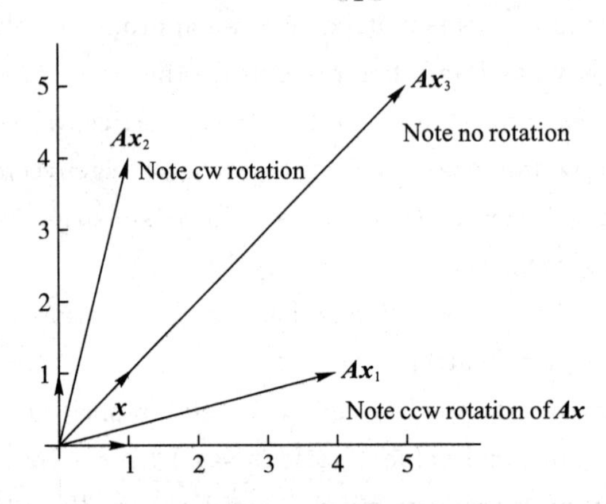

Fig. 1.2　Matrix-vector multiplication for various vectors

By comparing the vectors $\boldsymbol{x}_1$ and $\boldsymbol{A}\boldsymbol{x}_1$, we see that the product vector is scaled and rotated counter-clockwise with respect to $\boldsymbol{x}_1$.

Now consider the case where $\boldsymbol{x}_2 = [0, 1]^{\mathrm{T}}$. Then $\boldsymbol{A}\boldsymbol{x}_2 = [1, 4]^{\mathrm{T}}$. Here, we note a clockwise rotation of $\boldsymbol{A}\boldsymbol{x}_2 = [1, 4]^{\mathrm{T}}$ with respect to $\boldsymbol{x}_2$.

Now lets consider a more interesting case. Suppose $\boldsymbol{A}\boldsymbol{x}_3 = [1, 1]^{\mathrm{T}}$, Then $\boldsymbol{A}\boldsymbol{x}_3 = [5, 5]^{\mathrm{T}}$. Now the product vector points in the same direction as $\boldsymbol{x}_3$. The vector $\boldsymbol{A}\boldsymbol{x}_3$ is a scaled version of the vector $\boldsymbol{x}_3$. Because of this property, $\boldsymbol{x}_3 = [1, 1]^{\mathrm{T}}$ is an eigenvector of $\boldsymbol{A}$. The scale factor (which in this case is 5) is given the symbol λ and is referred to as an eigenvalue.

Note that $\boldsymbol{x} = [1, -1]^{\mathrm{T}}$ is also an eigenvector, because in this case, $\boldsymbol{A}\boldsymbol{x} = [1, 1]^{\mathrm{T}} = 3\boldsymbol{x}$. The corresponding eigenvalue is 3.

Thus we have, if $\boldsymbol{x}$ is an eigenvector of $\boldsymbol{A} \in \mathbb{R}^{n \times n}$,

$$\boldsymbol{A}\boldsymbol{x} = \lambda\boldsymbol{x}$$

$$\uparrow \text{scalar multiple} \tag{1.2.21}$$

$$(\text{eigenvalue})$$

i.e., the vector $\boldsymbol{A}\boldsymbol{x}$ is in the same direction as $\boldsymbol{x}$ but scaled by a factor λ.

Now that we have an understanding of the fundamental idea of an eigenvector, we proceed to develop the idea further. Equation (1.2.21) may be written in the form

$$(\boldsymbol{A}-\lambda\boldsymbol{I})\boldsymbol{x}=0 \tag{1.2.22}$$

Where $\boldsymbol{I}$ is the $n\times n$ identity matrix. Equation (1.2.22) is a homogeneous system of equations, and from fundamental linear algebra, we know that a nontrivial solution to exists if and only if

$$\det(\boldsymbol{A}-\lambda\boldsymbol{I})=0 \tag{1.2.23}$$

Where det(•) denotes determinant. Equation (1.2.23), when evaluated, becomes a polynomial in λ of degree n. For example, for the matrix $\boldsymbol{A}$ above we have

$$\begin{cases}\det\left[\begin{pmatrix}4 & 1\\ 1 & 4\end{pmatrix}-\lambda\begin{pmatrix}1 & 0\\ 0 & 1\end{pmatrix}\right]=0\\ \det\begin{bmatrix}4-\lambda & 1\\ 1 & 4-\lambda\end{bmatrix}=(4-\lambda)^2-1=\lambda^2-8\lambda+15=0\end{cases} \tag{1.2.24}$$

It is easily verified that the roots of this polynomial are (5, 3), which correspond to the eigenvalues indicated above.

Equation is referred to as the characteristic equation of $\boldsymbol{A}$, and the corresponding polynomial is the characteristic polynomial. The characteristic polynomial is of degree n.

More generally, if $\boldsymbol{A}$ is $n\times n$, then there are n solutions of, or n roots of the characteristic polynomial. Thus there are n eigenvalues of $\boldsymbol{A}$ satisfying; i. e.,

$$\boldsymbol{A}\boldsymbol{x}_i=\lambda_i\boldsymbol{x}_i \qquad i=1, 2, \cdots, n \tag{1.2.25}$$

If the eigenvalues are all distinct, there are n associated linearly-independent eigenvectors, whose directions are unique, which span an n-dimensional Euclidean space.

Repeated Eigenvalues: In the case where there are e. g., r repeated eigenvalues, then a linearly independent set of n eigenvectors exist, provided the rank of the matrix $(\boldsymbol{A}-\lambda\boldsymbol{I})$ in is rank $n-r$. Then, the directions of the r eigenvectors associated with the repeated eigenvalues are not unique.

In fact, consider a set of r linearly independent eigenvectors $\boldsymbol{v}_1, \boldsymbol{v}_2, \cdots, \boldsymbol{v}_r$ associated with the r repeated eigenvalues. Then, it may be shown that any vector in span $[v_1, v_2, \cdots, v_r]$ is also an eigenvector. This emphasizes the fact the eigenvectors are not unique in this case.

Example 1:

Consider the matrix given by

$$\begin{bmatrix}1 & 0 & 0\\ 0 & 0 & 0\\ 0 & 0 & 0\end{bmatrix}$$

It may be easily verified that any vector in span$[e_2, e_3]$ is an eigenvector associated with the zero repeated eigenvalue.

Example 2:

Consider the $n\times n$ identity matrix. It has n repeated eigenvalues equal to one. In this case, any n-dimensional vector is an eigenvector, and the eigenvectors span an n-dimensional space.

Equation (1. 2. 25) gives us a clue how to compute eigenvalues. We can formulate the characteristic polynomial and evaluate its roots to give the λ_i. Once the eigenvalues are available, it is possible to compute the corresponding eigenvectors $\boldsymbol{v}_i$ by evaluating the nullspace of the quantity$(\boldsymbol{A}-\boldsymbol{\lambda}_i\boldsymbol{I})$, for $i=1, 2, \cdots, n$. This approach is adequate for small systems, but for those of appreciable size, this⑤ method is prone to appreciable numerical error. Later, we consider various orthogonal transformations which lead to much more effective techniques for finding the eigenvalues.

We now present some very interesting properties of eigenvalues and eigenvectors, to aid in our understanding.

Property 1

If the eigenvalues of a Hermitian symmetric matrix are distinct, then the eigenvectors are orthogonal. ⑥

Proof

Let $\{\lambda_i\}$ and $\{\lambda_i\}$, $i=1, \cdots, n$ be the eigenvectors and corresponding eigenvalues respectively of $\boldsymbol{A}\in\mathbb{R}^{n\times n}$. Choose any $i, j\in[1, 2, \cdots, n]$, $i\neq j$. Then

$$\boldsymbol{A}\boldsymbol{v}_i = \lambda\boldsymbol{v}_i \tag{1.2.26}$$

and

$$\boldsymbol{A}\boldsymbol{v}_j = \lambda\boldsymbol{v}_j \tag{1.2.27}$$

Premultiply by $\boldsymbol{v}_j^{\mathrm{T}}$ and by $\boldsymbol{v}_i^{\mathrm{T}}$

$$\boldsymbol{v}_j^{\mathrm{T}}\boldsymbol{A}\boldsymbol{v}_i = \lambda_i\boldsymbol{v}_j^{\mathrm{T}}\boldsymbol{v}_i \tag{1.2.28}$$

$$\boldsymbol{v}_i^{\mathrm{T}}\boldsymbol{A}\boldsymbol{v}_j = \lambda_j\boldsymbol{v}_i^{\mathrm{T}}\boldsymbol{v}_j \tag{1.2.29}$$

The quantities on the left are equal when $\boldsymbol{A}$ is symmetric. We show this as follows. Since the left-hand side of $\boldsymbol{A}$ is a scalar, its transpose is equal to itself. Therefore, weget $\boldsymbol{v}_j^{\mathrm{T}}\boldsymbol{A}\boldsymbol{v}_i=\boldsymbol{v}_i^{\mathrm{T}}\boldsymbol{A}\boldsymbol{v}_j$. ⑦ But, since $\boldsymbol{A}$ is symmetric, $\boldsymbol{A}^{\mathrm{T}}=\boldsymbol{A}$. Thus, $\boldsymbol{v}_j^{\mathrm{T}}\boldsymbol{A}\boldsymbol{v}_i=\boldsymbol{v}_i^{\mathrm{T}}\boldsymbol{A}\boldsymbol{v}_j=\boldsymbol{v}_i^{\mathrm{T}}\boldsymbol{A}\boldsymbol{x}_j$, which was to be shown.

Subtracting from, we have

$$(\lambda_i-\lambda_j)\boldsymbol{v}_j^{\mathrm{T}}\boldsymbol{v}_i=0 \tag{1.2.30}$$

Where we have used the fact $\boldsymbol{v}_j^{\mathrm{T}}\boldsymbol{v}_i=\boldsymbol{v}_i^{\mathrm{T}}\boldsymbol{v}_j$. But by hypothesis $\lambda_i-\lambda_j\neq 0$. Therefore, is satisfied only if $\boldsymbol{v}_j^{\mathrm{T}}\boldsymbol{v}_i=0$, which means the vectors are orthogonal.

Here we have considered only the case where the eigenvalues are distinct.

---

⑤ A symmetric matrix is one where $\boldsymbol{A}=\boldsymbol{A}^{\mathrm{T}}$, where the superscript^{T} means transpose, i. e, for a symmetric matrix, an element $a_{ij}=a_{ji}$. A Hermitian symmetric (or just Hermitian) matrix is relevant only for the complex case, and is one where $\boldsymbol{A}=\boldsymbol{A}^{\mathrm{H}}$, where superscript^{H} denotes the Hermitian transpose. This means the matrix is transposed and complex conjugated. Thus for a Hermitian matrix, an element $a_{ij}=a_{ji}^{*}$.

⑥ In this course we will generally consider only real matrices. However, when complex matrices are considered, Hermitian symmetric is implied instead of symmetric.

⑦ Here, we have used the property that for matrices or vectors $\boldsymbol{A}$ and $\boldsymbol{B}$ of conformable size, $(\boldsymbol{AB})^{\mathrm{T}}=\boldsymbol{B}^{\mathrm{T}}\boldsymbol{A}^{\mathrm{T}}$.

If an eigenvalue $\tilde{\lambda}$ is repeated r times, and $\text{rank}(\boldsymbol{A}-\tilde{\lambda}\boldsymbol{I})=n-r$, then a mutually orthogonal set of n eigenvectors can still be found.

Another useful property of eigenvalues of symmetric matrices is as follows.

Property 2

The eigenvalues of a (Hermitian) symmetric matrix are real. ⑧

Proof

(By contradiction): First, we consider the case where $\boldsymbol{A}$ is real. Let λ be a non-zero complex eigenvalue of a symmetric matrix $\boldsymbol{A}$. Then, since the elements of $\boldsymbol{A}$ are real, λ^*, the complex-conjugate of λ, must also be an eigenvalue of $\boldsymbol{A}$, beacause the roots of the characteristic polynomial must occur in complex conjugate pairs. Also, if $\boldsymbol{v}$ is a nonzero eigenvector corresponding to λ, then an eigenvector corresponding λ^* must be $\boldsymbol{v}^*$, the complex conjugate of $\boldsymbol{v}$. But Property 1 requires that the eigenvectors be orthogonal; therefore $\boldsymbol{v}^{\mathrm{T}}\boldsymbol{v}^*=0$. But $\boldsymbol{v}^{\mathrm{T}}\boldsymbol{v}^*=(\boldsymbol{v}^H\boldsymbol{v})^*$, which is by definition the complex conjugate of the norm of $\boldsymbol{v}$. But the norm of a vector is a pure real number; hence, $\boldsymbol{v}^{\mathrm{T}}\boldsymbol{v}^*$ must be greater than zero, since $\boldsymbol{v}$ is by hypothesis nonzero. We therefore have a contradiction. It follows that the eigenvalues of a symmetric matrix cannot be complex; i. e., they are real.

While this proof considers only the real symmetric case, it is easily extended to the case where $\boldsymbol{A}$ is Hermitian symmetric.

Property 3

Let $\boldsymbol{A}$ be a matrix with eigenvalues $\lambda_i, i=1, 2, \cdots, n$ and eigenvectors $\boldsymbol{v}_i$. Then the eigenvalues of the matrix $\boldsymbol{A}+s\boldsymbol{I}$ are λ_i+s, with corresponding eigenvectors $\boldsymbol{v}_i$, where s is any real number.

Proof

From the definition of an eigenvector, we have $\boldsymbol{A}\boldsymbol{v}=\lambda\boldsymbol{v}$. Further, we have $s\boldsymbol{I}\boldsymbol{v}=s\boldsymbol{v}$. Adding, we have $(\boldsymbol{A}+s\boldsymbol{I})\boldsymbol{v}=(\lambda+s)\boldsymbol{v}$. This new eigenvector relation on the matrix $(\boldsymbol{A}+s\boldsymbol{I})$ shows the eigenvectors are unchanged, while the eigenvalues are displaced by s.

Property 4

Let $\boldsymbol{A}$ be an $n\times n$ matrix with eigenvalues λ_i, $i=1,2,\cdots,n$. Then

- The determinant $\det(\boldsymbol{A})=\prod_{i=1}^{n}\lambda_i$.
- The trace⑨ $\text{tr}(\boldsymbol{A})=\sum_{i=1}^{n}\lambda_i$

The proof is straightforward, but because it is easier using concepts presented later in the course, it is not given here.

---

⑧ From Lastman and Sinha, Microcomputer-based Numerical Methods for Science and Engineering.

⑨ The trace denoted tr(•) of a square matrix is the sum of its elements on the main diagonal (also called the "diagonal" elements).

Property 5

If $\boldsymbol{v}$ is an eigenvector of a matrix $\boldsymbol{A}$, then $c\boldsymbol{v}$ is also an eigenvector, where c is any real or complex constant.

The proof follows directly by substituting $c\boldsymbol{v}$ for $\boldsymbol{v}$ in $\boldsymbol{A}\boldsymbol{v}=\lambda\boldsymbol{v}$. This means that only the direction of an eigenvector can be unique; its norm is not unique.

1.2.7.2 Orthonormal Matrices

Before proceeding with the eigendecomposition of a matrix, we must develop the concept of an orthonormal matrix. This form of matrix has mutually orthogonal columns, each of unit norm. This implies that

$$\boldsymbol{q}_i^{\mathrm{T}}\boldsymbol{q}_j=\delta_{ij} \tag{1.2.31}$$

Where δ_{ij} is the Kronecker delta, and $\boldsymbol{q}_i$ and $\boldsymbol{q}_j$ are columns of the orthonormal matrix $\boldsymbol{Q}$. With in mind, we now consider the product $\boldsymbol{Q}^{\mathrm{T}}\boldsymbol{Q}$. The result may be visualized with the aid of the diagram below:

$$\boldsymbol{Q}^{\mathrm{T}}\boldsymbol{Q}=\begin{bmatrix}\boldsymbol{q}_1^{\mathrm{T}} & \rightarrow \\ \boldsymbol{q}_2^{\mathrm{T}} & \rightarrow \\ \vdots & \\ \boldsymbol{q}_N^{\mathrm{T}} & \rightarrow\end{bmatrix}\begin{bmatrix}\boldsymbol{q}_1, & \boldsymbol{q}_2, & \cdots, & \boldsymbol{q}_N \\ \downarrow & \downarrow & & \downarrow\end{bmatrix}=\boldsymbol{I} \tag{1.2.32}$$

(When $i=j$, the quantity $\boldsymbol{q}_i^{\mathrm{T}}\boldsymbol{q}_i$ defines the squared 2 norm of $\boldsymbol{q}_i$, which has been defined as unity. When $i\neq j$, $\boldsymbol{q}_i^{T}\boldsymbol{q}_j=0$, *due to the orthogonality of the* $\boldsymbol{q}_i$). Equation (1.2.32) is a fundamental property of an orthonormal matrix.

Thus, for an orthonormal matrix, implies the inverse may be computed simply by taking the transpose of the matrix, an operation which requires almost no computational effort.

Equation (1.2.32) follows directly from the fact $\boldsymbol{Q}$ has orthonormal columns. It is not so clear that the quantity $\boldsymbol{Q}\boldsymbol{Q}^{\mathrm{T}}$ should also equal the identity. We can resolve this question in the following way. Suppose that $\boldsymbol{A}$ and $\boldsymbol{B}$ are any two square invertible matrices such that $\boldsymbol{AB}=\boldsymbol{I}$. Then, $\boldsymbol{BAB}=\boldsymbol{B}$. By parsing this last expression, we have

$$(\boldsymbol{BA})\cdot\boldsymbol{B}=\boldsymbol{B}. \tag{1.2.33}$$

Clearly, if equation (1.2.33) is to hold, then the quantity $\boldsymbol{BA}$ must be the identity[10]; hence, if $\boldsymbol{AB}=\boldsymbol{I}$ then $\boldsymbol{BA}=\boldsymbol{I}$. Therefore, if $\boldsymbol{Q}^{\mathrm{T}}\boldsymbol{Q}=\boldsymbol{I}$, then also $\boldsymbol{Q}\boldsymbol{Q}^{\mathrm{T}}=\boldsymbol{I}$. From this fact, it follows that if a matrix has orthonormal columns, then it also must have orthonormal rows. We now develop a further useful property of orthonormal matrices:

Property 6

The vector 2-norm is invariant under an orthonormal transformation.

If $\boldsymbol{Q}$ is orthonormal, then

$$\|\boldsymbol{Qx}\|_2^2=\boldsymbol{x}^{\mathrm{T}}\boldsymbol{Q}^{\mathrm{T}}\boldsymbol{Qx}=\boldsymbol{x}^{\mathrm{T}}\boldsymbol{x}=\|\boldsymbol{x}\|_2^2 \tag{1.2.34}$$

---

[10] This only holds if $\boldsymbol{A}$ and $\boldsymbol{B}$ are square invertible.

Thus, because the norm does not change, an orthonormal transformation performs a rotation operation on a vector. We use this norm—invariance property later in our study of the least-squares problem.

Suppose we have a matrix $\boldsymbol{U} \in \mathbb{R}^{m\times n}$, where $m>n$, whose columns are orthonormal. We see in this case that $\boldsymbol{U}$ is a tall matrix, which can be formed by extracting only the first n columns of an arbitrary orthonormal matrix. (We reserve the term orthonormal matrix to refer to a complete $m\times m$ matrix). Because $\boldsymbol{U}$ has orthonormal columns, it follows that the quantity $\boldsymbol{U}^{\mathrm{T}}\boldsymbol{U}=\boldsymbol{I}_{n\times n}$. However, it is important to realize that the quantity $\boldsymbol{U}\boldsymbol{U}^{\mathrm{T}} \neq \boldsymbol{I}_{m\times m}$ in this case, in contrast to the situation when $m=n$. The latter relation follows from the fact that the m column vectors of $\boldsymbol{U}^{\mathrm{T}}$ of length n, $n<m$, cannot all be mutually orthogonal. In fact, we see later that $\boldsymbol{U}\boldsymbol{U}^{\mathrm{T}}$ is a projector onto the subspace $R(\boldsymbol{U})$.

Suppose we have a vector $\boldsymbol{b} \in \mathbb{R}^{m}$. Because it is easiest, by convention we represent b using the basis $[e_1, e_2, \cdots, e_m]$, where the e_i are the elementary vectors (all zeros except for a one in the i-th position). However it is often convenient to represent b in a basis formed from the columns of an orthonormal matrix $\boldsymbol{Q}$. In this case, the elements of the vector $\boldsymbol{c}=\boldsymbol{Q}^{\mathrm{T}}\boldsymbol{b}$ are the coefficients of $\boldsymbol{b}$ in the basis $\boldsymbol{Q}$. The orthonormal basis is convenient because we can restore b from c simply by taking $\boldsymbol{b}=\boldsymbol{Q}\boldsymbol{c}$.

An orthonormal matrix is sometimes referred to as a unitary matrix. This follows because the determinant of an orthonormal matrix is ± 1.

1.2.7.3 The Eigendecomposition (ED) of a Square Symmetric Matrix

Almost all matrices on which ED are performed (at least in signal processing) are symmetric. A good exampleis covariance matrices, which are discussed in some detail in the next section.

Let $\boldsymbol{A} \in \mathbb{R}^{n\times n}$ be symmetric. Then, for eigenvalues and eigenvectors $\boldsymbol{v}_i$, we have

$$\boldsymbol{A}\boldsymbol{v}_i=\lambda_i\boldsymbol{v}_i \quad i=1,2,\cdots,n \tag{1.2.35}$$

Let the eigenvectors be normalized to unit 2-norm. Then these n equations can be combined, or stacked side-by-side together, and represented in the following compact form:

$$\boldsymbol{AV}=\boldsymbol{V\Lambda} \tag{1.2.36}$$

Where $\boldsymbol{V}=[\boldsymbol{v}_1, \boldsymbol{v}_2, \cdots, \boldsymbol{v}_n]$ (i.e., each column of $\boldsymbol{V}$ is an eigenvector), and

$$\boldsymbol{\Lambda}=\begin{bmatrix} \lambda_1 & & & 0 \\ & \lambda_2 & & \\ & & \ddots & \\ 0 & & & \lambda_n \end{bmatrix}=\mathrm{diag}(\lambda_1, \lambda_2, \cdots, \lambda_n) \tag{1.2.37}$$

Corresponding columns from each side of represent one specific value of the index i in. Because we have assumed $\boldsymbol{A}$ is symmetric, from Property 1, the $\boldsymbol{v}_i$ are orthogonal. Furthermore, since we have assumed $\|\boldsymbol{v}_i\|_2=1$, $\boldsymbol{V}$ is an orthonormal matrix. Thus, post-multiplying both sides of by $\boldsymbol{V}^{\mathrm{T}}$ and using $\boldsymbol{V}\boldsymbol{V}^{\mathrm{T}}=\boldsymbol{I}$ we get

$$\boldsymbol{A}=\boldsymbol{V\Lambda}\boldsymbol{V}^{\mathrm{T}} \tag{1.2.38}$$

Equation (1.2.38) is called the eigendecomposition (ED) of $\boldsymbol{A}$. The columns of $\boldsymbol{V}$ are eigenvectors of $\boldsymbol{A}$, and the diagonal elements of $\boldsymbol{\Lambda}$ are the corresponding eigenvalues. Any symmetric matrix may be decomposed in this way. This form of decomposition, with $\boldsymbol{\Lambda}$ being diagonal, is of extreme interest and has many interesting consequences. It is this decomposition which leads directly to the Karhunen-Loeve expansion which we discuss shortly.

Note that from, equation (1.2.38) knowledge of the eigenvalues and eigenvectors of $\boldsymbol{A}$ is sufficient to completely specify $\boldsymbol{A}$. Note further that if the eigenvalues are distinct, then the ED is unique. There is only one orthonormal $\boldsymbol{\Lambda}$ and one diagonal $\boldsymbol{\Lambda}$ which satisfies.

Equation (1.2.38) can also be written as

$$\boldsymbol{V}^{\mathrm{T}}\boldsymbol{A}\boldsymbol{V}=\boldsymbol{\Lambda} \tag{1.2.39}$$

Since $\boldsymbol{\Lambda}$ is diagonal, we say that the unitary (orthonormal) matrix $\boldsymbol{V}$ of eigenvectors diagonalizes $\boldsymbol{A}$. No other orthonormal matrix can diagonalize $\boldsymbol{A}$. The fact that only $\boldsymbol{V}$ diagonalizes $\boldsymbol{A}$ is the fundamental property of eigenvectors. If you understand that the eigenvectors of a symmetric matrix diagonalize it, then you understand the "mystery" behind eigenvalues and eigenvectors. That's all there is to it. We look at the K-L expansion later in this lecture in order to solidify this interpretation, and to show some very important signal processing concepts which fall out of the K-L idea. But the K-L analysis is just a direct consequence of that fact that only the eigenvectors of a symmetric matrix diagonalize.

1.2.8 Matrix Norms

Now that we have some understanding of eigenvectors and eigenvalues, we can now present the matrix norm. The matrix norm is related to the vector norm: it is a function which maps $\mathbb{R}^{m\times n}$ into $\mathbb{R}$. A matrix norm must obey the same properties as a vector norm. Since a norm is only strictly defined for a vector quantity, a matrix norm is defined by mapping a matrix into a vector. This is accomplished by post multiplying the matrix by a suitable vector. Some useful matrix norms are now presented:

1.2.8.1 Matrix *p*-Norms

A matrix p-norm is defined in terms of a vector p-norm. The matrix p-norm of an arbitrary matrix $\boldsymbol{A}$, denoted $\|\boldsymbol{A}\|_p$, is defined as

$$\|\boldsymbol{A}\|_p=\sup_{x\neq 0}\frac{\|\boldsymbol{A}\boldsymbol{x}\|_p}{\|\boldsymbol{x}\|_p} \tag{1.2.40}$$

where "sup" means supremum; i.e., the largest value of the argument over all values of $x\neq 0$. Since a property of a vector norm is $\|c\boldsymbol{x}\|_p=|c|\,\|\boldsymbol{x}\|_p$ for any scalar c, we can choose c in equation (1.2.40) so that $\|\boldsymbol{x}\|_p=1$. Then, an equivalent statement to equation (1.2.40) is

$$\|\boldsymbol{A}\|_p=\max_{\|x\|_p=1}\|\boldsymbol{A}\boldsymbol{x}\|_p \tag{1.2.41}$$

We now provide some interpretation for the above definition for the specific case where $p=2$ and for $\boldsymbol{A}$ square and symmetric, in terms of the eigendecomposition of $\boldsymbol{A}$. To find the matrix 2-norm, we differentiate equation (1.2.41) and set the result to zero. Differentiating $\|\boldsymbol{Ax}\|_2$ directly is difficult.

However, we note that finding the $\boldsymbol{x}$ which maximizes $\|\boldsymbol{Ax}\|_2$ is equivalent to finding the $\boldsymbol{x}$ which maximizes $\|\boldsymbol{Ax}\|_2^2$ and the differentiation of the latter is much easier. In this case, we have $\|\boldsymbol{Ax}\|_2^2=\boldsymbol{x}^{\mathrm{T}}\boldsymbol{A}^{\mathrm{T}}\boldsymbol{Ax}$. To find the maximum, we use the method of Lagrange multipliers, since $\boldsymbol{x}$ is constrained by equation (1.2.41).

Therefore we differentiate the quantity

$$\boldsymbol{x}^{\mathrm{T}}\boldsymbol{A}^{\mathrm{T}}\boldsymbol{Ax}+\gamma(1-\boldsymbol{x}^{\mathrm{T}}\boldsymbol{x}) \tag{1.2.42}$$

and set the result to zero. The quantity γ above is the Lagrange multiplier. The interesting result of this process is that $\boldsymbol{x}$ must satisfy

$$\boldsymbol{A}^{\mathrm{T}}\boldsymbol{Ax}=\gamma\boldsymbol{x}, \quad \|\boldsymbol{x}\|_2=1 \tag{1.2.43}$$

Therefore the stationary points of equation (1.2.43) are the eigenvectors of $\boldsymbol{A}^{\mathrm{T}}\boldsymbol{A}$. When $\boldsymbol{A}$ is square and symmetric, the eigenvectors of $\boldsymbol{A}^{\mathrm{T}}\boldsymbol{A}$ are equivalent to those of $\boldsymbol{A}$. Therefore the stationary points of equation (1.2.41) are also the eigenvectors of $\boldsymbol{A}$. By substituting $\boldsymbol{x}=\boldsymbol{v}_1$ into equation (1.2.41) we find that $\|\boldsymbol{Ax}\|_2=\lambda_1$.

It then follows that the solution to equation (1.2.41) is given by the eigenvector corresponding to the largest eigenvalue of $\boldsymbol{A}$, and $\|\boldsymbol{Ax}\|_2$ is equal to the largest eigenvalue of $\boldsymbol{A}$.

More generally, it is shown in the next lecture for an arbitrary matrix $\boldsymbol{A}$ that

$$\|\boldsymbol{A}\|_2=\sigma_1 \tag{1.2.44}$$

Where σ_1 is the largest singular value of $\boldsymbol{A}$. This quantity results from the singular value decomposition, to be discussed next lecture.

Matrix norms for other values of p, for arbitrary $\boldsymbol{A}$, are given as

$$\|\boldsymbol{A}\|_1=\max_{1\leqslant j\leqslant n}\sum_{i=1}^{m}|a_{ij}| \ \text{(maximum column sum)} \tag{1.2.45}$$

and

$$\|\boldsymbol{A}\|_\infty=\max_{1\leqslant i\leqslant m}\sum_{j=1}^{n}|a_{ij}| \ \text{(maximum row sum)} \tag{1.2.46}$$

1.2.8.2 Frobenius Norm

The Frobenius norm is the 2-norm of the vector consisting of the 2-norms of the rows (or columns) of the matrix $\boldsymbol{A}$:

$$\|\boldsymbol{A}\|_F=\left[\sum_{i=1}^{m}\sum_{j=1}^{n}|a_{ij}|^2\right]^{1/2}$$

1.2.8.3 Properties of Matrix Norms

1. Consider the matrix $\boldsymbol{A}\in\mathbb{R}^{m\times n}$ and the vector $\boldsymbol{x}\in\mathbb{R}^n$. Then,

$$\|\boldsymbol{Ax}\|_p\leqslant\|\boldsymbol{A}\|_p\|\boldsymbol{x}\|_p \tag{1.2.47}$$

This property follows by dividing both sides of the above by $\|\boldsymbol{x}\|_p$, and applying.

2. If $\boldsymbol{Q}$ and $\boldsymbol{Z}$ are orthonormal matrices of appropriate size, then

$$\| \boldsymbol{QAZ} \|_2 = \| \boldsymbol{A} \|_2 \tag{1.2.48}$$

and

$$\| \boldsymbol{QAZ} \|_F = \| \boldsymbol{A} \|_F \tag{1.2.49}$$

Thus, we see that the matrix 2-norm and Frobenius norm are invariant to pre-and post-multiplication by an orthonormal matrix.

3. Further

$$\| \boldsymbol{A} \|_F^2 = \mathrm{tr}(\boldsymbol{A}^{\mathrm{T}}\boldsymbol{A}) \tag{1.2.50}$$

Where tr (•) denotes the trace of a matrix, which is the sum of its diagonal elements.

1.2.9 Covariance Matrices

Here, we investigate the concepts and properties of the covariance matrix $\boldsymbol{R}_{xx}$ corresponding to a stationary, discrete-time random process $x[n]$. We break the infinite sequence $x[n]$ into windows of length m, as shown in Fig. 1. 3. The windows generally overlap; in fact, they are typically displaced from one another by only one sample. The samples within the i-th window become an m-length vector $\boldsymbol{x}_i$, $i=1, 2, \cdots, n$. Hence, the vector corresponding to each window is a vector sample from the random process $x[n]$. Processing random signals in this way is the fundamental first step in many forms of electronic system which deal with real signals, such as process identification, control, or any form of communication system including telephones, radio, radar, sonar, etc.

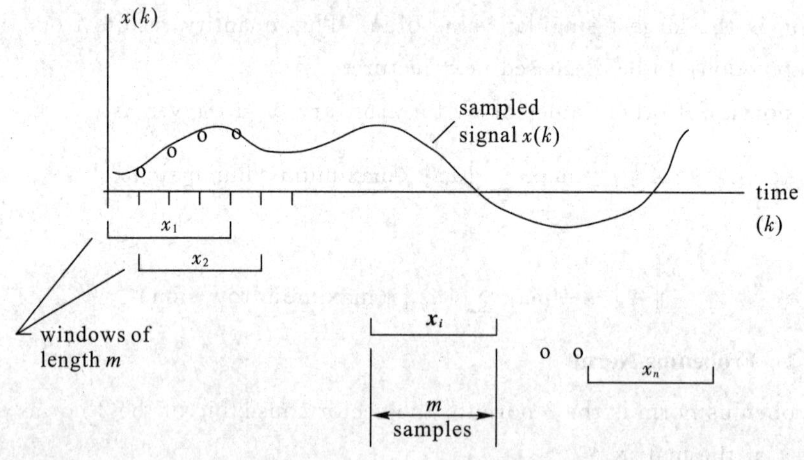

Fig. 1. 3 The received signal $x[n]$ is decomposed into windows of length m. The samples in the i-th window comprise the vector $\boldsymbol{x}_i$, $i=1, 2, 3, \cdots$

The word stationary as used above means the random process is one for which the corresponding joint m-dimensional probability density function describing the distribution of the vector sample $\boldsymbol{x}$ does not change with time. This means that all moments of the distribution (i. e., quantities such as the mean, the variance, and all cross-correlations, as well as all other higher-order statistical characterizations) are invariant with time. Here however, we deal with a weaker form of stationary referred to as wide-sense stationary (WSS). With these processes, only the first two moments (mean, variances and covariances) need be invariant with time. Strictly, the idea of a covariance matrix is only relevant for stationary or WSS processes, since expectations only have meaning if the underlying process is stationary.

The covariance matrix $\boldsymbol{R}_{xx} \in \mathbb{R}^{m \times m}$ corresponding to a stationary or WSS process $x[n]$ is defined as

$$\boldsymbol{R}_{xx} \triangleq \boldsymbol{E}[(\boldsymbol{x}-\mu)(\boldsymbol{x}-\mu)^{\mathrm{T}}] \tag{1.2.51}$$

Where μ is the vector mean of the process and $\boldsymbol{E}(\cdot)$ denotes the expectation operator over all possible windows of index i of length m in Fig. 1. 3. Often we deal with zero-mean processes, in which case we have

$$\boldsymbol{R}_{xx} = \boldsymbol{E}[\boldsymbol{x}_i \boldsymbol{x}_i^{\mathrm{T}}] = \boldsymbol{E}\left[\begin{pmatrix} x_1 \\ x_2 \\ \vdots \\ x_m \end{pmatrix} (x_1, x_2, \cdots, x_m)\right] = \boldsymbol{E}\begin{bmatrix} x_1 x_1 & x_1 x_2 & \cdots & x_1 x_m \\ x_2 x_1 & x_2 x_2 & \cdots & x_2 x_m \\ \vdots & \vdots & & \vdots \\ x_m x_1 & x_m x_2 & \cdots & x_m x_m \end{bmatrix} \tag{1.2.52}$$

where $(x_1, x_2, \cdots, x_m)^{\mathrm{T}} = \boldsymbol{x}_i$. Taking the expectation over all windows, equation (1. 2. 52) tells us that the element $r(1, 1)$ of $\boldsymbol{R}_{xx}$ is by definition $\boldsymbol{E}(\boldsymbol{x}_1^2)$, which is the mean-square value (the preferred term is variance, whose symbol is σ^2 of the first element $\boldsymbol{x}_1$ of all possible vector samples $\boldsymbol{x}_i$ of the process. $r(1, 1)=r(2, 2)=\cdots$, $r(m, m)$ are all equal to σ^2.) Thus all main diagonal elements of $\boldsymbol{R}_{xx}$ are equal to the variance of the process. The element $r(1, 2)=\boldsymbol{E}(x_1 x_2)$ is the cross-correlation between the first element of $\boldsymbol{x}_i$ and the second element. Taken over all possible windows, we see this quantity is the cross-correlation of the process and itself delayed by one sample. $r(1, 2)=r(2, 3)=\cdots=r(m-1, m)$, all elements on the first upper diagonal are equal to the cross correlation for a time-lag of one sample. Since multiplication is commutative, $r(2, 1)=r(1, 2)$, and therefore all elements on the first lower diagonal are also all equal to this same cross-correlation value. Using similar reasoning, all elements on the j-th upper or lower diagonal are all equal to the cross correlation value of the process for a time lag of j samples. Thus we see that the matrix $\boldsymbol{R}_{xx}$ is highly structured.

Let us compare the process shown in Fig. 1. 3 with that shown in Fig. 1. 4. In the former case, we see that the process is relatively slowly varying. Because we have assumed $x[n]$ to be zero mean, adjacent samples of the process in Fig. 1. 3 will have the same sign most of the time, and hence $\boldsymbol{E}(x_i x_{i+1})$ will be a positive number, coming close to the value $\boldsymbol{E}(x_i^2)$. The same can be said for $\boldsymbol{E}(x_i x_{i+2})$, except it is not so close to $\boldsymbol{E}(x_i^2)$. Thus, we see that for the process of Fig. 1. 3, the diagonals decay fairly slowly away from the main diagonal value.

However, for the process shown in Fig. 1. 4, adjacent samples are uncorrelated with each other. This means that adjacent samples are just as likely to have opposite signs as they are to have the same signs. On average, the terms with positive values have the same magnitude as those with negative values. Thus, when the expectations $\boldsymbol{E}(x_i x_{i+1})$, $\boldsymbol{E}(x_i x_{i+2})\cdots$ are taken, the resulting averages approach zero. In this case then, we see the covariance matrix concentrates around the main diagonal, and becomes equal to $\sigma^2\boldsymbol{I}$. We note that all the eigenvalues of $\boldsymbol{R}_{xx}$ are equal to the value σ^2. Because of this property, such processes are referred to as "white", in analogy to white light, whose spectral components are all of equal magnitude.

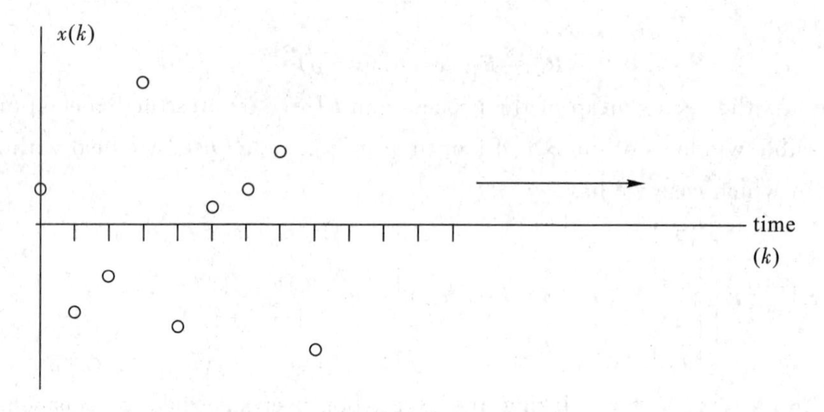

Fig 1.4　An uncorrelated discrete-time process

The sequence $\{r(1, 1), r(1, 2), \cdots, r(1, m)\}$ is equivalent to the autocorrelation function of the process, for lags 0 to $m-1$. The autocorrelation function of the process characterizes the random process $x[n]$ in terms of its variance, and how quickly the process varies over time.

In practice, it is impossible to evaluate the covariance matrix $\boldsymbol{R}_{xx}$ using expectations as in equation (1.2.51). Expectations cannot be evaluated in practice—they require an infinite amount of data which is never available, and furthermore, the data must be stationary over the observation interval, which is rarely the case. In practice, we evaluate an estimate $\hat{\boldsymbol{R}}_{xx}$ of $\boldsymbol{R}_{xx}$, based on an observation of finite length N of the process $x[n]$, by replacing the ensemble average (expectation) with a finite temporal average over the N available data points as follows:

$$\hat{\boldsymbol{R}}_{xx}=\frac{1}{N-m+1}\sum_{i=1}^{N-m+1}\boldsymbol{x}_i\boldsymbol{x}_i^{\mathrm{T}} \tag{1.2.53}$$

If equation (1.2.53) is used to evaluate $\hat{\boldsymbol{R}}$, then the process need only be stationary over the observation length. Thus, by using the covariance estimate given by, we can track slow changes in the true covariance matrix of the process with time, provided the change in the process is small over the observation interval N. Further properties and discussion covariance matrices are given in Haykin.

It is interesting to note that $\hat{\boldsymbol{R}}_{xx}$ can be formed in an alternate way from. Let $\boldsymbol{X}\in\mathbb{R}^{m\times(N-m+1)}$ be a matrix whose i-th column is the vector sample $\boldsymbol{x}_i\,(i=1,2,\cdots,N-m+1)$ of $x[n]$. Then $\hat{\boldsymbol{R}}_{xx}$ is also given as

$$\hat{\boldsymbol{R}}_{xx}=\frac{1}{N-m+1}\boldsymbol{X}\boldsymbol{X}^{\mathrm{T}} \tag{1.2.54}$$

Some Properties of $\boldsymbol{R}_{xx}$

1. $\boldsymbol{R}_{xx}$ is (Hermitian) symmetric i. e. $r_{ij}=r_{ji}^{*}$, where * denotes complex conjugation.

2. If $\boldsymbol{R}_{xx}$ is diagonal, then the elements of $\boldsymbol{x}$ are uncorrelated. If the magnitudes of the off-diagonal elements of $\boldsymbol{R}_{xx}$ are significant with respect to those on the main diagonal, the process is said to be highly correlated.

3. $\boldsymbol{R}$ is positive semi-definite. This implies that all the eigenvalues are greater than or equal to zero. We will discuss positive definiteness and positive semi-definiteness later.

4. If the stationary or WSS random process $\boldsymbol{x}$ has a Gaussian probability distribution, then the vector mean and the covariance matrix $\boldsymbol{R}_{xx}$ are enough to completely specify the statistical characteristics of the process.

1.3 Singular Value Decomposition (SVD)

In this lecture we learn about one of the most fundamental and important matrix decompositions of linear algebra: the SVD. It bears some similarity with the eigendecomposition (ED), but is more general. Usually, the ED is of interest only on symmetric square matrices, but the SVD may be applied to any matrix. The SVD gives us important information about the rank, the column and row spaces of the matrix, and leads to very useful solutions and interpretations of least squares problems. We also discuss the concept of matrix projectors, and their relationship with the SVD.

1.3.1 Orthogonalization

Before discussion of SVD, we first have a look at orthogonalization. For a square matrix with linearly independent eigenvalues, we have

$$\boldsymbol{A}=\boldsymbol{P}\boldsymbol{D}\boldsymbol{P}^{-1} \tag{1.3.1}$$

Where $\boldsymbol{P}$ is a matrix whose columns are the eigenvectors of $\boldsymbol{A}$, and $\boldsymbol{D}$ is a diagonal matrix with the eigenvalues of $\boldsymbol{A}$ along the diagonal.

Intuition to Orthogonalization

We can see equation (1. 3. 1) as a linear transformation from basis $\boldsymbol{A}$ to basis of eigenvalues $\boldsymbol{D}$ and $\boldsymbol{P}$ is the diagonalization is useful because it can easily to compute the exponential of a matrix:

$$\boldsymbol{A}^{n}=\boldsymbol{P}\boldsymbol{D}^{n}\boldsymbol{P}^{-1} \tag{1.3.2}$$

1.3.2 Existence Proof of the SVD

Consider two vectors $\boldsymbol{x}$ and $\boldsymbol{y}$ where $\|\boldsymbol{x}\|_2=\|\boldsymbol{y}\|_2=1$, s. t. $\boldsymbol{A}\boldsymbol{x}=\sigma\boldsymbol{y}$, where $\sigma=\|\boldsymbol{A}\|_2$. The fact that such vectors $\boldsymbol{x}$ and $\boldsymbol{y}$ can exist follows from the definition of the matrix 2-norm. We define orthonormal matrices $\boldsymbol{U}$ and $\boldsymbol{B}$ so that $\boldsymbol{x}$ and $\boldsymbol{y}$ form their first columns, as follows:

$$U=[y, U_1]$$
$$V=[x, V_1] \tag{1.3.3}$$

That is, U_1 consists of a set of non-unique orthonormal columns which are mutually orthogonal to themselves and to y; similarly for V_1.

We then define a matrix A_1 as

$$U^T AV = A_1 = \begin{bmatrix} y^T \\ U_1^T \end{bmatrix} A[x, V_1] \tag{1.3.4}$$

The matrix A_1 has the following structure:

$$\underbrace{\begin{bmatrix} y^T \\ U_1^T \end{bmatrix}}_{\text{orthonormal}} \quad A \quad \underbrace{[x \quad V_1]}_{\text{orthonormal}} = \begin{bmatrix} y^T \\ U_1^T \end{bmatrix} \quad [\sigma y \quad AV_1]$$

$$\begin{matrix} \sigma y^T y & y^T A V_1 \\ \downarrow & \downarrow \end{matrix}$$

$$\begin{matrix} \begin{bmatrix} \sigma & w^T \\ 0 & B \end{bmatrix} & \begin{matrix} 1 \\ m-1 \end{matrix} & \triangleq A_1 \\ \begin{matrix} 1 & \quad n-1 \end{matrix} & & \end{matrix} \tag{1.3.5}$$

Where $B \triangleq U_1^T A V_1$. The 0 in the equation, 1 block above follows from the fact that $U_1 \perp y$, because U is orthonormal.

Now, we post-multiply both sides of equation (1.3.5) by the vector $\begin{bmatrix} \sigma \\ w \end{bmatrix}$ and take 2-norm:

$$\left\| A_1 \begin{bmatrix} \sigma \\ w \end{bmatrix} \right\|_2^2 = \left\| \begin{bmatrix} \sigma & w^T \\ 0 & B \end{bmatrix} \begin{bmatrix} \sigma \\ W \end{bmatrix} \right\|_2^2 \geqslant (\sigma^2 + w^T w)^2 \tag{1.3.6}$$

This follows because the term on the extreme right is only the first element of the vector product of the middle term. But, as we have seen, matrix p-norms obey the following property:

$$\| Ax \|_2 \leqslant \| A \|_2 \; \| x \|_2 \tag{1.3.7}$$

Therefore using equation (1.3.6) and (1.3.7), we have

$$\| A_1 \|_2^2 \left\| \begin{bmatrix} \sigma \\ w \end{bmatrix} \right\|_2^2 \geqslant \left\| A_1 \begin{bmatrix} \sigma \\ w \end{bmatrix} \right\|_2^2 \geqslant (\sigma^2 + w^T w)^2 \tag{1.3.8}$$

Note that $\left\| \begin{bmatrix} \sigma \\ w \end{bmatrix} \right\|_2^2 = \sigma^2 + w^T w$. Dividing by this quantity, we obtain

$$\| A_1 \|_2^2 \geqslant \sigma^2 + w^T w \tag{1.3.9}$$

But, we defined $\sigma = \| A \|_2$. Therefore, the following must hold:

$$\sigma = \| A \|_2 = \| U^T AV \|_2 = \| A_1 \|_2 \tag{1.3.10}$$

Where the equality on the right follows, because the matrix 2-norm is invariant to matrix pre-and post-multiplication by an orthonormal matrix. By comparing equation (1.3.9) and (1.3.10), we have the result $w=0$.

Substituting this result back into equation (1.3.5), we now have

$$A_1=\begin{bmatrix}\sigma & 0\\ 0 & B\end{bmatrix} \tag{1.3.11}$$

The whole process repeats using only the component $\boldsymbol{B}$, until $\boldsymbol{A}_n$ becomes diagonal.

It is instructive to consider an alternative proof for the SVD. The following is useful because it is a constructive proof, which shows us how to form the components of the SVD.

Theorem 1

Let $\boldsymbol{A}\in\mathbb{R}^{m\times n}$ be a rank r matrix ($r\leqslant p=\min(m, n)$). Then there exist orthonormal matrices $\boldsymbol{U}$ and $\boldsymbol{V}$ such that

$$\boldsymbol{U}^{\mathrm{T}}\boldsymbol{A}\boldsymbol{V}=\begin{bmatrix}\tilde{\boldsymbol{\Sigma}} & 0\\ 0 & 0\end{bmatrix} \tag{1.3.12}$$

Where

$$\tilde{\boldsymbol{\Sigma}}=\operatorname{diag}(\sigma_1, \sigma_2, \cdots, \sigma_r) \tag{1.3.13}$$

Proof

Consider the square symmetric positive semi-definite matrix $\boldsymbol{A}^{\mathrm{T}}\boldsymbol{A}^1$. Let the eigenvalues greater than zero be $\sigma_1^2, \sigma_2^2, \cdots, \sigma_r^2$. Then, from our knowledge of the eigendecomposition, there exists an orthonormal matrix $\boldsymbol{V}\in\mathbb{R}^{n\times n}$ such that

$$\boldsymbol{V}^{\mathrm{T}}\boldsymbol{A}^{\mathrm{T}}\boldsymbol{A}\boldsymbol{V}=\begin{bmatrix}\tilde{\boldsymbol{\Sigma}}^2 & 0\\ 0 & 0\end{bmatrix} \tag{1.3.14}$$

Where $\tilde{\boldsymbol{\Sigma}}^2=\operatorname{diag}[\sigma_1^2, \sigma_2^2, \cdots, \sigma_r^2]$. We now partition $\boldsymbol{V}$ as $[\boldsymbol{V}_1 \quad \boldsymbol{V}_2]$, where $\boldsymbol{V}_1\in\mathbb{R}^{n\times r}$. Then equation (1.3.14) has the form

$$\begin{bmatrix}\boldsymbol{V}_1^{\mathrm{T}}\\ \boldsymbol{V}_2^{\mathrm{T}}\end{bmatrix}_n\boldsymbol{A}^{\mathrm{T}}\boldsymbol{A}\begin{bmatrix}\underset{r}{\boldsymbol{V}_1} & \underset{n-r}{\boldsymbol{V}_2}\end{bmatrix}=\begin{bmatrix}\tilde{\boldsymbol{\Sigma}}^2 & 0\\ 0 & 0\end{bmatrix} \tag{1.3.15}$$

Then by equating corresponding blocks in equation (1.3.15) we have

$$\boldsymbol{V}_1^{\mathrm{T}}\boldsymbol{A}^{\mathrm{T}}\boldsymbol{A}\boldsymbol{V}_1=\tilde{\boldsymbol{\Sigma}}^2 \quad (r\times r) \tag{1.3.16}$$

$$\boldsymbol{V}_2^{\mathrm{T}}\boldsymbol{A}^{\mathrm{T}}\boldsymbol{A}\boldsymbol{V}_2=0. \quad (n-r)\times(n-r) \tag{1.3.17}$$

From equation (1.3.16), we can write

$$\tilde{\boldsymbol{\Sigma}}^{-1}\boldsymbol{V}_1^{\mathrm{T}}\boldsymbol{A}^{\mathrm{T}}\boldsymbol{A}\boldsymbol{V}_1\tilde{\boldsymbol{\Sigma}}^{-1}=\boldsymbol{I} \tag{1.3.18}$$

Then, we define the matrix $\boldsymbol{U}_1\in\mathbb{R}^{m\times r}$ from equation (1.3.18) as

$$\boldsymbol{U}_1=\boldsymbol{A}\boldsymbol{V}_1\tilde{\boldsymbol{\Sigma}}^{-1} \tag{1.3.19}$$

Then from equation (1.3.18) we have $\boldsymbol{U}_1^{\mathrm{T}}\boldsymbol{U}_1=\boldsymbol{I}$ and it follows that

$$\boldsymbol{U}_1^{\mathrm{T}}\boldsymbol{A}\boldsymbol{V}_1=\tilde{\boldsymbol{\Sigma}} \tag{1.3.20}$$

From equation (1.3.17) we also have

$$\boldsymbol{A}\boldsymbol{V}_2=0 \tag{1.3.21}$$

We now choose a matrix $\boldsymbol{U}_2$ so that $\boldsymbol{U}=[\boldsymbol{U}_1 \ \boldsymbol{U}_2]$, where $\boldsymbol{U}_2 \in \mathbb{R}^{m\times(m-r)}$, is orthonormal. Then from equation (1.3.19) and because $\boldsymbol{U}_1 \perp \boldsymbol{U}_2$ we have

$$\boldsymbol{U}_2^{\mathrm{T}}\boldsymbol{U}_1=\boldsymbol{U}_2^{\mathrm{T}}\boldsymbol{A}\boldsymbol{V}_1\tilde{\boldsymbol{\Sigma}}^{-1}=0 \tag{1.3.22}$$

Therefore

$$\boldsymbol{U}_2^{\mathrm{T}}\boldsymbol{A}\boldsymbol{V}_1=0 \tag{1.3.23}$$

Combining equation (1.3.20), (1.3.21) and (1.3.23), we have

$$\boldsymbol{U}^{\mathrm{T}}\boldsymbol{A}\boldsymbol{V}=\begin{bmatrix}\boldsymbol{U}_1^{\mathrm{T}}\boldsymbol{A}\boldsymbol{V}_1 & \boldsymbol{U}_1^{\mathrm{T}}\boldsymbol{A}\boldsymbol{V}_2\\ \boldsymbol{U}_2^{\mathrm{T}}\boldsymbol{A}\boldsymbol{V}_1 & \boldsymbol{U}_2^{\mathrm{T}}\boldsymbol{A}\boldsymbol{V}_2\end{bmatrix}=\begin{bmatrix}\tilde{\boldsymbol{\Sigma}} & 0\\ 0 & 0\end{bmatrix} \tag{1.3.24}$$

The proof can be repeated using an eigendecomposition on the matrix $\boldsymbol{A}\boldsymbol{A}^{\mathrm{T}} \in \mathbb{R}^{m\times m}$ instead of on $\boldsymbol{A}^{\mathrm{T}}\boldsymbol{A}$. In this case, the roles of the orthonormal matrices $\boldsymbol{V}$ and $\boldsymbol{U}$ are interchanged.

The above proof is useful for several reasons:

• It is short and elegant.

• We can also identify which part of the SVD is not unique. Here, we assume that $\boldsymbol{A}^{\mathrm{T}}\boldsymbol{A}$ has norepeated non-zero eigenvalues. Because $\boldsymbol{V}_2$ are the eigenvectors corresponding to the zero eigenvalues of $\boldsymbol{A}^{\mathrm{T}}\boldsymbol{A}$, $\boldsymbol{V}_2$ is not unique when there are repeated zero eigenvalues. This happens when $m<n+1$, (i. e., $\boldsymbol{A}$ is sufficiently short) or when the nullity of $\boldsymbol{A}\geqslant 2$, or a combination of these conditions.

By its construction, the matrix $\boldsymbol{U}\in\mathbb{R}^{m\times m-r}$ is not unique whenever it consists of two or more columns. This happens when $m-2\geqslant r$.

It is left as an exercise to show that similar conclusions on the uniqueness of $\boldsymbol{U}$ and $\boldsymbol{V}$ can be made when the proof is developed using the matrix $\boldsymbol{A}\boldsymbol{A}^{\mathrm{T}}$.

1.3.3 Partitioning the SVD

Here we assume that $\boldsymbol{A}$ has $r\leqslant p$ non-zero singular values (and p-r zero singular values). Later, we see that $r=\mathrm{rank}(\boldsymbol{A})$. For convenience of notation, we arrange the singular values as:

$$\underbrace{\underset{\max}{\sigma_1}\geqslant\cdots\geqslant\underset{\substack{\min\\ \text{non-zero}\\ \text{s. v.}}}{\sigma_r}}_{r\text{ non-zero s. v's}}>\underbrace{\sigma_{r+1}=\cdots=\sigma_p=0}_{p\text{-}r\text{ zero s. v's}}$$

In the remainder of this lecture, we use the SVD partitioned in both $\boldsymbol{U}$ and $\boldsymbol{V}$. We can write the SVD of $\boldsymbol{A}$ in the form

$$\boldsymbol{A}=[\boldsymbol{U}_1 \quad \boldsymbol{U}_2]\begin{bmatrix}\tilde{\boldsymbol{\Sigma}} & 0\\ 0 & 0\end{bmatrix}\begin{bmatrix}\boldsymbol{V}_1^{\mathrm{T}}\\ \boldsymbol{V}_2^{\mathrm{T}}\end{bmatrix} \tag{1.3.25}$$

Where $\tilde{\boldsymbol{\Sigma}} \in \mathbb{R}^{r \times r} = \operatorname{diag}(\sigma_1, \sigma_2, \cdots, \sigma_r)$, and $\boldsymbol{U}$ is partitioned as

$$\boldsymbol{U} = \begin{bmatrix} \underset{r}{\boldsymbol{U}_1} & \underset{m-r}{\boldsymbol{U}_2} \end{bmatrix} \quad m \tag{1.3.26}$$

The columns of $\boldsymbol{U}_1$ are the left singular vectors associated with the r nonzero singular values, and the columns of $\boldsymbol{U}_2$ are the left singular vectors associated with the zero singular values. $\boldsymbol{V}$ is partitioned in an analogous manner:

$$\boldsymbol{U} = \begin{bmatrix} \underset{r}{\boldsymbol{V}_1} & \underset{n-r}{\boldsymbol{V}_2} \end{bmatrix} \quad n \tag{1.3.27}$$

1.3.4 Properties and Interpretations of the SVD

The above partition reveals many interesting properties of the SVD:

1.3.4.1 rank(*A*)=*r*

Using equation (1.3.25), we can write $\boldsymbol{A}$ as

$$\boldsymbol{A} = [\boldsymbol{U}_1 \quad \boldsymbol{U}_2] \begin{bmatrix} \tilde{\boldsymbol{\Sigma}} \boldsymbol{V}_1^{\mathrm{T}} \\ 0 \end{bmatrix} = \boldsymbol{U}_1 \tilde{\boldsymbol{\Sigma}} \boldsymbol{V}_1^{\mathrm{T}} = \boldsymbol{U}_1 \boldsymbol{B} \tag{1.3.28}$$

Where $\boldsymbol{B} \in \boldsymbol{R}^{r \times n} \triangleq \tilde{\boldsymbol{\Sigma}} \boldsymbol{V}_1^{\mathrm{T}}$. From equation (1.3.28) it is clear that the ith, $i = 1, 2, \cdots, n$, r column of $\boldsymbol{A}$ is a linear combination of the columns of $\boldsymbol{U}_1$, whose coefficients are given by the ith column of $\boldsymbol{B}$. But since there are $r \leqslant n$ columns in $\boldsymbol{U}_1$, there can only be r linearly independent columns in $\boldsymbol{A}$. It follows from the definition of rank that rank $(\boldsymbol{A}) = r$.

This point is analogous to the case previously considered, where we saw rank is equal to the number of non-zero eigenvalues, when $\boldsymbol{A}$ is a square symmetric matrix. In this case however, the result applies to any matrix. This is another example of how the SVD is a generalization of the eigendecomposition.

Determination of rank when $\sigma_1, \sigma_2, \cdots, \sigma_r$ are distinctly greater than zero, and when $\sigma_{r+1}, \cdots, \sigma_p$ are exactly zero is easy. But often in practice, due to finite precision arithmetic and fuzzy data, σ_r may be very small, and σ_{r+1} may be not quite zero. Hence, in practice, determination of rank is not so easy. A common method is to declare rank $\boldsymbol{A} = \boldsymbol{r}$ if $\sigma_{r+1} \leqslant \varepsilon$, where ε is a small number specific to the problem considered.

1.3.4.2 $N(A) = R(V_2)$

Recall the nullspace $N(\boldsymbol{A}) = \{\boldsymbol{x} \neq 0 \mid \boldsymbol{A}\boldsymbol{x} = 0\}$. So, we investigate the set $\{\boldsymbol{x}\}$ such that $\boldsymbol{A}\boldsymbol{x} = 0$. Let $\boldsymbol{x} \in \operatorname{span}(\boldsymbol{V}_2)$; i.e., $\boldsymbol{x} = \boldsymbol{V}_2 \boldsymbol{c}$, where $\boldsymbol{c} \in \mathbb{R}^{n-r}$. By substituting equation (1.3.25) for $\boldsymbol{A}$, by noting that $\boldsymbol{V}_1 \perp \boldsymbol{V}_2$ and that $\boldsymbol{V}_1^{\mathrm{T}} \boldsymbol{V}_1 = \boldsymbol{I}$, we have

$$\boldsymbol{A}\boldsymbol{x} = [\boldsymbol{U}_1 \quad \boldsymbol{U}_2] \begin{bmatrix} \tilde{\boldsymbol{\Sigma}} & 0 \\ 0 & 0 \end{bmatrix} \begin{bmatrix} 0 \\ C \end{bmatrix} = 0 \tag{1.3.29}$$

Thus, span $(\boldsymbol{V}_2)$ is at least a subspace of $N(\boldsymbol{A})$. However, if x contains any components of $\boldsymbol{V}_1$, then equation (1.3.29) will not be zero. But since $\boldsymbol{V} = [\boldsymbol{V}_1 \quad \boldsymbol{V}_2]$ is a complete basis in $\mathbb{R}^n$, we see that $\boldsymbol{V}_2$ alone is a basis for the nullspace of $\boldsymbol{A}$.

1.3.4.3　R(A) = R(U_1)

Recall that the definition of range $R(\boldsymbol{A})$ is $\{\boldsymbol{y} \mid \boldsymbol{y}=\boldsymbol{A}\boldsymbol{x},\ \boldsymbol{x}\in\mathbb{R}^n\}$. From equation (1.3.25),

$$\boldsymbol{A}x=[\boldsymbol{U}_1\quad \boldsymbol{U}_2]\begin{bmatrix}\tilde{\boldsymbol{\Sigma}} & 0\\ 0 & 0\end{bmatrix}\begin{bmatrix}\boldsymbol{V}_1^{\mathrm{T}}\\ \boldsymbol{V}_2^{\mathrm{T}}\end{bmatrix}x=[\boldsymbol{U}_1\quad \boldsymbol{U}_2]\begin{bmatrix}\tilde{\boldsymbol{\Sigma}} & 0\\ 0 & 0\end{bmatrix}\begin{bmatrix}d_1\\ d_2\end{bmatrix} \tag{1.3.30}$$

Where

$$\begin{matrix}r\\ n-r\end{matrix}\begin{bmatrix}d_1\\ d_2\end{bmatrix}=\begin{bmatrix}\boldsymbol{V}_1^{\mathrm{T}}\\ \boldsymbol{V}_2^{\mathrm{T}}\end{bmatrix}\boldsymbol{x} \tag{1.3.31}$$

From the above we have

$$\boldsymbol{A}\boldsymbol{x}=[\boldsymbol{U}_1\quad \boldsymbol{U}_2]\begin{bmatrix}\tilde{\boldsymbol{\Sigma}}\boldsymbol{d}_1\\ 0\end{bmatrix}=\boldsymbol{U}(\tilde{\boldsymbol{\Sigma}}\boldsymbol{d}_1) \tag{1.3.32}$$

We see that as $\boldsymbol{x}$ moves throughout $\mathbb{R}^n$, the quantity $\tilde{\boldsymbol{\Sigma}}\boldsymbol{d}_1$ moves throughout $\mathbb{R}^r$. Thus, the quantity $\boldsymbol{y}=\boldsymbol{A}\boldsymbol{x}$ in this context consists of all linear combinations of the columns of $\boldsymbol{U}_1$. Thus, an orthonormal basis for $R(\boldsymbol{A})$ is $\boldsymbol{U}_1$.

1.3.4.4　$R(A^{\mathrm{T}}) = R(V_1)$

Recall that $\boldsymbol{R}(\boldsymbol{A}^{\mathrm{T}})$is the set of all linear combinations of rows of **A**. Our property can be seen using a transposed version of the argument in Section 1.3.4.3 above. Thus, $\boldsymbol{V}_1$ is an orthonormal basis for the rows of **A**.

1.3.4.5　$R(A)_{\perp} = R(U_2)$

From Section 1.3.4.3, we see that $\boldsymbol{R}(\boldsymbol{A})=R(\boldsymbol{U}_1)$. Since from equation (1.3.25), $\boldsymbol{U}_1\perp\boldsymbol{U}_2$, then $\boldsymbol{U}_2$ is a basis for the orthogonal complement of $R(\boldsymbol{A})$. Hence the result.

1.3.4.6　$\|A\|_2=\sigma_1=\sigma_{\max}$

This is easy to see from the definition of the 2-norm and the ellipsoid example of Section 1.3.6.

1.3.4.7　Inverse of A

If the SVD of a square matrix **A** is given, it is easy to find the inverse. Of course, we must assume **A** is full rank, (which means $\sigma_i>0$) for the inverse to exist. The inverse of **A** is given from the svd, using the familiar rules, as

$$\boldsymbol{A}^{-1}=\boldsymbol{V}\boldsymbol{\Sigma}^{-1}\boldsymbol{U}^{\mathrm{T}} \tag{1.3.33}$$

The evaluation of $\boldsymbol{\Sigma}^{-1}$ is easy because $\boldsymbol{\Sigma}$ is square and diagonal. Note that this treatment indicates that the singular values of $\boldsymbol{A}^{-1}$ are $[\sigma_n^{-1}\sigma_{n-1}^{-1},\ \cdots,\ \sigma_1^{-1}]$.

1.3.4.8　The SVD Diagonalizes Any System of Equations

Consider the system of equations $\boldsymbol{A}\boldsymbol{x}=\boldsymbol{b}$, for an arbitrary matrix **A**. Using the SVD of **A**, we have

$$\boldsymbol{U}\boldsymbol{\Sigma}\boldsymbol{V}^{\mathrm{T}}\boldsymbol{x}=\boldsymbol{b} \tag{1.3.34}$$

Let us now represent b in the basis U, and x in the basis **V**, in the same way as in section 3.6. We therefore have

$$\boldsymbol{c}=\begin{matrix} r \\ m-r \end{matrix}\begin{bmatrix} c_1 \\ c_2 \end{bmatrix}=\begin{bmatrix} \boldsymbol{U}_1^{\mathrm{T}} \\ \boldsymbol{U}_2^{\mathrm{T}} \end{bmatrix}\boldsymbol{b} \tag{1.3.35}$$

And

$$\boldsymbol{d}=\begin{matrix} r \\ n-r \end{matrix}\begin{bmatrix} d_1 \\ d_2 \end{bmatrix}=\begin{bmatrix} \boldsymbol{V}_1^{\mathrm{T}} \\ \boldsymbol{V}_2^{\mathrm{T}} \end{bmatrix}\boldsymbol{x} \tag{1.3.36}$$

Substituting the above into equation (1.3.34), the system of equations becomes

$$\boldsymbol{\Sigma d}=\boldsymbol{c} \tag{1.3.37}$$

This shows that as long as we choose the correct bases, any system of equations can become diagonal. This property represents the power of the SVD; it allows us to transform arbitrary algebraic structures into their simplest forms.

If $m>n$ or if rank $r<\min(m, n)$, then the system of equations $\boldsymbol{Ax}=\boldsymbol{b}$ can only be satisfied if $\boldsymbol{b}\in R(\boldsymbol{U}_1)$. To see this, $\boldsymbol{\Sigma}$ above has an $(m-r)\times n$ block of zeros below the diagonal block of nonzero singular values. Thus, the lower $m-r$ elements of left-hand side of equation (1.3.37) are all zero. Then if the equality of equation (1.3.37) is to be satisfied, c_2 must also be zero. This means that $\boldsymbol{U}_2^{\mathrm{T}}\boldsymbol{b}=0$, or that $\boldsymbol{b}\in R(\boldsymbol{U}_1)$.

Further, if $n>m$, or if $r<\min(m, n)$, then, if x_0 is a solution to $\boldsymbol{Ax}=\boldsymbol{b}$, $\boldsymbol{x}+\boldsymbol{V}_2\boldsymbol{z}$ is also a solution, where $\boldsymbol{z}\in\mathbb{R}^{n-r}$. This follows because, as we have seen, $\boldsymbol{V}_2$ is a basis for $N(\boldsymbol{A})$; thus, the component $\boldsymbol{A}\,\boldsymbol{V}_2\boldsymbol{z}=0$, and $\boldsymbol{Ax}_0+\boldsymbol{A}\,\boldsymbol{V}_2\boldsymbol{z}=\boldsymbol{Ax}_0=\boldsymbol{b}$.

1.3.4.9 The "Rotation" Interpretation of the SVD

From the SVD relation $\boldsymbol{A}=\boldsymbol{USV}^{\mathrm{T}}$ we have

$$\boldsymbol{AV}=\boldsymbol{U\Sigma} \tag{1.3.38}$$

Note that since $\boldsymbol{\Sigma}$ is diagonal, the matrix on the right has orthogonal columns, whose 2-norm are equal to the corresponding singular value. We can therefore interpret the matrix $\boldsymbol{V}$ as an orthonormal matrix which rotates the rows of $\boldsymbol{A}$ so that the result is a matrix with orthogonal columns. Likewise, we have

$$\boldsymbol{U}^{\mathrm{T}}=\boldsymbol{\Sigma}\,\boldsymbol{V}^{\mathrm{T}} \tag{1.3.39}$$

The matrix $\boldsymbol{SV}^{\mathrm{T}}$ on the right has orthogonal rows with 2-norm equal to the corresponding singular value. Thus, the orthonormal matrix $\boldsymbol{U}^{\mathrm{T}}$ operates (rotates) the columns of $\boldsymbol{A}$ to produce a matrix with orthogonal rows.

In the case where $m > n$ ($\boldsymbol{A}$ is tall), then the matrix $\boldsymbol{\Sigma}$ is also tall, with zeros in the bottom $m-n$ rows. Then, only the first n columns of $\boldsymbol{U}$ are relevant in equation (1.3.38), and only the first n rows of $\boldsymbol{U}^{\mathrm{T}}$ are relevant in equation (1.3.39). When $m<n$, a corresponding transposed statement replacing $\boldsymbol{U}$ with $\boldsymbol{V}$ can be made.

1.3.5 Relationship between SVD and ED

It is clear that the eigendecomposition and the singular value decomposition share many properties in common. The price we pay for being able to perform a diagonal decomposition on an arbitrary matrix is that we need two orthonormal matrices instead of just one, as is the case for square symmetric matrices. In this section, we explore further relationships between the ED and the SVD.

Using equation (1. 3. 25), we can write

$$\boldsymbol{A}^{\mathrm{T}}\boldsymbol{A}=\boldsymbol{V}\begin{bmatrix}\tilde{\boldsymbol{\Sigma}} & 0\\ 0 & 0\end{bmatrix}\boldsymbol{U}^{\mathrm{T}}\boldsymbol{U}\begin{bmatrix}\tilde{\boldsymbol{\Sigma}} & 0\\ 0 & 0\end{bmatrix}\boldsymbol{V}^{\mathrm{T}}=\boldsymbol{V}\begin{bmatrix}\tilde{\boldsymbol{\Sigma}}^2 & 0\\ 0 & 0\end{bmatrix}\boldsymbol{V}^{\mathrm{T}} \tag{1.3.40}$$

Thus it is apparent, that the eigenvectors $\boldsymbol{V}$ of the matrix $\boldsymbol{A}^{\mathrm{T}}\boldsymbol{A}$ are the right singular vectors of $\boldsymbol{A}$, and that the singular values of $\boldsymbol{A}$ squared are the corresponding nonzero eigenvalues. Note that if $\boldsymbol{A}$ is short $(m<n)$ and full rank, the matrix $\boldsymbol{A}^{\mathrm{T}}\boldsymbol{A}$ will contain $n-m$ additional zero eigenvalues that are not included as singular values of $\boldsymbol{A}$. This follows because the rank of the matrix $\boldsymbol{A}^{\mathrm{T}}\boldsymbol{A}$ is m when A is full rank, yet the size of $\boldsymbol{A}^{\mathrm{T}}\boldsymbol{A}$ is $n\times n$.

As discussed in Golub and van Loan, the SVD is numerically more stable to compute than the ED. However, in the case where $n\gg m$, the matrix $\boldsymbol{V}$ of the SVD of $\boldsymbol{A}$ becomes large, which means the SVD on $\boldsymbol{A}$ becomes more costly to compute, relative to the eigendecomposition of $\boldsymbol{A}^{\mathrm{T}}\boldsymbol{A}$.

Further, we can also say, using the form $\boldsymbol{A}\boldsymbol{A}^{\mathrm{T}}$, that

$$\boldsymbol{A}\boldsymbol{A}^{\mathrm{T}}=\boldsymbol{U}\begin{bmatrix}\tilde{\boldsymbol{\Sigma}} & 0\\ 0 & 0\end{bmatrix}\boldsymbol{V}^{\mathrm{T}}\boldsymbol{V}\begin{bmatrix}\tilde{\boldsymbol{\Sigma}} & 0\\ 0 & 0\end{bmatrix}\boldsymbol{U}^{\mathrm{T}}=\boldsymbol{U}\begin{bmatrix}\tilde{\boldsymbol{\Sigma}}^2 & 0\\ 0 & 0\end{bmatrix}\boldsymbol{U}^{\mathrm{T}} \tag{1.3.41}$$

which indicates that the eigenvectors of $\boldsymbol{A}\boldsymbol{A}^{\mathrm{T}}$ are the left singular vectors $\boldsymbol{U}$ of $\boldsymbol{A}$, and the singular values of $\boldsymbol{A}$ squared are the nonzero eigenvalues of $\boldsymbol{A}\boldsymbol{A}^{\mathrm{T}}$. Notice that in this case, if $\boldsymbol{A}$ is tall and full rank, the matrix $\boldsymbol{A}\boldsymbol{A}^{\mathrm{T}}$ will contain $m\ n$ additional zero eigenvalues that are not included as singular values of $\boldsymbol{A}$.

We now compare the fundamental defining relationships for the ED and the SVD:

For the ED, if $\boldsymbol{A}$ is symmetric, we have:

$$\boldsymbol{A}=\boldsymbol{Q}\boldsymbol{\Lambda}\boldsymbol{Q}^{\mathrm{T}}\rightarrow\boldsymbol{A}\boldsymbol{Q}=\boldsymbol{Q}\boldsymbol{\Lambda}$$

Where $\boldsymbol{Q}$ is the matrix of eigenvectors, and $\boldsymbol{\Lambda}$ is the diagonal matrix of eigenvalues. Writing this relation column-by-column, we have the familiar eigenvector/eigenvalue relationship:

$$\boldsymbol{A}q_i=\lambda_i q_i \qquad i=1,2,\cdots,n \tag{1.3.42}$$

For the SVD, we have

$$\boldsymbol{A}=\boldsymbol{U}\boldsymbol{\Sigma}\boldsymbol{V}^{\mathrm{T}}\rightarrow\boldsymbol{A}\boldsymbol{V}=\boldsymbol{U}\boldsymbol{\Sigma}$$

or

$$\boldsymbol{A}\boldsymbol{u}_i=\boldsymbol{\sigma}_i\boldsymbol{u}_i \qquad i=1,2,\cdots,p \tag{1.3.43}$$

Where $p=\min(m, n)$. Also, since $\boldsymbol{A}^{\mathrm{T}}=\boldsymbol{V}\boldsymbol{\Sigma}\boldsymbol{U}^{\mathrm{T}}\rightarrow\boldsymbol{A}^{\mathrm{T}}\boldsymbol{U}=\boldsymbol{V}\boldsymbol{\Sigma}$, we have

$$\boldsymbol{A}^{\mathrm{T}}\boldsymbol{u}_i=\boldsymbol{\sigma}_i\boldsymbol{v}_i \qquad i=1,2,\cdots,p \tag{1.3.44}$$

Thus, by comparing equation (1. 3. 42), (1. 3. 43), and (1. 3. 44), we see the singular vectors and singular values obey a relation which is similar to that which defines the eigenvectors and eigenvalues. However, we note that in the SVD case, the fundamental relationship expresses left singular values in terms of right singular values, and vice-versa, whereas the eigenvectors are expressed in terms of themselves.

Exercise

Compare the ED and the SVD on a square symmetric matrix, when i) $\boldsymbol{A}$ is positive definite, and ii) when $\boldsymbol{A}$ has some positive and some negative eigenvalues.

1.3.6 Ellipsoidal Interpretation of the SVD

The singular values of $\boldsymbol{A}$, where $\boldsymbol{A} \in \mathbb{R}^{m\times n}$ are the lengths of the semi-axes of the hyperellipsoid $\boldsymbol{E}$ given by:

$$\boldsymbol{E}=\{\boldsymbol{y} \mid \boldsymbol{y}=\boldsymbol{A}\boldsymbol{x},\ \|\boldsymbol{x}\|_2=1\}$$

That is, $\boldsymbol{E}$ is the set of points mapped out as $\mathbf{x}$ takes on all possible values such that $\|\boldsymbol{x}\|_2=1$, as shown in Fig. 1. 5. To appreciate this point, let us look at the set of $\boldsymbol{y}$ corresponding to $\{\boldsymbol{x} \| \boldsymbol{x}\|_2=1\}$. We take

$$\boldsymbol{y}=\boldsymbol{A}\boldsymbol{x}=\boldsymbol{U}\boldsymbol{\Sigma}\boldsymbol{V}^{\mathrm{T}}\boldsymbol{x} \tag{1.3.45}$$

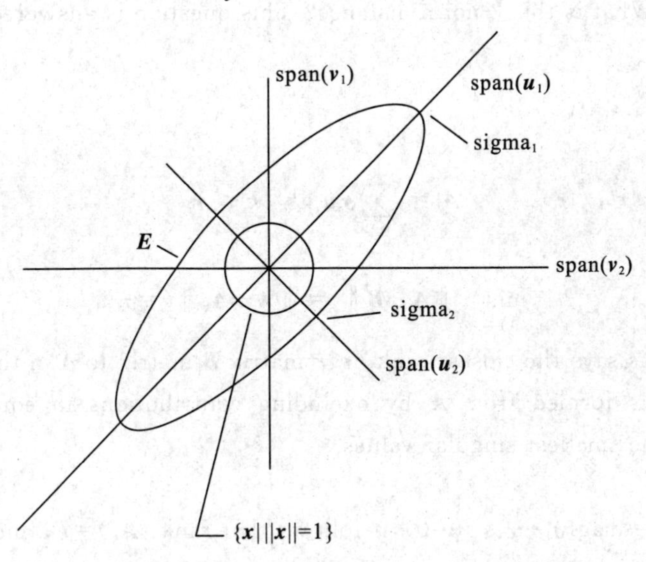

Fig. 1. 5 The ellipsoidal interpretation of the SVD

Let us change bases for both $\boldsymbol{x}$ and $\boldsymbol{y}$. Define

$$\boldsymbol{c}=\boldsymbol{U}^{\mathrm{T}}\boldsymbol{y}$$

$$\boldsymbol{d}=\boldsymbol{V}^{\mathrm{T}}\boldsymbol{x} \tag{1.3.46}$$

Then equation (1. 3. 45) becomes

$$\boldsymbol{c}=\boldsymbol{\Sigma}\boldsymbol{d} \tag{1.3.47}$$

The locus of points $\boldsymbol{E}=\{\boldsymbol{y} \mid \boldsymbol{y}=\boldsymbol{A}\boldsymbol{x},\ \|\boldsymbol{x}\|_2=1\}$ defines an ellipse. The principal axes of the ellipse are aligned along the left singular vectors u_i, with lengths equal to the corresponding singular value.

We note that $\|\boldsymbol{d}\|_2=1$ if $\|\boldsymbol{x}\|_2=1$. Thus, our problem is transformed into observing the set $\{c\}$ corresponding to the set $\{\boldsymbol{d} \mid \|\boldsymbol{d}\|_2 = 1\}$. The set $\{c\}$ can be determined by evaluating 2-norm on each side of equation (1.3.47):

$$\sum_{i=1}^{p}\left(\frac{c_i}{\sigma_i}\right)^2 = \sum_{i=1}^{p}(\boldsymbol{d}_i)^2 = 1 \tag{1.3.48}$$

We see that the set $\{c\}$ defined by equation (1.3.48) is indeed the canonical form of an ellipse in the basis $\boldsymbol{U}$. Thus, the principal axes of the ellipse are aligned along the columns u_i of $\boldsymbol{U}$, with lengths equal to the corresponding singular value σ_i. This interpretation of the SVD is useful later in our study of condition numbers.

1.3.7 An Interesting Theorem

First, we realize that the SVD of $\boldsymbol{A}$ provides a "sum of outer-products" representation:

$$\boldsymbol{A} = \boldsymbol{U}\boldsymbol{\Sigma}\boldsymbol{V}^{\mathrm{T}} = \sum_{\mathrm{i}}^{\mathrm{p}}\sigma_i u_i \boldsymbol{v}_i^{\mathrm{T}} \quad p = \min(m, n) \tag{1.3.49}$$

Given $\boldsymbol{A} \in \mathbb{R}^{m\times n}$ with rank r, then what is the matrix $\boldsymbol{B} \in \mathbb{R}^{m\times n}$ with rank $k < r$ closest to $\boldsymbol{A}$ in 2-norm. What is this 2-norm distance? This question is answered in the following theorem.

Theorem 2

Define

$$\boldsymbol{A}_k = \sum_{\mathrm{i}=1}^{\mathrm{k}}\boldsymbol{\sigma}_i \boldsymbol{u}_i \boldsymbol{v}_i^{\mathrm{T}} \quad k \leqslant r \tag{1.3.50}$$

then

$$\min_{\mathrm{rank}(\boldsymbol{B}) = k} \|\boldsymbol{A} - \boldsymbol{B}\|_2 = \|\boldsymbol{A} - \boldsymbol{A}_k\|_2 = \sigma_{k+1}$$

In words, this says the closest rank $k < r$ matrix $\boldsymbol{B}$ matrix to $\boldsymbol{A}$ in the 2-norm sense is given by $\boldsymbol{A}_k$. $\boldsymbol{A}_k$ is formed from $\boldsymbol{A}$ by excluding contributions in equation (1.3.49) associated with the smallest singular values.

Proof

Since $\boldsymbol{U}^{\mathrm{T}}\boldsymbol{A}_k\boldsymbol{V} = \mathrm{diag}(\sigma_1 \cdots \sigma_k, 0 \cdots 0)$, it follows that rank $(\boldsymbol{A}_k) = k$, and that

$$\begin{aligned}\|\boldsymbol{A} - \boldsymbol{A}_k\|_2 &= \|\boldsymbol{U}^{\mathrm{T}}(\boldsymbol{A} - \boldsymbol{A}_k)\boldsymbol{V}\|_2 \\ &= \|\mathrm{diag}(0 \cdots 0, \sigma_{k+1} \cdots \sigma_r, 0 \cdots 0)\|_2 \\ &= \sigma_{k+1}\end{aligned} \tag{1.3.51}$$

where the first line follows from the fact the 2-norm of a matrix is invariant to pre-and post-multiplication by an orthonormal matrix. Further, it may be shown that, for any matrix $\boldsymbol{B} \in \mathbb{R}^{m\times n}$ of rank $k < r$,

$$\|\boldsymbol{A} - \boldsymbol{B}\|_2 \geqslant \sigma_{k+1} \tag{1.3.52}$$

Comparing equation (1.3.51) and (1.3.52), we see the closest rank k matrix to $\boldsymbol{A}$ is $\boldsymbol{A}_k$ given by equation (1.3.50). This result is very useful when we wish to approximate a matrix by another of lower rank. For example, let us look at the Karhunen-Loeve expansion as discussed in Section 1.1. For a sample $\boldsymbol{x}_n$ of a random process $\boldsymbol{x} \in \mathbb{R}^m$, we express $\boldsymbol{x}$ as

$$\boldsymbol{x}_i = \boldsymbol{V}\theta_i \tag{1.3.53}$$

Where the columns of $\boldsymbol{V}$ are the eigenvectors of the covariance matrix $\boldsymbol{R}$. We saw in Section 1.2.7, that we may represent $\boldsymbol{x}_i$ with relatively few coefficients by setting the elements of θ associated with the smallest eigenvalues of $\boldsymbol{R}$ to zero. The idea was that the resulting distortion in $\boldsymbol{x}$ would have minimum energy.

This fact may now be seen in a different light with the aid of this theorem. Suppose we retain the $j=r$ elements of a given θ associated with the largest r eigenvalues. Let $\tilde{\theta} \triangleq [\theta_1, \theta_2, \cdots, \theta_r, 0, \cdots, 0]^{\mathrm{T}}$ and $\tilde{\boldsymbol{x}} = \boldsymbol{V}\theta$. Then

$$\tilde{\boldsymbol{R}} = \boldsymbol{E}(\tilde{\boldsymbol{x}}\,\tilde{\boldsymbol{x}}^{\mathrm{T}}) = \boldsymbol{E}(\boldsymbol{V}\tilde{\theta}\tilde{\theta}\boldsymbol{V})$$

$$= \boldsymbol{V}\begin{bmatrix} \boldsymbol{E}\,|\theta_1|^2 & & & & & \\ & \ddots & & & & \\ & & \boldsymbol{E}\,|\theta_r|^2 & & & \\ & & & 0 & & \\ & & & & \ddots & \\ & & & & & 0 \end{bmatrix}\boldsymbol{V}^{\mathrm{T}}$$

$$= \boldsymbol{V}\tilde{\boldsymbol{\Lambda}}\boldsymbol{V}^{\mathrm{T}} \tag{1.3.54}$$

Where $\tilde{\boldsymbol{\Lambda}} = \operatorname{diag}[\lambda_1, \lambda_2, \cdots, \lambda_r, 0, \cdots, 0]$. Since $\tilde{\boldsymbol{R}}$ is positive definite, square and symmetric, its eigendecomposition and singular value decomposition are identical; hence, $\lambda_i = \sigma_i$, $i = 1, 2, \cdots, r$. Thus from this theorem, and quation (1.3.54), we know that the covariance matrix $\tilde{\boldsymbol{R}}$ formed from truncating the K-L coefficients is the closest rank r matrix to the true covariance matrix $\boldsymbol{R}$ in the 2-norm sense.

1.4 The Quadratic Form

1.4.1 Quadratic Form Theory

We introduce the quadratic form by considering the idea of positive definiteness. A square matrix $\boldsymbol{A} \in \mathbb{R}^{n\times n}$ is positive definite if and only if for any

$$\boldsymbol{x}^{\mathrm{T}}\boldsymbol{A}\boldsymbol{x} > 0, \quad 0 \neq \boldsymbol{x} \in \mathbb{R}^n \tag{1.4.1}$$

The matrix $\boldsymbol{A}$ is positive semi-definite if and only if for any $0 \neq \boldsymbol{x}$ we have

$$\boldsymbol{x}^{\mathrm{T}}\boldsymbol{A}\boldsymbol{x} \geqslant 0 \tag{1.4.2}$$

It is only the symmetric part of $\boldsymbol{A}$ which is relevant in a quadratic form. This may be seen as follows, the symmetric part $\boldsymbol{T}$ of $\boldsymbol{A}$ is defined as $\mathbf{T}=\frac{1}{2}[\mathbf{A}+\mathbf{A}^{\mathrm{T}}]$, whereas the asymmetric part $\boldsymbol{S}$ of $\boldsymbol{A}$ is defined as $\boldsymbol{S}=\frac{1}{2}[\boldsymbol{A}-\boldsymbol{A}^{\mathrm{T}}]$. Then $\boldsymbol{A}=\boldsymbol{T}+\boldsymbol{S}$. It may be verified by direct multiplication that the quadratic form can also be expressed in the form

$$\boldsymbol{x}^{\mathrm{T}}\boldsymbol{a}\boldsymbol{x}=\sum_{i=1}^{n}\sum_{j=1}^{n}a_{ij}x_{i}x_{j} \tag{1.4.3}$$

Because $\boldsymbol{S}^{\mathrm{T}}=-\boldsymbol{S}$, the (i, j) term corresponding to the asymmetric part of exactly cancels that corresponding to the (j, i)-th term. Further, the terms corresponding to $i=j$ are zero for the asymmetric part. Thus the part of the quadratic form corresponding to the asymmetric part $\boldsymbol{S}$ is zero. Therefore, when considering quadratic forms, it suffices to consider only the symmetric part $\boldsymbol{T}$ of a matrix. Quadratic forms on positive definite matrices are used very frequently in least-squares and adaptive filtering applications.

Theorem 1

A matrix $\boldsymbol{A}$ is positive definite if and only if all eigenvalues of the symmetric part of $\boldsymbol{A}$ are positive.

Proof

Since only the symmetric part of $\boldsymbol{A}$ is relevant the quadratic form on $\boldsymbol{A}$ may be expressed as $\boldsymbol{x}^{\mathrm{T}}\boldsymbol{A}\boldsymbol{x}=\boldsymbol{x}^{\mathrm{T}}\boldsymbol{V}\boldsymbol{\Lambda}\ \boldsymbol{V}^{\mathrm{T}}\boldsymbol{x}$ where an eigendecomposition has been performed on the symmetric part of $\boldsymbol{A}$ Let us define $\boldsymbol{z}=\boldsymbol{V}^{\mathrm{T}}\boldsymbol{x}$. Thus we have

$$\boldsymbol{x}^{\mathrm{T}}\boldsymbol{A}\boldsymbol{x}=\boldsymbol{z}^{\mathrm{T}}\boldsymbol{\Lambda}\boldsymbol{z}=\sum_{i=1}^{n}z_{i}^{2}\lambda_{\mathrm{i}} \tag{1.4.4}$$

Thus equation (1.4.4) is greater than zero for arbitrary $\boldsymbol{x}$ if and only if $\lambda_i>0$, $i=1, 2, \cdots, n$. From equation (1.4.4), it is easy to verify that the equation $k=\sum_{i=1}^{n}z_i^2\lambda$, where k is a constant, defines a multi-dimensional ellipse where $\sqrt{\frac{k}{\lambda_i}}$ is the length of the i-th principal axis. Since $\boldsymbol{z}=\boldsymbol{V}^{\mathrm{T}}\boldsymbol{x}$ where $\boldsymbol{V}$ is orthonormal, $\boldsymbol{z}$ is a rotation transformation on $\boldsymbol{x}$, and the equation $k=\boldsymbol{x}^{\mathrm{T}}\boldsymbol{A}\boldsymbol{x}$ is a rotated version of equation (1.4.4). Thus $k=\boldsymbol{x}^{\mathrm{T}}\boldsymbol{A}\boldsymbol{x}$ is also an ellipse with principal axes given by $\sqrt{\frac{k}{\lambda_i}}$. In this case, the i-th principal axes of the ellipse lines up along the i-th eigenvector $\boldsymbol{v}_i$ of $\boldsymbol{A}$.

Positive definiteness of $\boldsymbol{A}$ in the quadratic form $\boldsymbol{x}^{\mathrm{T}}\boldsymbol{A}\boldsymbol{x}$ is the matrix analog to the scalar a being positive in the scalar expression $\boldsymbol{a}\boldsymbol{x}^2$. The scalar equation $\boldsymbol{y}=\boldsymbol{a}\boldsymbol{x}^2$ is a parabola which faces upwards if a is positive. Likewise, $\boldsymbol{y}=\boldsymbol{x}^{\mathrm{T}}\boldsymbol{A}\boldsymbol{x}$ is a multi-dimensional parabola which faces upwards in all directions if $\boldsymbol{A}$ is positive definite.

Example:

We now discuss an example to illustrate the above discussion. $\boldsymbol{A}$ three-dimensional plot of $\boldsymbol{y}=\boldsymbol{x}^{\mathrm{T}}\boldsymbol{A}\boldsymbol{x}$ is shown plotted in Fig. 1.6 for $\boldsymbol{A}$ given by

$$A=\begin{bmatrix}2 & 1\\ 1 & 2\end{bmatrix} \tag{1.4.5}$$

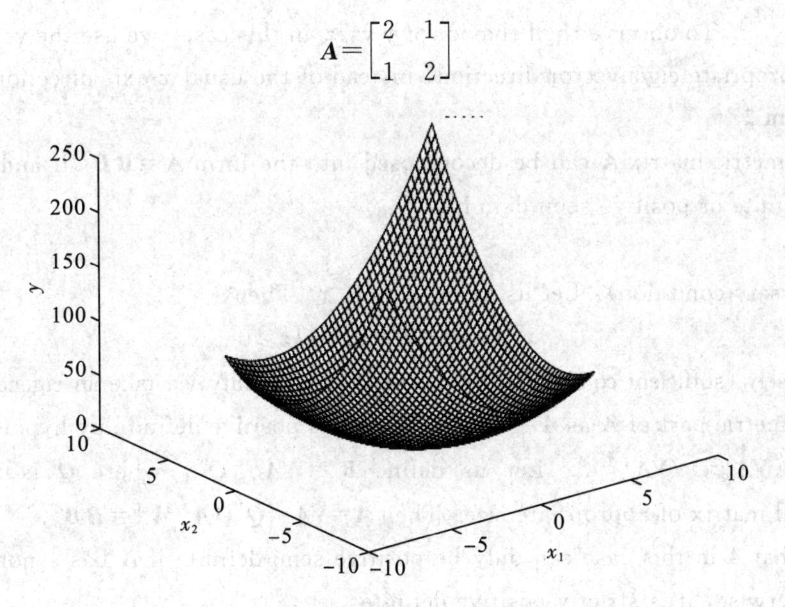

Fig. 1.6 Three-dimensional plot of quadratic form

The corresponding contour plot is plotted in Fig. 1.7 Note that this curve is elliptical in cross section in a plane as discussed above. It may be readily verified that the eigenvalues of $\boldsymbol{A}$ are 3, 1 with corresponding eigenvectors $[1, 1]^{\mathrm{T}}$ and $[1, -1]^{\mathrm{T}}$. For $y=k=1$ the lengths of the principal axes of the ellipse are then $\frac{1}{\sqrt{3}}$ and 1. It is seen from the figure these principal axes are indeed the lengths indicated and are lined up along the directions of the eigenvectors as required.

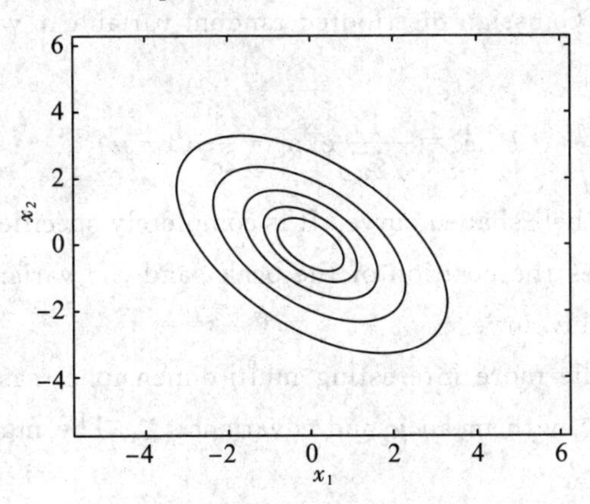

Fig. 1.7 Contour plot Innermost contour corresponds to $k=1$

We write the ellipse in the form

$$\boldsymbol{y}=\boldsymbol{x}^{\mathrm{T}}\boldsymbol{A}\boldsymbol{x}=\boldsymbol{z}^{\mathrm{T}}\boldsymbol{\Lambda}\boldsymbol{z}=\sum_{i=1}^{n} z_i^2\lambda_i \tag{1.4.6}$$

Where $\boldsymbol{z}=\boldsymbol{V}\boldsymbol{x}$ as before. It is seen from Fig. 1.7 that since $\boldsymbol{A}$ is positive definite, the curve defined by $\boldsymbol{y}=\boldsymbol{z}_i^2\lambda$, for all $\boldsymbol{z}_k$, $k\neq i$ held constant, is an upward-facing parabola for all

$i = 1,2, \cdots, n$. To observe the behavior of $\boldsymbol{y}$ vs. $\boldsymbol{z}_i$ in this case, we use the vertical axis $\boldsymbol{y}$, and the appropriate eigenvector direction, instead of the usual x-axis direction).

Theorem 2

A symmetric matrix $\boldsymbol{A}$ can be decomposed into the form $\boldsymbol{A}=\boldsymbol{B}\boldsymbol{B}^{\mathrm{T}}$ if and only if $\boldsymbol{A}$ is positive definite or positive semi-definite.

Proof

(Necessary condition): Let us define $\boldsymbol{z}$ as $\boldsymbol{B}^{\mathrm{T}}\boldsymbol{x}$. Then

$$\boldsymbol{x}^{\mathrm{T}}\boldsymbol{A}\boldsymbol{x} = \boldsymbol{x}^{\mathrm{T}}\boldsymbol{B}\boldsymbol{B}^{\mathrm{T}}\boldsymbol{x} = \boldsymbol{z}^{\mathrm{T}}\boldsymbol{z} \geqslant 0 \tag{1.4.7}$$

Conversely (sufficient condition) without loss of generality we take an eigendecomposition on the symmetric part of $\boldsymbol{A}$ as $\boldsymbol{A}=\boldsymbol{V}^{\mathrm{T}}\boldsymbol{\Lambda}\boldsymbol{V}$. Since $\boldsymbol{A}$ is positive definite by hypothesis we can write $\boldsymbol{A} = (\boldsymbol{V}\boldsymbol{\Lambda}^{1/2})(\boldsymbol{V}\boldsymbol{\Lambda}^{1/2})^{\mathrm{T}}$. Let us define $\boldsymbol{B} = \boldsymbol{V}\boldsymbol{\Lambda}^{1/2}\boldsymbol{Q}^{\mathrm{T}}$, where $\boldsymbol{Q}$ is an arbitrary orthonormal matrix of appropriate size. Then $\boldsymbol{A}=\boldsymbol{V}\boldsymbol{\Lambda}^{1/2}\boldsymbol{Q}^{\mathrm{T}}\boldsymbol{Q}\boldsymbol{\Lambda}^{1/2}\boldsymbol{V}^{\mathrm{T}}=\boldsymbol{B}\boldsymbol{B}^{\mathrm{T}}$.

Note that $\boldsymbol{A}$ in this case can only be positive semi-definite if $\boldsymbol{A}$ has a non-empty null space. Otherwise, it is strictly positive definite.

1.4.2 The Gaussian Multi-Variate Probability Density Function

Here, we very briefly introduce this topic so we can use this material for an example of the application of the Cholesky decomposition, and also in least squares analysis to follow shortly later. This topic is a good application of quadratic forms. More details are provided in several books.

First we consider the univariate case of the Gaussian probability distribution function pdf. The pdf $p(x)$ of a Gaussian distributed random variable x with mean μ and variance σ^2 is given a

$$P(x)=\frac{1}{\sqrt{2\pi}\sigma}\exp\left[-\frac{1}{2\sigma^2}(\boldsymbol{x}-\boldsymbol{\mu})^2\right] \tag{1.4.8}$$

This is the familiar bell-shaped curve. It is completely specified by two parameters the mean $\boldsymbol{\mu}$ which determines the position of the peak, and the variance $\boldsymbol{\sigma}^2$ which determines the width or spread of the curve.

We now consider the more interesting multi-dimensional case. Consider a Gaussian distributed vector $\boldsymbol{x}\in\mathbb{R}^n$ with mean $\boldsymbol{\mu}$ and covariance $\boldsymbol{\Sigma}$. The multivariate pdf describing the variation of $\boldsymbol{x}$ is

$$p(x)=(2\pi)^{-\frac{n}{2}}|\boldsymbol{\Sigma}|^{-\frac{1}{2}}\exp\left[-\frac{1}{2}(\boldsymbol{x}-\boldsymbol{\mu})\boldsymbol{\Sigma}^{-1}(\boldsymbol{x}-\boldsymbol{\mu})\right] \tag{1.4.9}$$

We can see that the multi-variate case collapses to the univariate case when the number of variables becomes one. A plot of $p(\boldsymbol{x})$ vs. $\boldsymbol{x}$ is shown in Fig. 1.8, for $\boldsymbol{\mu}=0$ and $\boldsymbol{\Sigma}$ and defined as

$$\boldsymbol{\Sigma}=\begin{bmatrix}2 & 1\\1 & 2\end{bmatrix} \tag{1.4.10}$$

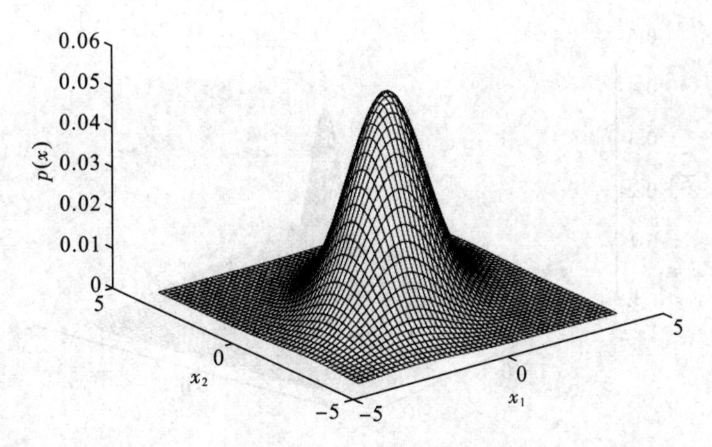

Fig. 1. 8 A Gaussian probability density function

Because the exponent in equation (1. 4. 9) is a quadratic form, the set of points satisfied by the equation$\left[\frac{1}{2}(\boldsymbol{x}-\boldsymbol{\mu})\boldsymbol{\Sigma}^{-1}(\boldsymbol{x}-\boldsymbol{\mu})\right]=k$ where k is a constant, is an ellipse. Therefore this ellipse defines a contour of equal probability density. The interior of this ellipse defines a region into which an observation will fall with a specified probability α which is dependent on k. This probability level α is given as

$$\alpha=\int_R(2\pi)^{\frac{n}{2}}\mid\boldsymbol{\Sigma}\mid^{-\frac{1}{2}}\exp\left[-\frac{1}{2}(\boldsymbol{x}-\boldsymbol{\mu})\boldsymbol{\Sigma}-1(\boldsymbol{x}-\boldsymbol{\mu})\right]\mathrm{d}x \tag{1.4.11}$$

Where R is the interior of the ellipse. Stated another way, an ellipse is the region in which any observation governed by the probability distribution (1. 4. 9) will fall with a specified probability level α. As k increases the ellipse gets larger and α increases. These ellipses are referred to as joint confidence regions at probability level α.

The covariance matrix $\boldsymbol{\Sigma}$ controls the shape of the ellipse. Because the quadratic form in this case involves $\boldsymbol{\Sigma}^{-1}$, the length of the i-th principal axis is $\sqrt{2k\lambda_i}$ instead of $\sqrt{2k/\lambda_i}$ as it would be if the quadratic form were in $\boldsymbol{\Sigma}$. Therefore as the eigenvalues of $\boldsymbol{\Sigma}$ increase, the size of the joint confidence regions increase (i. e., the spread of the distribution increases) for a given value of k. Now suppose we let $\boldsymbol{\Sigma}$ poorly conditioned in such a way that the variance (main diagonal elements of $\boldsymbol{\Sigma}$) remain constant. Then the ratio of the largest to smallest principal axes become large, and the ellipse becomes elongated. In this case, the pdf takes on more of the shape shown in Fig. 1. 9, which shows a multi-variate Gaussian pdf for $\mu=0$ for a relatively poorly conditioned given $\boldsymbol{\Sigma}$ as

$$\boldsymbol{\Sigma}=\begin{bmatrix}2 & 1.9\\ 1.9 & 2\end{bmatrix} \tag{1.4.12}$$

Here, because the ellipse describing the joint confidence region is elongated, we see that if one of the variables is known, the distribution of the other variable becomes more concentrated around the value of the first; i. e., knowledge of one variable tells us relatively more about the other. This implies the variables are more highly correlated with one another. But we have seen previously that if the variables in a vector random process

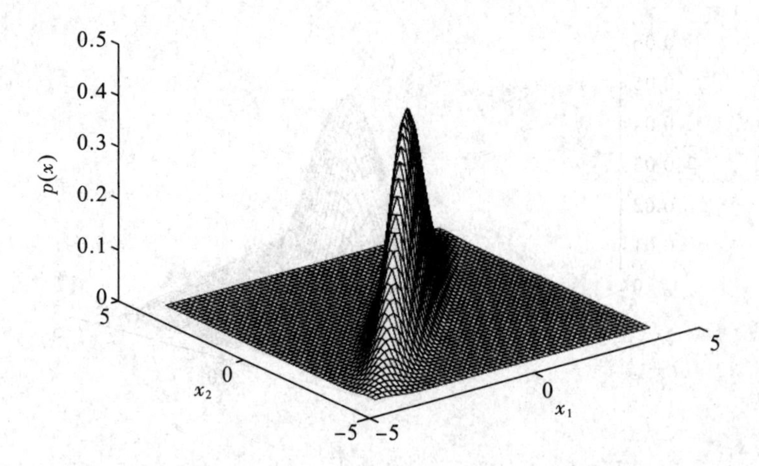

Fig. 1. 9　A Gaussian pdf with a more poorly conditioned covariance matrix

are highly correlated, then the off-diagonal elements of the covariance matrix become larger, which leads to their eigenvalues becoming more disparate; i. e., the condition number of the covariance matrix becomes worse. It is precisely this poorer condition number that causes the ellipse in Fig. 1. 9 to become elongated.

With this discussion, we now have gone full circle: a highly correlated system has large off-diagonal elements in its covariance matrix. This leads to a poorly conditioned covariance matrix. But a Gaussian distributed process with a poorly conditioned covariance matrix has a joint confidence region that is elongated. In turn an elongated joint confidence region means the system is highly correlated which takes us back to the beginning.

Understanding these relationships is a key element in the signal processing rigor.

1.4.3 The Rayleigh Quotient

The Rayleigh quotient is a simple mathematical structure that has a great deal of interesting uses. The Rayleigh quotient $r(\boldsymbol{x})$ is defined as

$$r(\boldsymbol{x})=\frac{\boldsymbol{X}^{\mathrm{T}}\boldsymbol{A}\boldsymbol{x}}{\boldsymbol{x}^{\mathrm{T}}\boldsymbol{x}} \tag{1. 4. 13}$$

It is easily verified that if $\boldsymbol{x}$ is the i-th eigenvector $\boldsymbol{v}_i$ of $\boldsymbol{A}$ (not necessarily normalized to unit norm), then $r(\boldsymbol{x})=\lambda_i$:

$$\frac{\boldsymbol{v}_i^{\mathrm{T}}\boldsymbol{A}\,\boldsymbol{v}_i}{\boldsymbol{v}_i^{\mathrm{T}}\,\boldsymbol{v}_i}=\frac{\lambda_i\,\boldsymbol{v}^{\mathrm{T}}\boldsymbol{v}}{\boldsymbol{v}^{\mathrm{T}}\boldsymbol{v}}=\lambda_i \tag{1. 4. 14}$$

In fact, it is easily shown by differentiating $r(\boldsymbol{x})$ with respect to $\boldsymbol{x}$, that $\boldsymbol{x}=\boldsymbol{v}_i$ is a stationary point of $r(\boldsymbol{x})$.

Further along this line of reasoning, let us define a subspace $\boldsymbol{S}_k$ as $\boldsymbol{S}_k=[\boldsymbol{v}_1,\boldsymbol{v}_2\cdots,\boldsymbol{v}_k]$, $k=1,2,\cdots,n$, where $\boldsymbol{v}_i$ is the i-th eigenvector of $\boldsymbol{A}\in\mathbb{R}^{n\times n}$, where $\boldsymbol{A}$ is symmetric. Then, a variation of the Courant Fischer minimax theorem says that

$$\lambda_k=\min_{0\neq\boldsymbol{x}\in S_k}\frac{\boldsymbol{x}^{\mathrm{T}}\boldsymbol{A}\boldsymbol{x}}{\boldsymbol{X}^{\mathrm{T}}\boldsymbol{x}} \tag{1. 4. 15}$$

Question

It is easily shown by differentiation that for $r(\boldsymbol{x})=\lambda_i$ minimizes $\|(\boldsymbol{A}=\lambda_i \boldsymbol{I})\boldsymbol{x}\|_2$. The perturbation theory of Golub and Van Loan says that if $\boldsymbol{x}$ in equation (1.4.13) is a good approximation to an eigenvector, then $r(\boldsymbol{x})$ is a good approximation to the corresponding eigenvalue, and vice versa. Starting with an initial estimate $\boldsymbol{x}_0$ with unit 2-norm, suggest an iteration using equation (1.4.13) which gives an improved estimate of the eigenvector. How can the eigenvalue be found?

This technique is referred to as the Rayleigh quotient iteration for computing an eigenvalue and eigenvector. In fact, this iteration is remarkably effective; it can be shown to have cubic convergence.

Chapter 2 The Solution of Least Squares Problems

In this section, we discuss the idea of linear least squares estimation of parameters. Least-squares (LS) analysis is the fundamental concept in adaptive systems linear prediction/signal encoding system identification and many other applications. The solution of least squares problems is very interesting from an algebraic perspective and also has several interesting statistical properties.

In thissection, we look at several applications of least squares, and then go on to develop the so-called normal equations for solving the LS problem. We then discuss several statistical properties of the LS solution and look at its performance in the presence of white and coloured noise.

In futuresection, we look at ways of solving the LS problem more efficiently, and deal with the case where the matrices involved are rank deficient.

2.1 Linear Least Squares Estimation

2.1.1 Example: Autoregressive Modelling

An autoregressive (AR) process is a random process which is the output of an all pole-filter when excited by white noise. The reason for this terminology is made apparent later. In this example, we deal in discrete time. An all-pole filter has a transfer function $\boldsymbol{H}(z)$ given by the expression

$$\boldsymbol{H}(z) = \frac{1}{\prod_{i=1}^{n}(1 - z_i z^{-1})} \equiv \frac{1}{1 - \sum_{i=1}^{n} h_i z^{-i}} \tag{2.1.1}$$

where z_i are the poles of the filter and h_i are the coefficients of the corresponding polynomial in z. Let $\boldsymbol{W}(z)$ and $\boldsymbol{Y}(z)$ denote the z-transforms of the input and output sequences, respectively. If $\boldsymbol{W}(z)=\sigma^2$ (corresponding to a white noise input) then

$$\boldsymbol{H}(z) = \frac{\boldsymbol{Y}(z)}{\boldsymbol{W}(z)} = \frac{1}{1 - \sum h_i z^{-i}} \tag{2.1.2}$$

or for this specific case,

$$\boldsymbol{Y}(z) = \frac{\sigma^2}{1 - \sum h_i z^{-i}}$$

Thus equation (2. 1. 2) may be expressed as

$$\boldsymbol{Y}(z)\left[1-\sum_{i=1}^{n}\boldsymbol{h}_i z^{-i}\right]=\sigma^2 \tag{2.1.3}$$

We now wish to transform this expression into the time domain. Each of the time-domain signals of equation (2. 1. 3) are given by the corresponding inverse z-transform relationship as

$$\boldsymbol{Z}^{-1}[\boldsymbol{Y}(z)]=[y_1, y_2, y_3, \cdots, y_n] \tag{2.1.4}$$

$$\boldsymbol{Z}^{-1}\left[1-\sum_{i=1}^{n}\boldsymbol{h}_i z^{-i}\right]=[1, -h_1, -h_2, \cdots, -h_n] \tag{2.1.5}$$

and the input sequence corresponding to the z-transform quantity σ^2 is

$$\boldsymbol{Z}^{-1}(\boldsymbol{\sigma}^2)=[w_1, w_2, \cdots, w_n] \tag{2.1.6}$$

where $\boldsymbol{w}_n$ is a white noise sequence with power $\boldsymbol{\sigma}^2$. The left-hand side of equation (2. 1. 3) is the product of z-transforms. Thus, the time-domain representation of the left-hand side of equation (2. 1. 3) is the convolution of the respective time-domain representations. Thus using equation (2. 1. 3) to (2. 1. 6) we have

$$y_i-\sum_{k=1}^{n}\boldsymbol{h}_k\boldsymbol{y}_{i-k}=\boldsymbol{w}_i \tag{2.1.7}$$

or

$$y_i=\sum_{k=1}^{n}\boldsymbol{h}_k\boldsymbol{y}_{i-k}+\boldsymbol{w}_i \tag{2.1.8}$$

Repeating this equation for m different values of the index i we have

$$\boldsymbol{y}_p=\boldsymbol{Yh}+\boldsymbol{w} \tag{2.1.9}$$

where the definitions of the matrix-vector quantities are apparent from previous discussions. From equation (2. 1. 8), we see that provided $\boldsymbol{\sigma}^2$ is suitably small, the present value of the output $\boldsymbol{y}$ is predicted from a linear combination of past values. Thus $\boldsymbol{y}_p$ is a vector of predicted values. (Hence the subscript: $p\equiv$"predicted"). If $\boldsymbol{\sigma}^2$ is small then the prediction has small error, and this is due to the "inertia" of the underlying all-pole system. In this case, the all-pole system is highly resonant with a relatively high Q-factor and the output process becomes highly predictable when driven by a white-noise input.

The mathematical model corresponding to equation (2. 1. 9) is sometimes referred to as a regression model. In the variables $\boldsymbol{y}$ are "regressed" onto themselves and hence the name "autoregressive".

So again, it makes sense to choose the h's in equation (2. 1. 5) so that the predicting term $\boldsymbol{Yh}$ is as close as possible to y_p in the 2-norm sense. Hence, as before, we choose $\boldsymbol{h}$ to satisfy

$$\min_h \|\boldsymbol{w}\|_2^2=\min_h(\boldsymbol{Yh}-\boldsymbol{y}_p)^{\mathrm{T}}(\boldsymbol{Yh}-\boldsymbol{y}_p) \tag{2.1.10}$$

Notice that if the parameters $\boldsymbol{h}$ are known the autoregressive process is completely characterized.

2.1.2 The Least-Squares Solution

We define our regression model corresponding to equation (2.1.11) as

$$\boldsymbol{b}=\boldsymbol{A}\boldsymbol{x}+\boldsymbol{n} \tag{2.1.11}$$

and we wish to determine the value $\boldsymbol{x}_{\mathrm{LS}}$ which solves

$$\boldsymbol{x}_{\mathrm{LS}}=\arg\min_{x}\|\boldsymbol{A}\boldsymbol{x}-\boldsymbol{b}\|_2^2 \tag{2.1.12}$$

where $\boldsymbol{A}\in\mathbb{R}^{m\times n}$, $m>n$, $\boldsymbol{b}\in\mathbb{R}^m$. The matrix $\boldsymbol{A}$ is assumed full rank.

In this general context, we note that $\boldsymbol{b}$ is a vector of observations, which correspond to a linear model of the form $\boldsymbol{A}\boldsymbol{x}$, contaminated by a noise contribution, $\boldsymbol{n}$. The matrix $\boldsymbol{A}$ is a constant. In determining $\boldsymbol{x}_{\mathrm{LS}}$, we find that value of $\boldsymbol{x}$ which provides the best fit of the observations to the model, in the 2-norm sense.

We now discuss a few relevant points concerning the LS problem:

• The system equation (2.1.12) is overdetermined and hence no solution exists in the general case for which $\boldsymbol{A}\boldsymbol{x}=\boldsymbol{b}$ exactly.

• Of all commonly used values of p for the norm $\|\cdot\|_p$ in equation (2.1.12), $p=2$ is the only one for which the norm is differentiable for all values of $\boldsymbol{x}$. Thus, for any other value of p, the optimal solution in not obtainable by differentiation.

• Note that for $\boldsymbol{Q}$ orthonormal, we have (only for $p=2$)

$$\|\boldsymbol{A}\boldsymbol{x}-\boldsymbol{b}\|_2^2=\|\boldsymbol{Q}^{\mathrm{T}}\boldsymbol{A}\boldsymbol{x}-\boldsymbol{Q}^{\mathrm{T}}\boldsymbol{b}\|_2^2 \tag{2.1.13}$$

This fact is used to advantage later on.

• We define the minimum sum of squares of the residual $\|\boldsymbol{A}\boldsymbol{x}_{\mathrm{LS}}-\boldsymbol{b}\|_2^2$ as ρ_{LS}^2.

• If $r=\mathrm{rank}(\boldsymbol{A})<n$, then there is no unique $\boldsymbol{x}_{\mathrm{LS}}$ which minimizes $\|\boldsymbol{A}\boldsymbol{x}-\boldsymbol{b}\|_2$. However, the solution can be made unique by considering only that element of set $\{x_{\mathrm{LS}}\in\mathbb{R}^n \mid \|\boldsymbol{A}\boldsymbol{x}_{\mathrm{LS}}-\boldsymbol{b}\|_2=\min\}$ which has minimum norm.

We wish to estimate the parameters x by solving equation (2.1.12). The method we choose to solve equation (2.1.12) is to differentiate the quantity $\|\boldsymbol{A}\boldsymbol{x}_{\mathrm{LS}}-\boldsymbol{b}\|_2^2$ with respect to x and set the result to zero. Thus, the remaining portion of this section is devoted to this differentiation. The result is a closed form expression for the solution of equation (2.1.12).

The expression $\|\boldsymbol{A}\boldsymbol{x}-\boldsymbol{b}\|_2^2$ can be written as

$$\|\boldsymbol{A}\boldsymbol{x}-\boldsymbol{b}\|_2^2=(\boldsymbol{A}\boldsymbol{x}-\boldsymbol{b})^{\mathrm{T}}(\boldsymbol{A}\boldsymbol{x}-\boldsymbol{b})=\boldsymbol{b}^{\mathrm{T}}\boldsymbol{b}-\boldsymbol{x}^{\mathrm{T}}\boldsymbol{A}^{\mathrm{T}}\boldsymbol{b}-\boldsymbol{b}^{\mathrm{T}}\boldsymbol{A}\boldsymbol{x}+\boldsymbol{x}^{\mathrm{T}}\boldsymbol{A}^{\mathrm{T}}\boldsymbol{A}\boldsymbol{x} \tag{2.1.14}$$

The solution $\boldsymbol{x}_{\mathrm{LS}}$ is that value of x which satisfies

$$\frac{\mathrm{d}}{\mathrm{d}x}[\boldsymbol{b}^{\mathrm{T}}\boldsymbol{b}-\boldsymbol{x}^{\mathrm{T}}\boldsymbol{A}^{\mathrm{T}}\boldsymbol{b}-\boldsymbol{b}^{\mathrm{T}}\boldsymbol{A}\boldsymbol{x}+\boldsymbol{x}^{\mathrm{T}}\boldsymbol{A}^{\mathrm{T}}\boldsymbol{A}\boldsymbol{x}]=0 \tag{2.1.15}$$

Define each term in the square brackets above as $t_1(x)$, …, $t_4(x)$, respectively.

$$\frac{\mathrm{d}}{\mathrm{d}x}[t_2(x)+t_3(x)+t_4(x)]=0 \tag{2.1.16}$$

where we have noted that the derivative $\frac{\mathrm{d}}{\mathrm{d}\boldsymbol{x}}t_1=0$, since $\boldsymbol{b}$ is independent of $\boldsymbol{x}$.

We see that every term of equation (2.1.16) is a scalar. To differentiate equation (2.1.16) with respect to the vector x, we differentiate each term of equation (2.1.16) with respect to each element of x and then assemble all the results back into a vector. We now discuss the differentiation of each term of equation (2.1.16).

2.1.2.1 Differentiation of $t_2(x)$ and $t_3(x)$ with Respect to x

Let us define the quantity $\boldsymbol{c} \triangleq \boldsymbol{A}^{\mathrm{T}}\boldsymbol{b}$, This implies that the component c_k of c is $\boldsymbol{a}_k^{\mathrm{T}}\boldsymbol{b}$, $k=1,2,\cdots,n$, where $\boldsymbol{a}_k^{\mathrm{T}}$ is the transpose of the k-th column of $\boldsymbol{A}$.

Thus, $t_2(x)=-\boldsymbol{x}^{\mathrm{T}}\boldsymbol{c}$. Therefore,

$$\frac{\mathrm{d}}{\mathrm{d}x_k}t_2(x)=\frac{\mathrm{d}}{\mathrm{d}x_k}(-\boldsymbol{x}^{\mathrm{T}}\boldsymbol{c})=-c_k=-\boldsymbol{a}_k^{\mathrm{T}}\boldsymbol{b} \quad k=1,2,\cdots,n \tag{2.1.17}$$

Combining these results for $k=1,2,\cdots,n$ back into a column vector, we get

$$\frac{\mathrm{d}}{\mathrm{d}x}t_2(x)=\frac{\mathrm{d}}{\mathrm{d}x}(-\boldsymbol{x}^{\mathrm{T}}\boldsymbol{A}\boldsymbol{b})=-\boldsymbol{A}^{\mathrm{T}}\boldsymbol{b} \tag{2.1.18}$$

Since Term 3 of equation (2.1.16) is the transpose of term 2 and both are scalars, the terms are equal. Hence,

$$\frac{\mathrm{d}}{\mathrm{d}x}t_3(x)=-\boldsymbol{A}^{\mathrm{T}}\boldsymbol{b} \tag{2.1.19}$$

2.1.2.2 Differentiation of $t_4(x)$ with Respect to x

For notational convenience, let us define the matrix $\boldsymbol{R} \triangleq \boldsymbol{A}^{\mathrm{T}}\boldsymbol{A}$. Thus, the quadratic form $t_4(x)$ may be expressed as

$$t_4(x) = \boldsymbol{x}^{\mathrm{T}}\boldsymbol{R}\boldsymbol{x} = \sum_{i=1}^{n}\sum_{j=1}^{n}x_i x_j r_{ij} \tag{2.1.20}$$

When differentiating the above with respect to a particular element x_k, we need only consider the terms when either index i or j equals k. Therefore,

$$\frac{\mathrm{d}}{\mathrm{d}x_k}t_4(x) = \frac{\mathrm{d}}{\mathrm{d}x_k}\left[\sum_{i=1}^{n}x_k x_j r_{kj} + \sum_{j\neq k}^{n}x_i x_k r_{ik} + x_k^2 r_{kk}\right] \tag{2.1.21}$$

where the first term of equation (2.1.21) corresponds to holding i constant at the value k. Care must be taken to include the term $x_k^2\boldsymbol{r}_{kk}$ corresponding to $i=j=k$ only once; therefore, it is excluded in the first two terms and added in separately. Equation (2.1.21) evaluates to

$$\frac{\mathrm{d}}{\mathrm{d}x_k}t_4(x) = \sum_{j\neq k}x_j r_{kj} + \sum_{i\neq k}x_i r_{ik} + 2x_k r_{kk} = \sum_{j}x_j r_{kj} + \sum_{i}x_i r_{ik} = 2(\boldsymbol{R}\boldsymbol{x})_k$$

Where $(\cdot)_k$ denotes k-th element of the corresponding vector. In the above, we have used the fact that $\boldsymbol{R}$ is symmetric; hence, $r_{ki}=r_{ik}$. Assembling all these components for $k=1,2,\cdots,n$ together into a column vector, we get

$$\frac{\mathrm{d}}{\mathrm{d}x}t_4(x)=\frac{\mathrm{d}}{\mathrm{d}x}(\boldsymbol{x}^{\mathrm{T}}\boldsymbol{A}^{\mathrm{T}}\boldsymbol{A}\boldsymbol{x})=2\boldsymbol{A}^{\mathrm{T}}\boldsymbol{A}\boldsymbol{x} \tag{2.1.22}$$

Substituting equation (2.1.18), (2.1.19) and (2.1.22) into (2.1.16) we get the important desired result:

$$\boldsymbol{A}^{\mathrm{T}}\boldsymbol{A}\boldsymbol{x}=\boldsymbol{A}^{\mathrm{T}}\boldsymbol{b} \tag{2.1.23}$$

The value $\boldsymbol{x}_{\mathrm{LS}}$ of $\boldsymbol{x}$ which solves equation (2. 1. 23) is the least-squares solution corresponding to equation (2. 1. 12). Equations (2. 1. 23) is called the normal equation. The reason for this terminology is discussed in the next section.

2.1.3 Interpretation of the Normal Equations

Equation (2. 1. 23) can be written in the form

$$\boldsymbol{A}^{\mathrm{T}}(\boldsymbol{b}-\boldsymbol{A}\boldsymbol{x}_{\mathrm{LS}})=0$$

or

$$\boldsymbol{A}^{\mathrm{T}}\boldsymbol{r}_{\mathrm{LS}}=0 \tag{2.1.24}$$

where

$$\boldsymbol{r}_{\mathrm{LS}}\triangleq\boldsymbol{b}-\boldsymbol{A}\boldsymbol{x}_{LS} \tag{2.1.25}$$

is the least squares error vector between $\boldsymbol{A}\boldsymbol{x}_{LS}$ and $\boldsymbol{b}$, $\boldsymbol{r}_{\mathrm{LS}}$ must be orthogonal to $R(\boldsymbol{A})$ for the LS solution $\boldsymbol{x}_{LS}$. Hence, the name "normal equations". This fact gives an important interpretation to least-squares estimation, which we now illustrate for the 3×2 case. Equation (2. 1. 11) may be expressed as

$$\boldsymbol{b}=[\boldsymbol{a}_1, \boldsymbol{a}_2]\begin{bmatrix}x_1\\x_2\end{bmatrix}+\boldsymbol{n}$$

We see from equation (2. 1. 24) that the point $\boldsymbol{A}\boldsymbol{x}_{\mathrm{LS}}$ is at the foot of a perpendicular dropped from $\boldsymbol{b}$ into $\mathrm{R}(\boldsymbol{A})$. The solution $\boldsymbol{x}_{\mathrm{LS}}$ are the weights of the linear combination of columns of $\boldsymbol{A}$ which equal the "foot vector".

This interpretation may be augmented as follows. From we see that

$$\boldsymbol{x}_{\mathrm{LS}}=(\boldsymbol{A}^{\mathrm{T}}\boldsymbol{A})^{-1}\boldsymbol{A}^{\mathrm{T}}\boldsymbol{b} \tag{2.1.26}$$

Hence the point $\boldsymbol{A}\boldsymbol{x}_{\mathrm{LS}}$ which is in $\mathrm{R}(\boldsymbol{A})$ is given by

$$\boldsymbol{A}\boldsymbol{x}_{\mathrm{LS}}=\boldsymbol{A}(\boldsymbol{A}^{\mathrm{T}}\boldsymbol{A})^{-1}\boldsymbol{A}^{\mathrm{T}}\boldsymbol{b}\equiv\boldsymbol{P}\boldsymbol{b}$$

Where $\boldsymbol{P}$ is the projector onto $\mathrm{R}(\boldsymbol{A})$. Thus, we see from another point of view that the least-squares solution is the result of projecting $\boldsymbol{b}$ (the observation) onto $R(\boldsymbol{A})$.

It is seen from equation (2. 1. 11) that in the noise-free case, the vector $\boldsymbol{b}$ is equal to the vector $\boldsymbol{A}\boldsymbol{x}_{\mathrm{LS}}$. The fact that $\boldsymbol{A}\boldsymbol{x}_{\mathrm{LS}}$ should be at the foot of a perpendicular from $\boldsymbol{b}$ into $R(\boldsymbol{A})$ makes intuitive sense, because a perpendicular is the shortest distance from $\boldsymbol{b}$ into $R(\boldsymbol{A})$. This, after all, is the objective of the LS problem as expressed by equation (2. 1. 12).

There is a further point we wish to address in the interpretation of the normal equations. Substituting equation (2. 1. 26) into (2. 1. 25) we have

$$\boldsymbol{r}_{\mathrm{LS}}=\boldsymbol{b}-\boldsymbol{A}(\boldsymbol{A}^{\mathrm{T}}\boldsymbol{A})^{-1}\boldsymbol{A}^{\mathrm{T}}\boldsymbol{b}=(\boldsymbol{I}-\boldsymbol{P})\boldsymbol{b}=\boldsymbol{P}_{\perp}\boldsymbol{b} \tag{2.1.27}$$

Thus, $\boldsymbol{r}_{\mathrm{LS}}$ is the projection of $\boldsymbol{b}$ onto $\mathrm{R}(\boldsymbol{A})_{\perp}$. We can now determine the value ρ_{LS}^2, which is the squared 2-norm of the LS residual:

$$\rho_{\mathrm{LS}}\triangleq\|\boldsymbol{r}_{\mathrm{LS}}\|_2^2=\|\boldsymbol{P}_{\perp}\boldsymbol{b}\|_2^2 \tag{2.1.28}$$

The fact that $\boldsymbol{r}_{\mathrm{LS}}$ is orthogonal to $\mathrm{R}(\boldsymbol{A})$ is of fundamental importance. In fact, it is easy to show that choosing $\boldsymbol{x}$ so that $\boldsymbol{r}_{\mathrm{LS}}\perp R(\boldsymbol{A})$ is a sufficient condition for the least-squares solution. Often in analysis $\boldsymbol{x}_{\mathrm{LS}}$ is determined this way, instead of through the normal equations. This concept is referred to as the principle of orthogonality.

2.1.4 Properties of the LS Estimate

Here we consider the regression equation (2.1.11) again. It is reproduced below for convenience.

$$\boldsymbol{b}=\boldsymbol{A}\boldsymbol{x}_0+\boldsymbol{n} \tag{2.1.29}$$

We now briefly clarify various aspects of notation in equation (2.1.29). We use the quantity $\boldsymbol{x}$ to mean an unknown vector of parameters; as such it is considered a variable whose true value $\boldsymbol{x}_0$ we wish to estimate. The quantity $\boldsymbol{x}_{LS}$ is the estimate of $\boldsymbol{x}$ obtained specifically by the least squares procedure. Along the same lines $\boldsymbol{r}_{LS}$ in equation (2.1.25) is the error between the model and the observation for the specific value $\boldsymbol{x}=\boldsymbol{x}_{LS}$. On the other hand, $\boldsymbol{n}$ in equation (2.1.29) is the true (unknown) value of error between the true model ($\boldsymbol{A}\boldsymbol{x}_0$) and the observation. The residual $\boldsymbol{r}_{LS}$ is not equal to $\boldsymbol{n}$ unless $\boldsymbol{x}_{LS}=\boldsymbol{x}_0$, which is very unlikely.

For this section we view the left hand side as a vector $\boldsymbol{b}$ of observations generated from a known constant matrix $\boldsymbol{A}$ and a vector of parameters $\boldsymbol{x}$ whose true values are $\boldsymbol{x}_0$. The observation is contaminated by noise, $\boldsymbol{n}$. Thus, $\boldsymbol{b}$ is a random variable described by the pdf of $\boldsymbol{n}$. But from equation (2.1.26), we see that $\boldsymbol{x}_{LS}$ is a linear transform of $\boldsymbol{b}$; therefore $\boldsymbol{x}_{LS}$ is also a random variable. We now study its properties.

In order to discuss useful and interesting properties of the LS estimate we make the following assumptions:

A1: $\boldsymbol{n}$ is a zero mean random vector with uncorrelated elements; i.e., $E(\boldsymbol{n}\boldsymbol{n}^T)=\sigma^2\boldsymbol{I}$.

A2: $\boldsymbol{A}$ is a constant matrix which is known with negligible error. That is, there is no uncertainty in $\boldsymbol{A}$.

Under A1 and A2, we have the following properties of the LS estimate given by equation (2.1.26).

2.1.4.1 X_{LS} is an Unbiased Estimate of X_0 the True Value

To show this, we have from equation (2.1.26)

$$\boldsymbol{x}_{LS}=(\boldsymbol{A}^T\boldsymbol{A})^{-1}\boldsymbol{A}^T\boldsymbol{b} \tag{2.1.30}$$

But from the regression equation (2.1.29), we realize that the observed data $\boldsymbol{b}$ are generated from the true values $\boldsymbol{x}_0$ of $\boldsymbol{x}$. Hence from equation (2.1.29)

$$\boldsymbol{x}_{LS}=(\boldsymbol{A}^T\boldsymbol{A})^{-1}\boldsymbol{A}^T(\boldsymbol{A}\boldsymbol{x}_0+\boldsymbol{n})=\boldsymbol{x}_0+(\boldsymbol{A}^T\boldsymbol{A})^{-1}\boldsymbol{A}^T\boldsymbol{n} \tag{2.1.31}$$

Therefore the $\boldsymbol{E}(\boldsymbol{x})$ is given as

$$\boldsymbol{E}(\boldsymbol{x})=\boldsymbol{x}_0+(\boldsymbol{A}^T\boldsymbol{A})^{-1}\boldsymbol{A}^T\boldsymbol{n}=\boldsymbol{x}_0 \tag{2.1.32}$$

which follows because $\boldsymbol{n}$ is zero mean from assumption A1. Therefore the expectation of $\boldsymbol{x}$ is its true value and $\boldsymbol{x}_{LS}$ is unbiased.

2.1.4.2 Covariance Matrix of $\boldsymbol{x}_{LS}$

The definition of the covariance matrix *cov* ($\boldsymbol{x}_{LS}$) of the non-zero mean process $\boldsymbol{x}_{LS}$ is:

$$\operatorname{cov}(\boldsymbol{x}_{LS})=\boldsymbol{E}[(\boldsymbol{x}_{LS}-E(\boldsymbol{x}_{LS}))(\boldsymbol{x}_{LS}-E(\boldsymbol{x}_{LS}))^T] \tag{2.1.33}$$

For these purposes we define $E(\boldsymbol{x}_{\mathrm{LS}})$ as

$$\boldsymbol{E}(\boldsymbol{x}_{\mathrm{LS}})=(\boldsymbol{A}^{\mathrm{T}}\boldsymbol{A})^{-1}\boldsymbol{A}^{\mathrm{T}}\boldsymbol{E}(\boldsymbol{b})=(\boldsymbol{A}^{\mathrm{T}}\boldsymbol{A})^{-1}\boldsymbol{A}^{\mathrm{T}}(\boldsymbol{A}\boldsymbol{x}_0) \tag{2.1.34}$$

Substituting equation (2.1.34) and (2.1.26) in (2.1.33), we have

$$\mathrm{cov}(\boldsymbol{x}_{\mathrm{LS}})=E[(\boldsymbol{A}^{\mathrm{T}}\boldsymbol{A})^{-1}\boldsymbol{A}^{\mathrm{T}}(\boldsymbol{b}-\boldsymbol{A}\boldsymbol{x}_0)(\boldsymbol{b}-\boldsymbol{A}\boldsymbol{x}_0)^{\mathrm{T}}\boldsymbol{A}\ (\boldsymbol{A}^{\mathrm{T}}\boldsymbol{A})^{-1}] \tag{2.1.35}$$

From assumption A2 we can move the expectation operator inside. Therefore,

$$\begin{aligned}\mathrm{cov}(\boldsymbol{x}_{\mathrm{LS}})&=[(\boldsymbol{A}^{\mathrm{T}}\boldsymbol{A})^{-1}\boldsymbol{A}^{\mathrm{T}}\ \underbrace{\boldsymbol{E}(\boldsymbol{b}-\boldsymbol{A}\boldsymbol{x}_0)(\boldsymbol{b}-\boldsymbol{A}\boldsymbol{x}_0)^{\mathrm{T}}}_{\sigma^2\boldsymbol{I}}\boldsymbol{A}\ (\boldsymbol{A}^{\mathrm{T}}\boldsymbol{A})^{-1}]\\&=(\boldsymbol{A}^{\mathrm{T}}\boldsymbol{A})^{-1}\boldsymbol{A}^{\mathrm{T}}(\sigma^2\boldsymbol{I})\boldsymbol{A}(\boldsymbol{A}^{\mathrm{T}}\ \boldsymbol{A})^{-1}\\&=\sigma^2(\boldsymbol{A}^{\mathrm{T}}\boldsymbol{A})^{-1}\end{aligned} \tag{2.1.36}$$

where we have used the result $\boldsymbol{b}-\boldsymbol{A}\boldsymbol{x}_0$ is the noise vector $\boldsymbol{n}$ and that $\mathrm{cov}(\boldsymbol{n})=\sigma^2\boldsymbol{I}$ from A1.

It is desirable to for the variances of the estimates $\boldsymbol{x}_{\mathrm{LS}}$ to be as small as possible. How small does equation (2.1.36) say they are? We see that if σ^2 is large then the variances which are the diagonal elements of cov ($\boldsymbol{x}_{\mathrm{LS}}$) are also large. This makes sense because if the variances of the elements of $\boldsymbol{n}$ are large, then the variances of the elements of $\boldsymbol{x}_{\mathrm{LS}}$ could also be expected to be large. But more importantly, equation (2.1.36) also says that if $\boldsymbol{A}^{\mathrm{T}}\boldsymbol{A}$ is "big" in some norm sense, then $\mathrm{cov}(\boldsymbol{x}_{\mathrm{LS}})$ is "small" which is desirable . We discuss this point in further detail shortly. We can also infer that if $\boldsymbol{A}$ is rank deficient then $\boldsymbol{A}^{\mathrm{T}}\boldsymbol{A}$ is rank deficient, and the variances of each component of $\boldsymbol{x}$ approach infinity which implies the results are meaningless.

2.1.4.3 x_{LS} is a BLUE

According to equation (2.1.26), we see that $\boldsymbol{x}_{\mathrm{LS}}$ is a linear estimate since it is a linear transformation of $\boldsymbol{b}$, where the transformation matrix is $(\boldsymbol{A}^{\mathrm{T}}\boldsymbol{A})^{-1}\boldsymbol{A}^{\mathrm{T}}$. Further from Section 2.1.6.1 we see that $\boldsymbol{x}_{\mathrm{LS}}$ is unbiased. With the following theorem, we show that $\boldsymbol{x}_{\mathrm{LS}}$ is the best linear unbiased estimator (BLUE).

Theorem 1

Consider any linear unbiased estimate $\bar{\boldsymbol{x}}$ of $\boldsymbol{x}$, defined by

$$\bar{\boldsymbol{x}}=\boldsymbol{B}\boldsymbol{b} \tag{2.1.37}$$

Where $\boldsymbol{B}\in\mathbb{R}^{m\times n}$ is an estimator, or transformation matrix. Then under A1 and A2, $\boldsymbol{x}_{\mathrm{LS}}$ is a BLUE.

Proof

Substituting equation (2.1.29) into (2.1.37) we have

$$\bar{\boldsymbol{x}}=\boldsymbol{B}\boldsymbol{A}\boldsymbol{x}_0+\boldsymbol{B}\boldsymbol{n} \tag{2.1.38}$$

because $\boldsymbol{n}$ has zero mean A1,

$$E(\bar{\boldsymbol{x}})=\boldsymbol{B}\boldsymbol{A}\boldsymbol{x}_0$$

For $\bar{\boldsymbol{x}}$ to be unbiased, we therefore require

$$\boldsymbol{B}\boldsymbol{A}=\boldsymbol{I} \tag{2.1.39}$$

We can now write equation (2.1.38) as

$$\bar{\boldsymbol{x}}=\boldsymbol{x}_0+\boldsymbol{B}\boldsymbol{n}$$

The covariance matrix of $\bar{x}$ is then

$$\mathrm{cov}(\boldsymbol{x}_{\mathrm{LS}})=E[(\bar{\boldsymbol{x}}-\boldsymbol{x}_0)(\bar{\boldsymbol{x}}-\boldsymbol{x}_0)^{\mathrm{T}}]=\mathrm{E}[\boldsymbol{B}\boldsymbol{n}\boldsymbol{n}^{\mathrm{T}}\boldsymbol{B}^{\mathrm{T}}]=\sigma^2\boldsymbol{B}\boldsymbol{B}^{\mathrm{T}} \qquad (2.1.40)$$

where we have used A1 in the last line.

We now consider a matrix $\boldsymbol{\Psi}$ defined as the difference of the estimator matrix $\boldsymbol{B}$ and the least-squares estimator matrix $(\boldsymbol{A}^{\mathrm{T}}\boldsymbol{A})^{-1}\boldsymbol{A}^{\mathrm{T}}$:

$$\boldsymbol{\Psi}=\boldsymbol{B}-(\boldsymbol{A}^{\mathrm{T}}\boldsymbol{A})^{-1}\boldsymbol{A}^{\mathrm{T}}$$

No using equation (2.1.39) we form the matrix product $\boldsymbol{\Psi}\boldsymbol{\Psi}^{\mathrm{T}}$:

$$\begin{aligned}\boldsymbol{\Psi}\boldsymbol{\Psi}^{\mathrm{T}} &=[\boldsymbol{B}-(\boldsymbol{A}^{\mathrm{T}}\boldsymbol{A})^{-1}\boldsymbol{A}^{\mathrm{T}}][\boldsymbol{B}^{\mathrm{T}}-\boldsymbol{A}(\boldsymbol{A}^{\mathrm{T}}\boldsymbol{A})^{-1}]\\ &=\boldsymbol{B}\boldsymbol{B}^{\mathrm{T}}-\boldsymbol{B}\boldsymbol{A}(\boldsymbol{A}^{\mathrm{T}}\boldsymbol{A})^{-1}-(\boldsymbol{A}^{\mathrm{T}}\boldsymbol{A})^{-1}\boldsymbol{A}^{\mathrm{T}}\boldsymbol{B}^{\mathrm{T}}+(\boldsymbol{A}^{\mathrm{T}}\boldsymbol{A})^{-1}\\ &=\boldsymbol{B}\boldsymbol{B}^{\mathrm{T}}-(\boldsymbol{A}^{\mathrm{T}}\boldsymbol{A})^{-1}\end{aligned} \qquad (2.1.41)$$

We note that the i-th diagonal element of $\boldsymbol{\Psi}\boldsymbol{\Psi}^{\mathrm{T}}$ is the squared 2-norm of the i-th row of $\boldsymbol{\Psi}$; hence $(\boldsymbol{\Psi}\boldsymbol{\Psi}^{\mathrm{T}})_{ii}\geqslant 0^5$. Hence from equation (2.1.41) we have

$$\sigma^2(\boldsymbol{B}\boldsymbol{B}^{\mathrm{T}})_{ii}\geqslant\sigma^2(\boldsymbol{A}^{\mathrm{T}}\boldsymbol{A})_{ii}^{-1} \quad i=1,2,\cdots,n \qquad (2.1.42)$$

We note that the diagonal elements of a covariance matrix are the variances of the individual element. But from equation (2.1.40) and (2.1.36) we see that $\sigma^2\boldsymbol{B}\boldsymbol{B}^{\mathrm{T}}$ and $\sigma^2(\boldsymbol{A}^{\mathrm{T}}\boldsymbol{A})^{-1}$ are the covariance matrices of $\bar{\boldsymbol{x}}$ and $\boldsymbol{x}_{\mathrm{LS}}$ respectively. Therefore, equation (2.1.42) tells us that the variances of the elements of $\bar{\boldsymbol{x}}$ are never better than those of $\boldsymbol{x}_{\mathrm{LS}}$. Thus within the class of linear unbiased estimators and under assumptions A1 and A2, no other estimator has smaller variance than the L-S estimate.

A3: For the following properties, we further assume $\boldsymbol{n}$ is jointly Gaussian distributed, with mean 0, covariance $\sigma^2\boldsymbol{I}$.

2.1.4.4 Probability Density Function of x_{LS}

It is a fundamental property of Gaussian-distributed random variables that any linear transformation of a Gaussian distributed quantity is also Gaussian. From equation (2.1.26) we see that $\boldsymbol{x}_{\mathrm{LS}}$ is a linear transformation of $\boldsymbol{b}$, which is Gaussian by hypothesis. Since the Gaussian pdf is completely specified from the expectation and covariance, given respectively by equation (2.1.32) and (2.1.36), then $\boldsymbol{x}_{\mathrm{LS}}$ has the Gaussian pdf given by

$$p(\boldsymbol{x}_{\mathrm{LS}})=(2\pi)^{-\frac{n}{2}}|\sigma^{-2}\boldsymbol{A}^{\mathrm{T}}\boldsymbol{A}|^{\frac{1}{2}}\exp\left[-\frac{1}{2\sigma^2}(\boldsymbol{x}_{\mathrm{LS}}-\boldsymbol{x}_0)^{\mathrm{T}}\boldsymbol{A}^{\mathrm{T}}\boldsymbol{A}(\boldsymbol{x}_{\mathrm{LS}}-\boldsymbol{x}_0)\right] \qquad (2.1.43)$$

We see that the elliptical joint confidence region of $\boldsymbol{x}_{\mathrm{LS}}$ is the set of points $\boldsymbol{\psi}$ defined as

$$\boldsymbol{\psi}=\left\{\boldsymbol{x}_{\mathrm{LS}}\,\middle|\,-\frac{1}{2\sigma^2}(\boldsymbol{x}_{\mathrm{LS}}-\boldsymbol{x}_0)^{\mathrm{T}}\boldsymbol{A}^{\mathrm{T}}\boldsymbol{A}(\boldsymbol{x}_{\mathrm{LS}}-\boldsymbol{x}_0)=k\right\} \qquad (2.1.44)$$

where k is some constant which determines the probability level that an observation will fall within $\boldsymbol{\psi}$. Note that if the joint confidence region becomes elongated in any direction, then the variance of the associated components of $\boldsymbol{x}_{\mathrm{LS}}$ become large. Let us rewrite the quadratic form in equation (2.1.44) as

$$-\frac{1}{2\sigma^2}z^{\mathrm{T}}\boldsymbol{\Lambda}^{-1}z$$

where $z\triangleq\boldsymbol{V}^{\mathrm{T}}(\boldsymbol{x}_{\mathrm{LS}}-\boldsymbol{x}_0)$ and $\boldsymbol{V}\boldsymbol{\Lambda}\boldsymbol{V}^{\mathrm{T}}$ length of the i-th principal axis of the associated ellipse is

$1/\sqrt{\lambda_i}$. This means that if a particular eigenvalue is small, then the length of the corresponding axis is large, and z has large variance in the direction of the corresponding eigenvector $\boldsymbol{v}_i$. If $\boldsymbol{v}_i$ has significant components along any component of $\boldsymbol{x}_{\mathrm{LS}}$, then these components of $\boldsymbol{x}_{\mathrm{LS}}$ have large variances too. On the other hand, if all the eigenvalues are large then the variances of z, and hence $\boldsymbol{x}_{\mathrm{LS}}$, are low in all directions.

Generalizing to multiple dimensions, we see that if all components of $\boldsymbol{x}_{\mathrm{LS}}$ are to have small variance, then all eigenvalues of $\boldsymbol{A}^{\mathrm{T}}\boldsymbol{A}$ must be large. Thus for desirable variance properties of $\boldsymbol{x}_{\mathrm{LS}}$, the matrix $\boldsymbol{A}^{\mathrm{T}}\boldsymbol{A}$ must be well-conditioned. This is the "sense" referred to earlier in which the matrix $\boldsymbol{A}^{\mathrm{T}}\boldsymbol{A}$ must be "big" in order for the variances to be small.

From the above we see that one small eigenvalue has the ability to make the variances of all components of $\boldsymbol{x}_{\mathrm{LS}}$ large. In the next lecture, we will present the pseudo inverse, which can mitigate the effect of a small eigenvalue destroying the desirable variance properties of $\boldsymbol{x}_{\mathrm{LS}}$.

Let us examine this concept from a different perspective. We go back to the regression equation

$$\boldsymbol{b}=\boldsymbol{A}\boldsymbol{x}+\boldsymbol{n}$$

where we see that observations $\boldsymbol{b}$ are perturbed by the noise $\boldsymbol{n}$ which has a given noise power σ^2. If the eigenvalues of $\boldsymbol{A}^{\mathrm{T}}\boldsymbol{A}$ are relatively large then the singular values of $\boldsymbol{A}$ are large. Then the variation in x to compensate for changes in $\boldsymbol{n}$ so that $\|\boldsymbol{b}-\boldsymbol{A}\boldsymbol{x}\|_2^2$ remains minimum is relatively small. On the other hand, if the singular values of $\boldsymbol{A}$ are relatively small, larger changes in x are required to compensate changes in $\boldsymbol{b}$ due to the noise. The variance of an estimate x is the amount of variation in the population of the estimates of x against all possible noise samples. So we see again if all singular eigenvalues are large then the variance of x is small and vice-versa.

We summarize the preceding discussion in the following theorem, which has already been justified.

Theorem 2

The least squares estimate $\boldsymbol{x}_{\mathrm{LS}}$ will have large variances if at least one of the eigenvalues of $\boldsymbol{A}^{\mathrm{T}}\boldsymbol{A}$ is small where the associated eigenvectors have significant components along the x-axes.

2.1.4.5 Maximum-Likelihood Property

In this vein, the least-squares estimate $\boldsymbol{x}_{\mathrm{LS}}$ is the maximum likelihood estimate of $\boldsymbol{x}_0$. To show this property, we first investigate the probability density function of $\boldsymbol{n}=\boldsymbol{A}\boldsymbol{x}-\boldsymbol{b}$, given for the more general case wherecov$(\boldsymbol{n})=\boldsymbol{\Sigma}$:

$$\begin{aligned}p(\boldsymbol{n})&=p(\boldsymbol{b}\,|\,\boldsymbol{x}=\boldsymbol{x}_0)\\&=(2\pi)^{-\frac{n}{2}}\,|\boldsymbol{\Sigma}|^{-\frac{1}{2}}\exp\left[-\frac{1}{2}(\boldsymbol{A}\boldsymbol{x}_o-\boldsymbol{b})^{\mathrm{T}}\boldsymbol{\Sigma}^{-1}(\boldsymbol{A}\boldsymbol{x}_0-\boldsymbol{b})\right]\end{aligned}\tag{2.1.45}$$

The conditional pdf $p(\boldsymbol{b}|\boldsymbol{x})$ describes the variation in the observation $\boldsymbol{b}$ as a result of the noise, assuming that $\boldsymbol{A}$ is a known constant and that $\boldsymbol{x}$ is assigned its true value.

The physical mechanism or process which generates the observations $\boldsymbol{b}$ takes random noise samples distributed according to equation (2.1.45) and adds them to the value $\boldsymbol{A}\boldsymbol{x}_0$. The generating process considers the value $\boldsymbol{A}\boldsymbol{x}_0$, or just $\boldsymbol{x}_0$ constant, but considers $\boldsymbol{n}$ and therefore $\boldsymbol{b}$ as a random variable.

But now lets consider process which observes or receives, $\boldsymbol{b}$. The observation $\boldsymbol{b}$ is now a constant, since it is a measured quantity. Now, $\boldsymbol{x}_0$ is not known but is to be estimated. Therefore we consider the associated quantity $\boldsymbol{x}$ as variable. This is exactly the opposite situation to the generating process.

In order to estimate the value of $\boldsymbol{x}_0$ based on an observation $\boldsymbol{b}$, we use a simple but very elegant trick. We choose the value of $\boldsymbol{x}$ which is most likely to have given rise to the observation $\boldsymbol{b}$. This is the value of $\boldsymbol{x}$ for which the pdf given by equation (2.1.45) is maximum with respect to variation $\boldsymbol{x}$ with $\boldsymbol{b}$ held constant at the value which was observed not with $\boldsymbol{x}$ constant and $\boldsymbol{b}$ variable as it was for the generator. The value of $\boldsymbol{x}$ which maximizes equation (2.1.45) for $\boldsymbol{b}$ constant at its observed value is referred to as the maximum likelihood estimate of $\boldsymbol{x}$. It is a very powerful estimation technique and has many desirable properties discussed in several texts.

Note from equation (2.1.45) that if $\boldsymbol{\Sigma}=\sigma^2\boldsymbol{I}$, then the value $\boldsymbol{x}$ which maximizes the probability $p(\boldsymbol{b}|\boldsymbol{x})$ as a function of $\boldsymbol{x}$ is $\boldsymbol{x}_{\mathrm{LS}}$. This follows because $\boldsymbol{x}_{\mathrm{LS}}$ is by definition that value of $\boldsymbol{x}$ which minimizes the quadratic form of the exponent in equation (2.1.45) for the case $\boldsymbol{\Sigma}=\sigma^2\boldsymbol{I}$. It is also the $\boldsymbol{x}$ for which $p(\boldsymbol{b}|\boldsymbol{x})$ in equation (2.1.45) is maximum. Thus, $\boldsymbol{x}_{\mathrm{LS}}$ is the maximum likelihood estimate of $\boldsymbol{x}$.

2.1.5 Linear Least-Squares Estimation and the Cramer Rao Lower Bound

In this section we discuss the relationship between the cramer rao lower bound (CRLB) and the linear least-squares estimate. We first discuss the CRLB itself, and then go on to discuss the relationship between the CRLB and linear least squares estimation in white and coloured noise.

2.1.5.1 The Cramer Rao Lower Bound

Here we assume that the observed data $\boldsymbol{b}$ is generated from the model (2.1.29), for the specific case when the noise $\boldsymbol{n}$ is a joint Gaussian zero mean process. In order to address the CRLB, we consider a matrix $\boldsymbol{J}$ defined by

$$(\boldsymbol{J})_{ij}=-E\frac{\partial^2 \ln p(\boldsymbol{b}|\boldsymbol{x})}{\partial x_i \partial y_i} \tag{2.1.46}$$

In our case, $\boldsymbol{J}$ is defined as a matrix of second derivatives related to equation (2.1.45). The constant terms preceding the exponent in are not functions of $\boldsymbol{x}$, and so are not relevant with regard to the differentiation. Thus we need to consider only the exponential term of equation (2.1.45). Because of the ln(•) operation reduces to the second derivative matrix of the quadratic form in the exponent. This second derivative matrix is referred to as the Hessian. The expectation operator of equation (2.1.46) is redundant in our specific case because all the second derivative quantities are constant. Thus,

$$(\boldsymbol{J})_{ij}=\frac{\partial^2}{\partial x_i \partial y_i}\left[\frac{1}{2\sigma^2}(\boldsymbol{x}-\boldsymbol{x}_0)^{\mathrm{T}}(\boldsymbol{A}^{\mathrm{T}}\boldsymbol{A})(\boldsymbol{x}-\boldsymbol{x}_0)\right] \tag{2.1.47}$$

Using the analysis of Section 2.1.4.1 and 2.1.4.2, it is easy to show that

$$\boldsymbol{J}=\frac{1}{\sigma^2}(\boldsymbol{A}^{\mathrm{T}}\boldsymbol{A}) \tag{2.1.48}$$

The matrix $\boldsymbol{J}$ defined by equation (2.1.48) is referred to as the Fisher information matrix. Now consider a matrix $\boldsymbol{U}$ which is the covariance matrix of parameter estimates, obtained by some arbitrary estimation process; i.e., $\mathrm{cov}(\bar{\boldsymbol{x}})=\boldsymbol{U}$, where $\bar{\boldsymbol{x}}$ is some estimate of $\boldsymbol{x}$ obtained by some arbitrary estimator.

Then,

$$u_{ij}\geqslant j^{ii} \tag{2.1.49}$$

where j^{ii} denotes the (i, i)-th element of the inverse of $\boldsymbol{J}$. Because the diagonal elements of a covariance matrix are the variances of the individual elements, tells us that the individual variances of the estimates $\bar{\boldsymbol{x}}_i$ obtained by some arbitrary estimator are greater than or equal to the corresponding diagonal term of $\boldsymbol{J}^{-1}$. The CRLB thus puts a lower bound on how low the variances can be, regardless of how good the estimation procedure is.

2.1.5.2 Least-Squares Estimation and the CRLB for White Noise

Using equation (2.1.45), we now evaluate the CRLB for data generated according to the linear regression model of (2.1.11), for the specific case of white noise where $\boldsymbol{\Sigma}=\sigma^2\boldsymbol{I}$. That is, if we observe data which obey the model (2.1.11), what is the lowest possible variance on the estimates given by equation (2.1.26) from (2.1.48),

$$\boldsymbol{J}^{-1}=\sigma^2(\boldsymbol{A}^{\mathrm{T}}\boldsymbol{A})^{-1} \tag{2.1.50}$$

Comparing equation (2.1.50) and (2.1.36), we see that the covariance matrix of the LS estimates satisfies the equality of the CRLB. Thus, no other estimator can do better than the LS estimator under A1 to A3. This is a very important feature of linear least-squares estimates. For this reason, we refer to the LS estimator as a minimum variance unbiased estimator (MVUB).

2.1.5.3 Least-Squares Estimation and the CRLB for Coloured Noise

In this case, we consider $\boldsymbol{\Sigma}$ to be an arbitrary covariance matrix, i.e., $\boldsymbol{E}(\boldsymbol{n}\boldsymbol{n}^{\mathrm{T}})=\boldsymbol{\Sigma}$. By substituting equation (2.1.45) and evaluating, we can easily show that the Fisher information matrix $\boldsymbol{J}$ for this case is given by

$$\boldsymbol{J}=\boldsymbol{A}^{\mathrm{T}}\boldsymbol{\Sigma}^{-1}\boldsymbol{A} \tag{2.1.51}$$

We now develop the version of the covariance matrix of the LS estimate corresponding to equation (2.1.36) for the coloured noise case. Suppose we use the normal equation (2.1.23) to produce the estimate $\boldsymbol{x}_{\mathrm{LS}}$ for this coloured noise case. Using the same analysis as in Section 2.1.1.4, except using $E(\boldsymbol{b}-\boldsymbol{A}\boldsymbol{x}_0)(\boldsymbol{b}-\boldsymbol{A}\boldsymbol{x}_0)^{\mathrm{T}}=\boldsymbol{\Sigma}$ instead of $\sigma^2\boldsymbol{I}$ as before, we get:

$$\mathrm{cov}(\boldsymbol{x}_{LS})=(\boldsymbol{A}^T\boldsymbol{A})^{-1}\boldsymbol{A}^T\boldsymbol{\Sigma}\boldsymbol{A}(\boldsymbol{A}^T\boldsymbol{A})^{-1} \tag{2.1.52}$$

Note that the covariance matrix of the estimate in this case is not equal to $\boldsymbol{J}^{-1}$ from equation (2.1.51). In this case, $\boldsymbol{J}^{-1}$ from equation (2.1.51) expresses the CRLB, which is the minimum possible variance on the elements of $\boldsymbol{x}_{LS}$, Because equation (2.1.52) is not equal to the corresponding CRLB expression, the variances on the elements of $\boldsymbol{x}_{LS}$ when the ordinary normal equation (2.1.23) are used in coloured noise are not the minimum possible. Therefore, the ordinary normal equations do not give a MVUE in coloured noise.

We now show however, that if $\boldsymbol{\Sigma}$ is known we may improve the situation by pre-whitening the noise. Let $\boldsymbol{\Sigma}=\boldsymbol{G}\boldsymbol{G}^T$, where $\boldsymbol{G}$ is the Choleski factor. Then multiplying both sides of equation (2.1.11) by $\boldsymbol{G}^{-1}$, the noise is whitened, and we have

$$\boldsymbol{G}^{-1}\boldsymbol{b}=\boldsymbol{G}^{-1}\boldsymbol{A}\boldsymbol{x}+\boldsymbol{G}^{-1}\boldsymbol{n} \tag{2.1.53}$$

Using the above as the regression model, and substituting $\boldsymbol{G}^{-1}\boldsymbol{A}$ for $\boldsymbol{A}$, and $\boldsymbol{G}^{-1}\boldsymbol{b}$ for $\boldsymbol{b}$ in equation (2.1.23), we get

$$\boldsymbol{x}_{LS}=(\boldsymbol{A}^T\boldsymbol{\Sigma}^{-1}\boldsymbol{A})^{-1}\boldsymbol{A}^T\boldsymbol{\Sigma}^{-1}\boldsymbol{b} \tag{2.1.54}$$

The covariance matrix corresponding to this estimate is found as follows. We can write

$$\boldsymbol{E}(\boldsymbol{x}_{LS})=(\boldsymbol{A}^T\boldsymbol{\Sigma}^{-1}\boldsymbol{A})^{-1}\boldsymbol{A}^T\boldsymbol{\Sigma}^{-1}\boldsymbol{A}\boldsymbol{x}_0 \tag{2.1.55}$$

Substituting equation (2.1.56) and (2.1.54) into (2.1.33) we get

$$\begin{aligned}\mathrm{cov}(\boldsymbol{x}_{LS})&=\boldsymbol{E}(\boldsymbol{A}^T\boldsymbol{\Sigma}^{-1}\boldsymbol{A})^{-1}\boldsymbol{A}^T\boldsymbol{\Sigma}^{-1}(\boldsymbol{b}-\boldsymbol{A}\boldsymbol{x}_0)(\boldsymbol{b}-\boldsymbol{A}\boldsymbol{x}_0)^T\boldsymbol{\Sigma}^{-1}\boldsymbol{A}(\boldsymbol{A}^T\boldsymbol{\Sigma}^{-1}\boldsymbol{A})^-1\\&=(\boldsymbol{A}^T\boldsymbol{\Sigma}^{-1}\boldsymbol{A})^{-1}\boldsymbol{A}^T\boldsymbol{\Sigma}^{-1}\underbrace{\boldsymbol{E}(\boldsymbol{b}-\boldsymbol{A}\boldsymbol{x}_0)(\boldsymbol{b}-\boldsymbol{A}\boldsymbol{x}_0)^T)}_{\boldsymbol{\Sigma}}\boldsymbol{\Sigma}^{-1}\boldsymbol{A}(\boldsymbol{A}^T\boldsymbol{\Sigma}^{-1}\boldsymbol{A})^-1\\&=(\boldsymbol{A}^T\boldsymbol{\Sigma}^{-1}\boldsymbol{A})^{-1}\end{aligned} \tag{2.1.56}$$

Notice that in the coloured noise case when the noise is pre-whitened as in equation (2.1.53), the resulting matrix $\mathrm{cov}(\boldsymbol{x}_{LS})$ is equivalent to $\boldsymbol{J}^{-1}$ in equation (2.1.51) which is the corresponding form of the CRLB; i.e., the equality of the bound is now satisfied, provided the noise is pre-whiten.

Hence, in the presence of coloured noise with known covariance matrix, pre-whitening the noise before applying the linear least-squares estimation procedure also results in a MVUE of $\boldsymbol{x}$. We have seen this is not the case when the noise is not pre-whitened.

2.2 A Generalized "Pseudo-Inverse" Approach to Solving the Least-squares Problem

In this lecture we discuss the solution of the least squares problem using the pseudo inverse of a matrix. We first develop its structure and then look at its properties.

2.2.1 Least Squares Solution Using the SVD

Previously we have seen that the LS problem may be posed as

$$\min_{x} \| \boldsymbol{Ax} - \boldsymbol{b} \|_2^2 \tag{2.2.1}$$

where the observation $\boldsymbol{b}$ is generated from the regression model $\boldsymbol{b}=\boldsymbol{Ax}_0+\boldsymbol{n}$. For the case where $\boldsymbol{A}$ is full rank we saw that the solution $\boldsymbol{x}_{\text{LS}}$ which solves is given by the normal equation

$$\boldsymbol{A}^{\mathrm{T}}\boldsymbol{Ax}=\boldsymbol{A}^{\mathrm{T}}\boldsymbol{b} \tag{2.2.2}$$

If the matrix $\boldsymbol{A}$ is rank deficient, then a unique solution to the normal equations does not exist. There is an infinity of solutions which minimize with respect to $\boldsymbol{x}$. However, we can generate a unique solution if, amongst the set of $\boldsymbol{x}$ satisfying, we choose that value of $\boldsymbol{x}$ which itself has minimum norm. Thus, when $\boldsymbol{A}$ is rank deficient, $\boldsymbol{x}_{\text{LS}}$ is the result of two 2-norm minimizing procedures. The first determines a set $\{\boldsymbol{x}\}$ which minimizes $\| \boldsymbol{Ax}-\boldsymbol{b} \|_2^2$, and the second determines $\boldsymbol{x}_{\text{LS}}$ as that element of $\{\boldsymbol{x}\}$ for which $\| \boldsymbol{x} \|_2$ is minimum.

In this respect we develop the pseudo-inverse of a matrix. The pseudo-inverse gives the solution $\boldsymbol{x}_{\text{LS}}$ which is the solution to both these 2-norm minimizing procedures. It is useful for solving least-squares problems when the matrix $\boldsymbol{A}$ is full-rank or rank-deficient. The procedure which follows opens up some very interesting aspects of least-squares analysis which we explore later.

We are given $\boldsymbol{A}\in\mathbb{R}^{m\times n}$, $m>n$ and $\text{rank}(\boldsymbol{A})=r\leqslant n$. If the svd of $\boldsymbol{A}$ is given as $\boldsymbol{U\Sigma V}^{\mathrm{T}}$, then we define $\boldsymbol{A}^{+}$ as the pseudo-inverse of $\boldsymbol{A}$, defined by

$$\boldsymbol{A}^{+}=\boldsymbol{V\Sigma}^{+}\boldsymbol{U}^{\mathrm{T}} \tag{2.2.3}$$

The matrix $\boldsymbol{\Sigma}^{+}$ is related to $\boldsymbol{\Sigma}$ in the following way. If

$$\boldsymbol{\Sigma}=\text{diag}(\sigma_1, \sigma_2, \cdots, \sigma_r, 0, \cdots, 0)$$

then

$$\boldsymbol{\Sigma}^{+}=\text{diag}(\sigma_1^{-1}, \sigma_2^{-1}, \cdots, \sigma_r^{-1}, 0, \cdots, 0) \tag{2.2.4}$$

where $\boldsymbol{\Sigma}$ and $\boldsymbol{\Sigma}^{+}$ are padded with zeros in an appropriate manner to maintain dimensional consistency.

Theorem

When $\boldsymbol{A}$ is rank deficient the unique solution $\boldsymbol{x}_{\text{LS}}$ minimizing such that $\| \boldsymbol{x} \|_2$ is minimum is given by

$$\boldsymbol{x}_{\text{LS}}=\boldsymbol{A}^{+}\boldsymbol{b} \tag{2.2.5}$$

where $\boldsymbol{A}^{+}$ is defined by equation (2.2.3). Further, we have

$$\rho_{\text{LS}}^2 \triangleq \| \boldsymbol{r}_{\text{LS}} \| = \sum_{i=r+1}^{m} (\boldsymbol{u}_i^{\mathrm{T}}\boldsymbol{b})^2 \tag{2.2.6}$$

Proof

For any $\boldsymbol{x}\in\mathbb{R}^n$ we have

$$\| \boldsymbol{Ax}-\boldsymbol{b} \|_2^2 = \| \boldsymbol{U}^{\mathrm{T}}\boldsymbol{AV}(\boldsymbol{V}^{\mathrm{T}}\boldsymbol{x})-\boldsymbol{U}^{\mathrm{T}}\boldsymbol{b} \|_2^2 \tag{2.2.7}$$

$$= \left\| \begin{bmatrix} \boldsymbol{\Sigma}' & 0 \\ 0 & 0 \end{bmatrix} \begin{bmatrix} \boldsymbol{w}_1 \\ \boldsymbol{w}_2 \end{bmatrix} - \begin{bmatrix} \boldsymbol{c}_1 \\ \boldsymbol{c}_2 \end{bmatrix} \right\|_2^2 \tag{2.2.8}$$

where

$$\boldsymbol{w}=\begin{bmatrix}\boldsymbol{w}_1\\ \boldsymbol{w}_2\end{bmatrix}=\begin{matrix}r\\ n-r\end{matrix}\overset{n}{\begin{bmatrix}\boldsymbol{V}_1^{\mathrm{T}}\\ \boldsymbol{V}_2^{\mathrm{T}}\end{bmatrix}}\boldsymbol{x}=\boldsymbol{V}^{\mathrm{T}}\boldsymbol{x} \tag{2.2.9}$$

and

$$\boldsymbol{c}=\begin{bmatrix}\boldsymbol{c}_1\\ \boldsymbol{c}_2\end{bmatrix}=\begin{matrix}r\\ m-r\end{matrix}\overset{m}{\begin{bmatrix}\boldsymbol{U}_1^{\mathrm{T}}\\ \boldsymbol{U}_2^{\mathrm{T}}\end{bmatrix}}\boldsymbol{b}=\boldsymbol{U}^{\mathrm{T}}\boldsymbol{b} \tag{2.2.10}$$

and

$$\boldsymbol{\Sigma}'=\mathrm{diag}[\sigma_1, \sigma_2, \cdots, \sigma_r] \tag{2.2.11}$$

Note that we can write the quantity $\|\boldsymbol{Ax}-\boldsymbol{b}\|_2^2$ in the form of equation (2.2.7), since the 2-norm is invariant to the orthonormal transformation $\boldsymbol{U}^{\mathrm{T}}$, and the quantity $\boldsymbol{VV}^{\mathrm{T}}$ which is inserted between $\boldsymbol{A}$ and $\boldsymbol{x}$ is identical to $\boldsymbol{I}$.

From equation (2.2.8) we can make several immediate conclusions, as follows:

1. Because of the zero blocks in the right column of the matrix in equation (2.2.8), we see that the solution $\boldsymbol{w}$ is independent of $\boldsymbol{w}_2$. Therefore, $\boldsymbol{w}_2$ is arbitrary.

2. Since the argument of the left-hand side of equation (2.2.8) is a vector, it may be expressed as

$$\|\boldsymbol{Ax}-\boldsymbol{b}\|_2^2=\|\boldsymbol{\Sigma}'\boldsymbol{w}_1-\boldsymbol{c}_1\|_2^2+\|\boldsymbol{c}_2\|_2^2 \tag{2.2.12}$$

Therefore, equation (2.2.8) is minimized by choosing $\boldsymbol{w}_1$ to satisfy

$$\boldsymbol{\Sigma}'\boldsymbol{w}_1=\boldsymbol{c}_1 \tag{2.2.13}$$

Note that this fact is immediately apparent, without having to resort to tedious differentiations, which is because the svd reveals so much about the structure of the underlying problem.

3. From equation (2.2.9) we have $\boldsymbol{x}=\boldsymbol{Vw}$. Therefore,

$$\|\boldsymbol{x}\|_2^2=\|\boldsymbol{w}\|_2^2=\|\boldsymbol{w}_1\|_2^2+\|\boldsymbol{w}_2\|_2^2 \tag{2.2.14}$$

Clearly $\|\boldsymbol{x}\|_2^2$ is minimum when $\boldsymbol{w}_2=0$.

4. Combining our definitions for $\boldsymbol{w}_1$ and $\boldsymbol{w}_2$ together, we have

$$\boldsymbol{w}=\begin{bmatrix}\boldsymbol{w}_1\\ \boldsymbol{w}_2\end{bmatrix}=\begin{bmatrix}(\boldsymbol{\Sigma}')^{-1} & 0\\ 0 & 0\end{bmatrix}\boldsymbol{c}=\boldsymbol{\Sigma}^{+}\boldsymbol{c} \tag{2.2.15}$$

This can be written as

$$\boldsymbol{V}^{\mathrm{T}}\boldsymbol{x}_{\mathrm{LS}}=\boldsymbol{\Sigma}^{+}\boldsymbol{U}^{\mathrm{T}}\boldsymbol{b}$$

or

$$\boldsymbol{x}_{\mathrm{LS}}=\boldsymbol{V\Sigma}^{+}\boldsymbol{U}^{\mathrm{T}}\boldsymbol{b}=\boldsymbol{A}^{+}\boldsymbol{b} \tag{2.2.16}$$

which was to be shown. Furthermore, we can say from the bottom half of equation (2.2.8) that

$$\rho_{\mathrm{LS}}^2=\|\boldsymbol{c}_2\|_2^2=\|(\boldsymbol{u}_{r+1}, \cdots, \boldsymbol{u}_m)^{\mathrm{T}}\boldsymbol{b}\|_2^2=\sum_{i=r+1}^{m}(\boldsymbol{u}_i^{\mathrm{T}}\boldsymbol{b})^2 \tag{2.2.17}$$

Note that $\boldsymbol{A}^{+}$ is defined even if $\boldsymbol{A}$ is singular.

This preceding analysis brings out relevant advantages of the singular value decomposition. The svd immediately reveals a great deal about the structure of the matrix

$\boldsymbol{A}$ and as a result, allows a relatively simple development of the pseudo inverse solution. For example in using the svd in least-squares analysis we can see in the rank deficient case, that we can add any vector in span$[\boldsymbol{v}_{r+1}, \cdots, \boldsymbol{v}_n]$ to $\boldsymbol{x}_{\mathrm{LS}}$ without changing ρ_{LS}. Also, it is easy to determine that the residual vector $\boldsymbol{r}_{\mathrm{LS}}=\boldsymbol{A}\boldsymbol{x}_{\mathrm{LS}}-\boldsymbol{b}$ lies in the space $\boldsymbol{U}_2$.

2.2.2 Interpretation of the Pseudo-Inverse

2.2.2.1 Geometrical Interpretation

Let us now take another look at the geometry of least squares. It shows a simple LS problem for the case $\boldsymbol{A}\in\mathbb{R}^{2\times 1}$. We again see that $\boldsymbol{x}_{\mathrm{LS}}$ is the solution which corresponds to projecting $\boldsymbol{b}$ onto $\boldsymbol{R}(\boldsymbol{A})$. In fact, substituting into the expression $\boldsymbol{A}\boldsymbol{x}_{\mathrm{LS}}$, we get

$$\boldsymbol{A}\boldsymbol{x}_{\mathrm{LS}}=\boldsymbol{A}\boldsymbol{A}^{+}\boldsymbol{b} \tag{2.2.18}$$

But, for the specific case where $m>n$, we know from our previous discussion on linear least squares, that

$$\boldsymbol{A}\boldsymbol{x}_{\mathrm{LS}}=\boldsymbol{P}\boldsymbol{b} \tag{2.2.19}$$

where $\boldsymbol{P}$ is the projector onto $\boldsymbol{R}(\boldsymbol{A})$. Comparing equation (2.2.18) and (2.2.19), and noting the projector is unique, we have

$$\boldsymbol{P}=\boldsymbol{A}\boldsymbol{A}^{+} \tag{2.2.20}$$

Thus, the matrix $\boldsymbol{A}\boldsymbol{A}^{+}$ is a projector onto $\boldsymbol{R}(\boldsymbol{A})$.

This may also be seen in a different way as follows Using the definition of $\boldsymbol{A}^{+}$, we have

$$\boldsymbol{A}\boldsymbol{A}^{+}=\boldsymbol{U}\boldsymbol{\Sigma}\boldsymbol{V}^{\mathrm{T}}\boldsymbol{V}\boldsymbol{\Sigma}^{+}\boldsymbol{U}^{\mathrm{T}}=\boldsymbol{U}\begin{bmatrix}\boldsymbol{I}_r & 0\\ 0 & 0\end{bmatrix}\boldsymbol{U}^{\mathrm{T}}=\boldsymbol{U}_r\boldsymbol{U}_r^{\mathrm{T}} \tag{2.2.21}$$

Where $\boldsymbol{I}_r$ is the $r\times r$ identity and $\boldsymbol{U}_r=[\boldsymbol{u}_1, \cdots, \boldsymbol{u}_r]$. From our discussion on projectors, we know $\boldsymbol{U}_r\boldsymbol{U}_r^{\mathrm{T}}$ is also a projector onto $\boldsymbol{R}(\boldsymbol{A})$ which is the same as the column space of $\boldsymbol{A}$.

We also note that it is just as easy to show that for the case $m<n$, the matrix $\boldsymbol{A}^{+}\boldsymbol{A}$ is a projector onto the row space of $\boldsymbol{A}$.

2.2.2.2 Relationship of Pseudo-Inverse Solution to Normal Equations

Suppose $\boldsymbol{A}\in\mathbb{R}^{m\times n}$, $m>n$, the normal equations give us

$$\boldsymbol{x}_{\mathrm{LS}}=(\boldsymbol{A}^{\mathrm{T}}\boldsymbol{A})^{-1}\boldsymbol{A}^{\mathrm{T}}\boldsymbol{b} \tag{2.2.22}$$

but the pseudo-inverse gives:

$$\boldsymbol{x}_{\mathrm{LS}}=\boldsymbol{A}^{+}\boldsymbol{b} \tag{2.2.23}$$

In the full rank case, these two quantities must be equal. We can indeed show this is the case as follows:

We let

$$\boldsymbol{A}^{\mathrm{T}}\boldsymbol{A}=\boldsymbol{V}\boldsymbol{\Sigma}^2\boldsymbol{V}^{\mathrm{T}} \tag{2.2.24}$$

be the ED of $\boldsymbol{A}^{\mathrm{T}}\boldsymbol{A}$ and we let the SVD of $\boldsymbol{A}^{\mathrm{T}}$ be defined as

$$\boldsymbol{A}^{\mathrm{T}}=\boldsymbol{V}\boldsymbol{\Sigma}\boldsymbol{U}^{\mathrm{T}} \tag{2.2.25}$$

Using these relations we have

$$(\boldsymbol{A}^{\mathrm{T}}\boldsymbol{A})^{-1}\boldsymbol{A}^{\mathrm{T}}=(\boldsymbol{V}\boldsymbol{\Sigma}^{2}\boldsymbol{V}^{\mathrm{T}})\boldsymbol{V}\boldsymbol{\Sigma}\boldsymbol{U}^{\mathrm{T}}=\boldsymbol{V}\boldsymbol{\Sigma}^{-1}\boldsymbol{U}^{\mathrm{T}}=\boldsymbol{A}^{+} \tag{2.2.26}$$

as desired, where the last line follows from . Thus, for the full-rank case for $m>n$, $\boldsymbol{A}^{+}=(\boldsymbol{A}^{\mathrm{T}}\boldsymbol{A})^{-1}\boldsymbol{A}^{\mathrm{T}}$. In a similar way, we can also show that $\boldsymbol{A}^{+}=\boldsymbol{A}(\boldsymbol{A}^{\mathrm{T}}\boldsymbol{A})^{-1}$ for the case $m<n$.

2.2.2.3 The Pseudo-Inverse as a Generalized Linear System Solver

If we are willing to accept the least squares solution when the ordinary solution to a system of linear equations does not exist (e.g. when the system is over-determined), and if we can accept the definition of a unique solution as that which has minimum 2-norm then $\boldsymbol{x}=\boldsymbol{A}^{+}\boldsymbol{b}$ solves the system $\boldsymbol{A}\boldsymbol{x}=\boldsymbol{b}$ regardless of whether $\boldsymbol{A}$ is full rank or rank deficient, when the system is either over determined square or under determined.

A generalized inverse $\boldsymbol{X}\in\mathbb{R}^{n\times m}$ of $\boldsymbol{A}\in\mathbb{R}^{m\times n}$ (note dimensions are transposes of each other) must satisfy the following 4 Moore-Penrose conditions:

(1) $\boldsymbol{AXA}=\boldsymbol{A}$

(2) $\boldsymbol{XAX}=\boldsymbol{X}$

(3) $(\boldsymbol{AX})^{\mathrm{T}}=\boldsymbol{AX}$

(4) $(\boldsymbol{XA})^{\mathrm{T}}=\boldsymbol{XA}$

It is easily shown that $\boldsymbol{A}^{+}$ defined by equation (2.2.3) indeed satisfies these conditions.

The four Moore-Penrose conditions are equivalent to the matrices $\boldsymbol{AX}$ and $\boldsymbol{XA}$ being projectors onto the column space (for $m>n$) and row space (for $m<n$) of $\boldsymbol{A}$ respectively. We illustrate this loosely as follows. Recall that a projector matrix must have three specific properties. From condition (1) above we have $\boldsymbol{AXA}=\boldsymbol{PA}=\boldsymbol{A}$ which means that $\boldsymbol{P}$ spans the column of $\boldsymbol{A}$, as required which is one of the required properties of a projector. Conditions (3) and (4) lead directly to the symmetry property of the projector. The idempotent property follows from pre-multiplying or post-multiplying (1) or (2) by $\boldsymbol{X}$ or $\boldsymbol{A}$ as appropriate, to obtain $\boldsymbol{P}^{2}=\boldsymbol{P}$.

Chapter 3 Principal Component Analysis

3.1 Introductory Example

To set the stage for this paper, we will start with a small example where principal component analysis (PCA) can be useful. Red wines, 44 samples, produced from the same grape (*Cabernet Sauvignon*) were collected. Six of these were from Argentina, fifteen from Chile, twelve from Australia and eleven from South Africa. A Foss WineScan instrument was used to measure 14 characteristic parameters of the wines such as the ethanol content, pH, etc. (Table 3.1).

Table 3.1 Chemical Parameters Determined on the Wine Samples

(data from http://www.models.life.ku.dk/Wine_GCMS_FTIR)

Ethanol (vol%)
Total acid (g L^{-1})
Volatile acid (g L^{-1})
Malic acid (g L^{-1})
PH
Lactic acid (g L^{-1})
Rest Sugar (Glucose + Fructose) (g L^{-1})
Citric acid (mg L^{-1})
CO_2 (g L^{-1})
Density (g mL^{-1})
Total polyphenol index
Glycerol (g L^{-1})
Methanol (vol%)
Tartaric acid (g L^{-1})

Hence, a dataset is obtained which consists of 44 samples and 14 variables. The actual measurements can be arranged in a table or a matrix of size 44×14. A portion of this table is shown in Fig. 3.1.

	A	Ethanol	TotalAcid	VolatileA	MalicAcid	pH	Lactic Acid	ReSugar	CitricAcid	CO2	Density	FolinC	Glycerol	Methanol	TartaricA
2	ARG-BNS1	13,62	3,54	0,29	0,89	3,71	0,78	1,46	0,31	85,61	0,99	60,92	9,72	0,16	1,74
3	ARG-DDA1	14,06	3,74	0,59	0,24	3,73	1,25	2,42	0,18	175,20	1,00	70,64	10,05	0,20	1,58
4	ARG-FFL1	13,74	3,2					1,52	0,39	513,74	0,99	63,59	10,92	0,18	1,24
5	ARG-FLM1	13,95	[illegible]					4,17	0,41	379,40	1,00	73,30	9,69	0,23	2,26
6	ARG-ICR1	14,47	3,6					1,25	0,14	154,88	0,99	71,69	10,81	0,20	1,22
7	ARG-SAL1	14,61	3,4					1,40	0,10	156,30	0,99	71,79	10,19	0,19	0,90
8	AUS-CAV1	13,65	4,3					3,80	0,24	462,62	1,00	59,60	10,66	0,25	1,81
9	AUS-EAG1	14,12	3,8					4,32	0,32	244,15	1,00	59,50	11,07	0,25	1,65
10	AUS-HAR1	13,13	3,8					3,99	0,34	212,00	1,00	59,42	8,89	0,23	2,12
11	AUS-IB41	13,49	3,6					6,40	0,13	419,38	1,00	63,86	10,35	0,26	1,81
12	AUS-KIL1	15,09	3,9					1,06	-0,04	48,02	0,99	70,10	11,43	0,19	1,47
13	AUS-KIR1	14,63	4,7								1,00	72,37	11,64	0,28	2,12
14	AUS-NUG1	13,63	4,6								1,00	55,07	9,59	0,25	1,36
15	AUS-SOC1	13,67	3,8								0,99	63,04	11,28	0,14	1,01
16	AUS-TGH1	14,43	4,5								1,00	63,52	10,93	0,30	1,81
17	AUS-VAF1	13,45	4,3								0,99	62,69	9,46	0,18	2,13
18	AUS-WBL1	13,83	4,22	0,33	0,45						0,99	59,08	11,10	0,22	1,55
19	AUS-WES1	13,85	4,16	0,36	[illegible]						1,00	83,51	10,45	0,24	2,47
20	CHI-CDD1	13,97	3,54	0,29	0,48						0,99	64,31	10,58	0,18	1,72
21	CHI-CDM1	12,84	3,22	0,34	0,42						1,00	53,10	8,80	0,17	1,85
22	CHI-CMO1	14,19	3,40	0,35	0,46						0,99	66,82	10,11	0,18	1,48
23	CHI-CSU1	14,13	[illegible]	0,33	0,31						0,99	64,83	9,85	0,22	1,83
24	CHI-GNE1	13,66	3,08	0,28	0,42						1,00	52,16	9,54	0,18	1,38
25	CHI-IND1	14,27	3,43	0,44	0,45	3,76	0,79	1,52	0,06	247,29	0,99	63,75	9,93	0,21	1,48
26	CHI-LIO1	13,84	3,05	0,26	0,47	3,71	0,80	2,08	0,21	399,07	0,99	56,55	9,48	0,19	1,66

Fig. 3. 1 A subset of the wine dataset

With 44 samples and 14 columns, it is quite complicated to get an overview of what kind of information is available in the data. A good starting point is to plot individual variables or samples. Three of the variables are shown in Fig. 3. 2. It can be seen that total acid as well as methanol tends to be higher in samples from Australia and South Africa whereas there are less pronounced regional differences in the ethanol content.

Fig. 3. 2

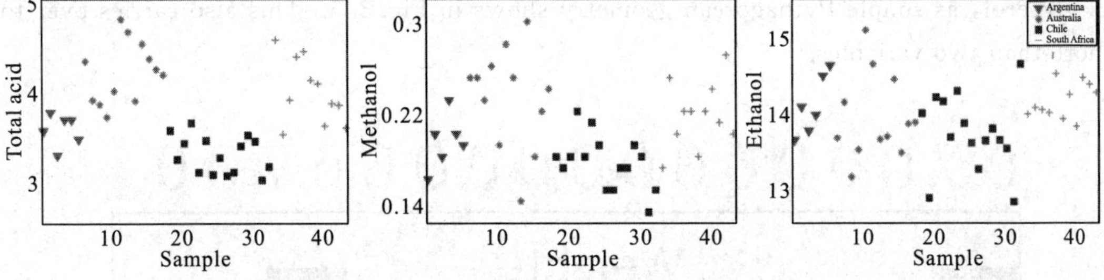

Fig. 3. 2 Three variables coloured according to the region

Even though Fig. 3. 2 may suggest that there is little relevant regional information in ethanol, it is dangerous to rely too much on univariate analysis. In univariate analysis, any co-variation with other variables is explicitly neglected and this may lead to important features being ignored. For example, plotting ethanol versus glycerol (Fig. 3. 3) shows an interesting correlation between the two. This is difficult to deduce from plots of the individual variables. If glycerol and ethanol were completely correlated, it would, in fact, be possible to simply use e. g. the average or the sum of the two as one new variable that could replace the two original ones. No information would be lost as it would always be possible to go from e. g. the average to the two original variables.

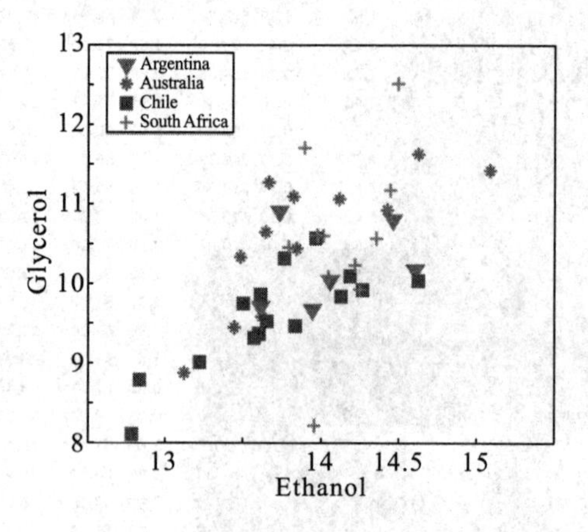

Fig. 3. 3　A plot of ethanol versus glycerol

This concept of using suitable linear combinations of the original variables will turn out to be essential in PCA and is explained in a bit more detail and a slightly unusual way here. The new variable, say, the average of the two original ones, can be defined as a weighted average of all 14 variables; only the other variables will have weight zero. These 14 weights are shown in Fig. 3. 4. Rather than having the weights of ethanol and glycerol to be 0. 5 as they would in an ordinary average, they are chosen as 0. 7 to make the whole 14-vector of weights scaled to be a unit vector. When the original variables ethanol and glycerol are taken to be of length one (unit length) then it is convenient to also have the linear combination of those to be of length one. This then defines the unit on the combined variable. To achieve this it is necessary to take 0. 7 ($\sqrt{2}/2$ to be exact) of ethanol and 0. 7 of glycerol, as simple Pythagorean geometry shows in Fig. 3. 5. This also carries over to more than two variables.

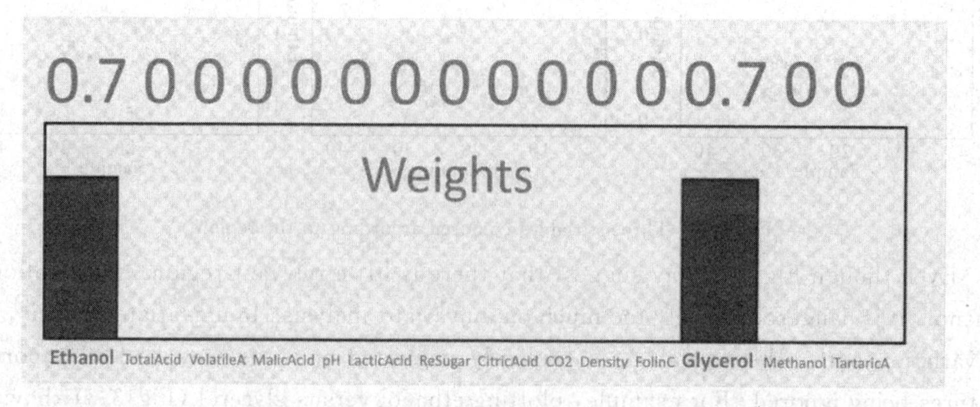

Fig. 3. 4　Defining the weights for a variable that includes only ethanol and glycerol information

Using a unit weight vector has certain advantages. The most important one is that the

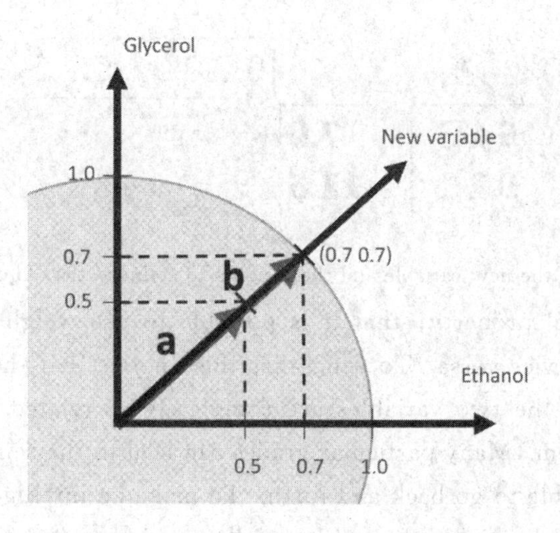

Fig. 3. 5 The concept of a unit vector

unit vector preserves the size of the variation. Imagine there are ten variables rather than two that are being averaged. Assume, for simplicity that all ten have the value five.

Regardless of whether the average is calculated from two or ten variables, the average remains five. Using the unit vector, though, will provide a measure of the number of variables showing variation. In fact, the variance of the original variables and this newly calculated one will be the same, if the original variables are all correlated. Thus, using the unit vector preserves the variation in the data and this is an attractive property. One of the reasons is that it allows for going back and forth between the space of the original variables (say glycerol-ethanol) and the new variable. With this deftnition of weights, it is now possible to calculate the new variable, the ' average', for any sample, as indicated in Fig. 3. 6.

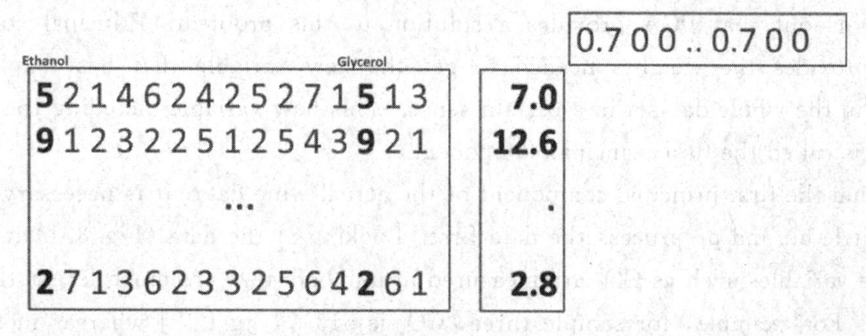

Fig. 3. 6 Using defined weights to calculate a new variable that is a scaled average of ethanol and glycerol (arbitrary numbers used here). The average is calculated as the inner product of the 14 measurements of a sample and the weight vector. Some didactical rounding has been used in the example

As mentioned above, it is possible to go back and forth between the original two variables and the new variable. Multiplying the new variable with the weights provides an estimation of the original variables (Fig. 3. 7).

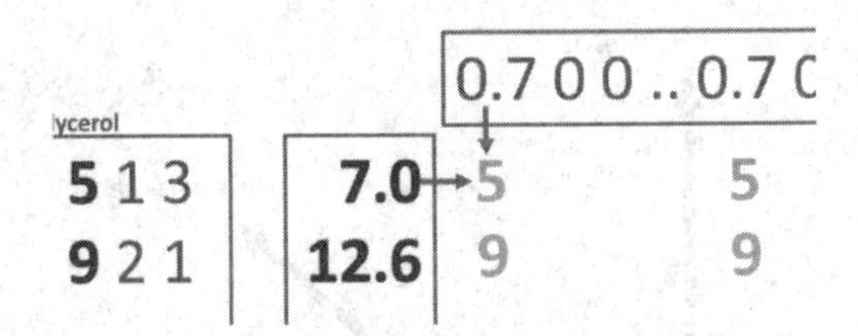

Fig. 3. 7 Using the new variable and the weights to estimate the old original variables

This is a powerful property; that it is possible to use weights to condense several variables into one and vice versa. To generalize this, notice that the current concept only works perfectly when the two variables are completely correlated. Think of an average grade in a school system. Many particular grades can lead to the same average grade, so it is not in general possible to go back and forth. To make an intelligent new variable, it is natural to ask for a new variable that will actually provide a nice model of the data. That is, a new variable which, when multiplied with the weights, will describe as much as possible the whole matrix (Fig. 3. 8). Such a variable will be an optimal representative of the whole data in the sense that no other weighted average simultaneously describes as much of the information in the matrix.

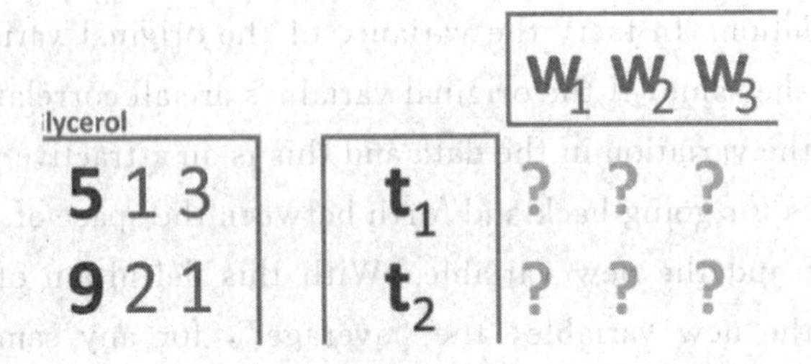

Fig. 3. 8 Defining weights (w's) that will give a new variable which leads to a good model of the data

It turns out that PCA provides a solution to this problem. Principal component analysis provides the weights needed to get the new variable that best explains the variation in the whole dataset in a certain sense. This new variable including the defining weights, is called the first principal component.

To find the first principal component of the actual wine data, it is necessary to jump ahead a little bit and preprocess the data first. Looking at the data (Fig. 3. 1) it is seen, that some variables such as CO_2 are measured in numbers that are much larger than e. g. methanol. For example, for sample three, CO_2 is 513. 74 [g L^{-1}] whereas methanol is 0. 18 [vol%]. If this difference in scale and possibly offset is not handled, then the PCA model will only focus on variables measured in large numbers. It is desired to model all variables, and there is a preprocessing tool called autoscaling which will make each column have the same "size" so that all variables have an equal opportunity of being modelled. Autoscaling means that from each variable, the mean value is subtracted and then the variable is divided by its standard deviation. Autoscaling will be described in more detail, but for now, it is just important to note that each variable is transformed to equal size and in the process, each variable will have negative as well as positive values because the mean of it has been subtracted. Note that an average sample now corresponds to all zeroes. Hence, zero is no longer absence of a "signal" but instead indicates an average "signal".

With this pre-processing of the data, PCA can be performed. The technical details of how to do that will follow, but the first principal component is shown in Fig. 3. 9. In the lower plot, the weights are shown. Instead of the quite sparse weights in Fig. 3. 4, these weights are non-zero for all variables. This first component does not explain all the variation, but it does explain 25% of what is happening in the data. As there are 14 variables, it would be expected that if every variable showed variation independent of the other, then each original variable would explain 100%/14=7% of the variation. Hence, this first component is wrapping up information, which can be said to correspond to approximately 3 - 4 variables.

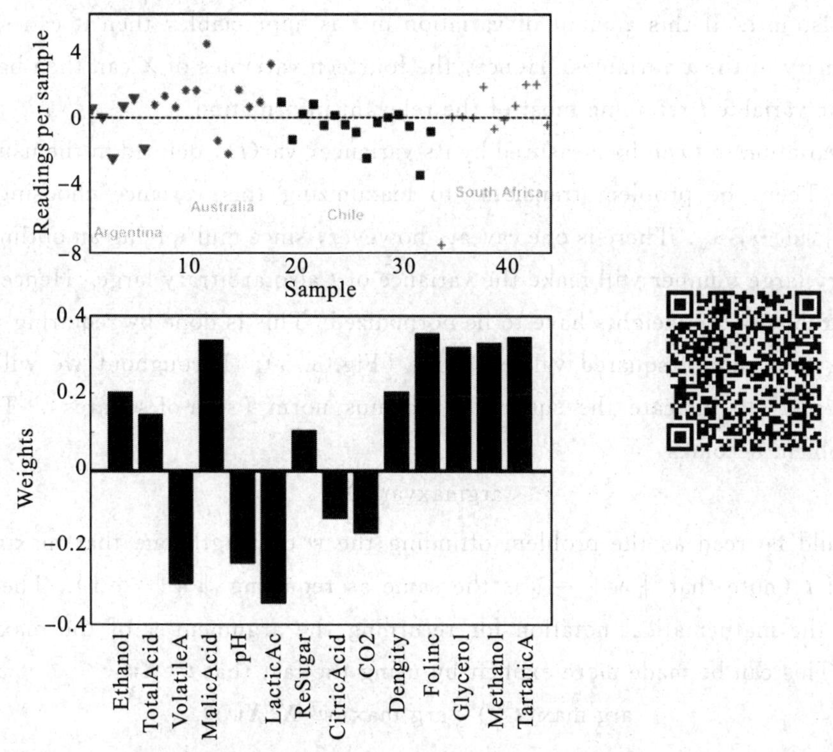

Fig. 3. 9 The first principal component of the wine data. The lower plot shows the weights and the upper plot shows the weighted averages obtained with those weights

Just like the average of ethanol and glycerol or the average school grade, the new variable can be interpreted as "just a variable". The weights define how the variable is determined and how many scores each sample has of this linear combination. For example, it is seen that most of the South African samples have positive scores and hence, will have fairly high values on variables that have positive weights such as for example methanol. This is confirmed in Fig. 3. 3.

3.2 Theory

3.2.1 Taking Linear Combinations

The data are collected in a matrix $\boldsymbol{X}$ with I rows ($i=1,2,\cdots,I$; usually samples/objects) and J columns ($j=1,2,\cdots,J$; usually variables), hence of size $I\times J$. The individual variables (columns) of $\boldsymbol{X}$ are denoted by $\boldsymbol{x}_j(j=1,2,\cdots,J)$ and are all vectors in the I-dimensional space. A linear combination of those $\boldsymbol{x}$ variables can be written as $\boldsymbol{t}=w_1\times\boldsymbol{x}_1+\cdots w_J\times\boldsymbol{x}_J$, where $\boldsymbol{t}$ is now a new vector in the same space as the $\boldsymbol{x}$ variables (because it is a linear combination of these). In matrix notation, this becomes $\boldsymbol{t}=\boldsymbol{X}\boldsymbol{w}$, with $\boldsymbol{w}$ being the vector with elements $w_j(j=1,2,\cdots,J)$. Since the matrix $\boldsymbol{X}$ contains variation relevant to the problem, it seems reasonable to have as much as possible of that variation also in $\boldsymbol{t}$. If this amount of variation in $\boldsymbol{t}$ is appreciable, then it can serve as a good summary of the $\boldsymbol{x}$ variables. Hence, the fourteen variables of $\boldsymbol{X}$ can then be replaced by only one variable $\boldsymbol{t}$ retaining most of the relevant information.

The variation in $\boldsymbol{t}$ can be measured by its variance, var($\boldsymbol{t}$), defined in the usual way in statistics. Then the problem translates to maximizing this variance choosing optimal weights $w_1, w_2\cdots, w_J$. There is one caveat, however, since multiplying an optimal $\boldsymbol{w}$ with an arbitrary large number will make the variance of $\boldsymbol{t}$ also arbitrary large. Hence, to have a proper problem, the weights have to be normalized. This is done by requiring that their norm, i. e. the sum-of-squared values is one (Fig. 3. 5). Throughout we will use the symbol $\|\cdot\|^2$ to indicate the squared Frobenius norm (sum-of-squares). Thus, the formal problem becomes

$$\underset{\|\boldsymbol{w}\|=1}{\mathrm{argmax}}\,\mathrm{var}(\boldsymbol{t}) \tag{3.2.1}$$

which should be read as the problem offinding the $\boldsymbol{w}$ of length one that maximizes the variance of $\boldsymbol{t}$ (note that $\|\boldsymbol{w}\|=1$ is the same as requiring $\|\boldsymbol{w}\|^2=1$). The function argmax is the mathematical notation for returning the argument $\boldsymbol{w}$ of the maximization function. This can be made more explicit by using the fact that $\boldsymbol{t}=\boldsymbol{X}\boldsymbol{w}$:

$$\underset{\|\boldsymbol{w}\|=1}{\arg\max}(\boldsymbol{t}^{\mathrm{T}}\boldsymbol{t})=\underset{\|\boldsymbol{w}\|=1}{\arg\max}(\boldsymbol{w}^{\mathrm{T}}\boldsymbol{X}^{\mathrm{T}}\boldsymbol{X}\boldsymbol{w}) \tag{3.2.2}$$

where it is assumed that the matrix $\boldsymbol{X}$ is mean-centered (then all linear combinations are also mean-centered). The latter problem is a standard problem in linear algebra and the optimal $\boldsymbol{w}$ is the (standardized) first eigenvector (i. e. the eigenvector with the largest value) of the covariance matrix $\boldsymbol{X}^{\mathrm{T}}\boldsymbol{X}/(n-1)$ or the corresponding cross-product matrix $\boldsymbol{X}^{\mathrm{T}}\boldsymbol{X}$.

3.2.2 Explained Variation

The variance of $\boldsymbol{t}$ can now be calculated but a more meaningful assessment of the summarizing capability of $\boldsymbol{t}$ is obtained by calculating how representative $\boldsymbol{t}$ is in terms of replacing $\boldsymbol{X}$. This can be done by projecting the columns of $\boldsymbol{X}$ on $\boldsymbol{t}$ and calculating the residuals of that projection. This is performed by regressing all variables of $\boldsymbol{X}$ on $\boldsymbol{t}$ using the ordinary regression equation

$$\boldsymbol{X}=\boldsymbol{t}\boldsymbol{p}^{\mathrm{T}}+\boldsymbol{E} \tag{3.2.3}$$

where $\boldsymbol{p}$ is the vector of regression coefficients and $\boldsymbol{E}$ is the matrix of residuals. Interestingly, $\boldsymbol{p}$ equals $\boldsymbol{w}$ and the whole machinery of regression can be used to judge the quality of the summarizer $\boldsymbol{t}$. Traditionally, this is done by calculating

$$\frac{\|\boldsymbol{X}\|^{2}-\|\boldsymbol{E}\|^{2}}{\|\boldsymbol{X}\|^{2}}100\% \tag{3.2.4}$$

which is referred to as the percentage of explained variation of $\boldsymbol{t}$.

In Fig. 3. 10, it is illustrated how the explained variation is calculated as also explained around equation (3. 2. 4).

$$100\left(1-\frac{\sum E^{2}}{\sum X^{2}}\right)$$

Unexplained fraction

Explained fraction of variation

Percentage variation explained

Fig. 3. 10 Exemplifying how explained variation is calculated using the data and the residuals

Note that, the measures above are called variations rather than variances. In order to talk about variances, it is necessary to correct for the degrees of freedom consumed by the model and this is not a simple task. Due to the non-linear nature of the PCA model, degrees of freedom are not as simple to define as for linear models such as in linear regression or analysis of variance. Hence, throughout this paper, the magnitude of variation will simply be expressed in terms of sums of squares.

3.2.3 PCA as a Model

Equation (3. 2. 3) highlights an important interpretation of PCA: it can be seen as a modelling activity (Fig. 3. 11). By rewriting equation (3. 2. 3) as

$$\boldsymbol{X}=\boldsymbol{t}\boldsymbol{p}^{\mathrm{T}}+\boldsymbol{E}=\hat{\boldsymbol{X}}+\boldsymbol{E} \tag{3.2.5}$$

shows that the (outer-) product $\boldsymbol{t}\boldsymbol{p}^{\mathrm{T}}$ serves as a model of $\boldsymbol{X}$ (indicated with a hat). In this equation, vector $\boldsymbol{t}$ was a fixed regressor and vector $\boldsymbol{p}$ the regression coefficient to be found. It can be shown that actually both $\boldsymbol{t}$ and $\boldsymbol{p}$ can be established from such an equation by solving

$$\underset{t,p}{\arg\min}\|\boldsymbol{X}-\boldsymbol{t}\boldsymbol{p}^{\mathrm{T}}\|^{2} \tag{3.2.6}$$

which is also a standard problem in linear algebra and has the same solution as equation

(3.2.2). Note that the solution does not change if $\boldsymbol{t}$ is pre-multiplied by $a \neq 0$ and simultaneously $\boldsymbol{p}$ is divided by that same value. This property is called the scaling ambiguity and it can be solved in different ways. In chemometrics, the vector $\boldsymbol{p}$ is normalized to length one ($\|\boldsymbol{p}\|=1$). The vector $\boldsymbol{t}$ is usually referred to as the score vector (or scores in shorthand) and the vector $\boldsymbol{p}$ is called the loading vector (or loadings in shorthand). The term principal component is not clearly defined and can mean either the score vector or the loading vector or the combination. Since the score and loading vectors are closely tied together it seems logical to reserve the term principal component for the pair $\boldsymbol{t}$ and $\boldsymbol{p}$.

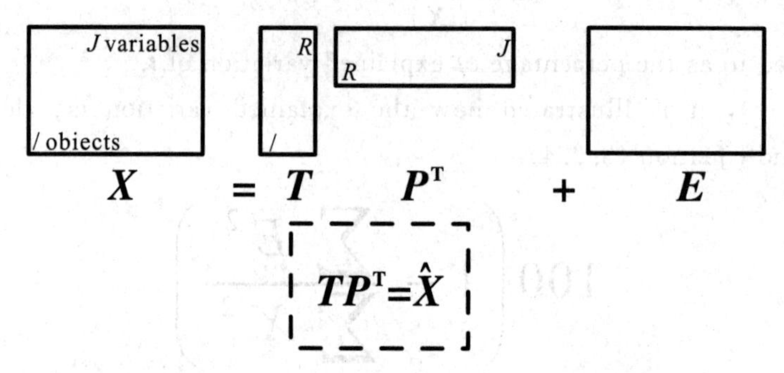

Fig. 3.11 The structure of a PCA model. Note that residuals ($\boldsymbol{E}$) have the same structure as the data and so does the model approximation of the data ($\boldsymbol{TP}^{\mathrm{T}}$)

3.2.4 Taking More Components

If the percentage of explained variation of equation (3.2.4) is too small, then the $\boldsymbol{t}$, $\boldsymbol{p}$ combination is not a sufficiently good summarizes of the data. Equation (3.2.5) suggests an extension by writing

$$\boldsymbol{X}=\boldsymbol{TP}^{\mathrm{T}}+\boldsymbol{E}=\boldsymbol{t}_1\boldsymbol{p}_1{}^{\mathrm{T}}+\cdots+\boldsymbol{t}_R\boldsymbol{p}_R^{\mathrm{T}}=\hat{X}+\boldsymbol{E} \tag{3.2.7}$$

where $\boldsymbol{T}=[\boldsymbol{t}_1, \boldsymbol{t}_2, \cdots, \boldsymbol{t}_R](I\times R)$ and $\boldsymbol{P}=[\boldsymbol{p}_1, \boldsymbol{p}_2, \cdots \boldsymbol{p}_R](J\times R)$ are now matrices containing, respectively, R score vectors and R loading vectors. If R is (much) smaller than J, then $\boldsymbol{T}$ and $\boldsymbol{P}$ still amount to a considerably more parsimonious description of the variation in $\boldsymbol{X}$. To identify the solution, $\boldsymbol{P}$ can be taken such that $\boldsymbol{P}^{\mathrm{T}}\boldsymbol{P}=\boldsymbol{I}$ and $\boldsymbol{T}$ can be taken such that $\boldsymbol{T}^{\mathrm{T}}\boldsymbol{T}$ is a diagonal matrix. This corresponds to the normalization of the loadings mentioned above. Each loading vector, thus has norm one and is orthogonal to other loading vectors (an orthogonal basis). The constraint on $\boldsymbol{T}$ implies that the score vectors are orthogonal to each other. This is the usual way to perform PCA in chemometrics. Due to the orthogonality in $\boldsymbol{P}$, the R components have independent contributions to the overall explained variation

$$\|\boldsymbol{X}\|^2=\|\boldsymbol{t}_1\boldsymbol{p}_1^{\mathrm{T}}\|^2+\cdots+\|\boldsymbol{t}_R\boldsymbol{p}_R^{\mathrm{T}}\|^2+\|\boldsymbol{E}\|^2 \tag{3.2.8}$$

and the term "explained variation per component" can be used, similarly as in equation (3.2.4).

3.3 History of PCA

PCA has been (re-)invented several times. The earliest presentation was in terms of equation (3.2.6). This interpretation stresses the modelling properties of PCA and is very much rooted in regression-thinking: variation explained by the principal components (Pearson's view). Later, in the thirties, the idea of taking linear combinations of variables was introduced and the variation of the principal components was stressed (equation (3.2.1); Hotelling's view). This is a more multivariate statistical approach. Later, it was realized that the two approaches were very similar. Similar, but not the same. There is a fundamental conceptual difference between the two approaches, which is important to understand. In the Hotelling approach, the principal components are taken seriously in their specific direction. The first component explains the most variation, the second component the second most, etc. This is called the principal axis property: the principal components define new axes which should be taken seriously and have a meaning. PCA finds these principal axes. In contrast, in the Pearson approach it is the subspace, which is important, not the axes as such. The axes merely serve as a basis for this subspace. In the Hotelling approach, rotating the principal components destroys the interpretation of these components whereas in the Pearson conceptual model rotations merely generate a different basis for the (optimal) subspace.

3.4 Practical Aspects

3.4.1 Preprocessing

Often a PCA performed on the raw data is not very meaningful. In regression analysis, often an intercept or offset is included since it is the deviation from such an offset, which represents the interesting variation. In terms of the prototypical example, the absolute levels of the pH is not that interesting but the variation in pH of the different Cabernets is relevant. For PCA to focus on this type of variation it is necessary to mean-center the data. This is simply performed by subtracting from every variable in $\boldsymbol{X}$ the corresponding mean-level.

Sometimes it is also necessary to think about the scales of the data. In the wine example, there were measurements of concentrations and of pH. These are not on the same scales (not even in the same units) and to make the variables more comparable, the variables are scaled by dividing them by the corresponding standard deviations. The combined process of centering and scaling in this way is often called autoscaling. For a more detailed account of centering and scaling, see the references[1][2].

Centering and scaling are the two most common types of preprocessing and they normally always have to be decided upon. There are many other types of preprocessing methods available though. The appropriate preprocessing typically depends on the nature of the data investigated[3-7].

3.4.2 Choosing the Number of Components

A basic rationale in PCA is that the informative rank of the data is less than the number of original variables. Hence, it is possible to replace the original J variables with $R(R \ll J)$ components and gain a number of benefits. The influence of noise is minimized as the original variables are replaced with weighted averages[8], and the interpretation and visualization is greatly aided by having a simpler (fewer variables) view to all the variations. Furthermore, the compression of the variation into fewer components can yield statistical benefits in further modelling with the data. Hence, there are many good reasons to use PCA. In order to use PCA, though, it is necessary to be able to decide on how many components to use. The answer to that problem depends a little bit on the purpose of the analysis, which is why the following three sections will provide different answers to that question.

3.4.2.1 Exploratory Studies

In exploratory studies, there is no quantitatively well-defined purpose with the analysis. Rather, the aim is often to just "have a look at the data". The short answer to how many components to use then is "just use the first few components". A slightly more involved answer is that in exploratory studies, it is quite common not to fix the number of components very accurately. Often, the interest is in looking at the main variation and per definition, the first components provide information on that. As e. g. component one and three do not change regardless of whether component six or seven is included, it is often not too critical to establish the exact number of components. Components are looked at and interpreted from the first component and downwards. Each extra component is less and less interesting as the variation explained is smaller and smaller, so often a gradual decline of interest is attached to components. Note that this approach for assessing the importance of components is not to be taken too literally. There may well be reasons why smaller variations are important for a specific dataset[9].

If outliers are to be diagnosed with appropriate statistics (see next section), then, however, it is more important to establish the number of components to use. For example, the residual will change depending on how many components are used, so in order to be able to assess residuals, a reasonable number of components must be used. There are several ad hoc approaches that can be used to determine the number of components. A selection of methods is offered below, but note that these methods seldom provide clear-cut and definitive answers. Instead, they are often used in a combined way to get an impression on the effective rank of the data.

3.4.2.2 Eigenvalues and Their Relation to PCA

Before the methods are described, it is necessary to explain the relation between PCA and eigenvalues. An eigenvector of a (square) matrix $\mathbf{A}$ is defined as the nonzero vector z with the following property:

$$\mathbf{A}z=\lambda z \tag{3.4.1}$$

Where z is called the eigenvector. If matrix $\mathbf{A}$ is symmetric (semi-) positive definite, then the full eigenvalue decomposition of $\mathbf{A}$ becomes:

$$\mathbf{A}=\mathbf{Z}\boldsymbol{\Lambda}\mathbf{Z}^{\mathrm{T}} \tag{3.4.2}$$

Where $\mathbf{Z}$ is an orthogonal matrix and $\boldsymbol{\Lambda}$ is a nonzero diagonal matrix, and it is customary to work with co-variance or correlation matrices and these are symmetric (semi) positive definite. Hence, equation (3.4.2) describes their eigenvalue decomposition. Since all eigenvalues of such matrices are nonnegative, it is customary to order them from high to low; and refer to the first eigenvalue as the largest one.

The singular value decomposition of $\mathbf{X}(I\times J)$ is given by

$$\mathbf{X}=\mathbf{U}\mathbf{S}\mathbf{V}^{\mathrm{T}} \tag{3.4.3}$$

Where $\mathbf{U}$ is an $(I\times J)$ orthogonal matrix $\mathbf{U}^{\mathrm{T}}\mathbf{U}=\mathbf{I}$; $\mathbf{S}(I\times J)$ is a diagonal matrix with the nonzero singular values on its diagonal and $\mathbf{V}$ is an $(J\times J)$ orthogonal matrix ($\mathbf{V}^{\mathrm{T}}\mathbf{V}=\mathbf{V}\mathbf{V}^{\mathrm{T}}=\mathbf{I}$). This is for the case of $I>J$, but the other cases follow similarly. Considering $\mathbf{X}^{\mathrm{T}}\mathbf{X}$ and upon using equation (3.4.3) and it follows:

$$\mathbf{X}^{\mathrm{T}}\mathbf{X}=\mathbf{V}\mathbf{S}^{\mathrm{T}}\mathbf{U}^{\mathrm{T}}\mathbf{U}\mathbf{S}\mathbf{V}^{\mathrm{T}}=\mathbf{U}\mathbf{S}^{2}\mathbf{V}^{\mathrm{T}}=\mathbf{Z}\boldsymbol{\Lambda}\mathbf{Z}^{\mathrm{T}} \tag{3.4.4}$$

This shows the relationship between the singular values and the eigenvalues. The eigenvalue corresponding to a component is the same as the squared singular value which again is the variation of the particular component.

3.4.2.3 Scree Test

The scree test was developed by R. B. Cattell in 1966[10]. It is based on the assumption that relevant information is larger than random noise and that the magnitude of the variation of random noise seems to level off quite linearly with the number of components. Traditionally, the eigenvalues of the cross-product of the preprocessed data, are plotted as a function of the number of components, and when only noise is modelled, it is assumed that the eigenvalues are small and decline gradually. In practice, it may be difficult to see this in the plot of eigenvalues due to the huge eigenvalues and often the logarithm of the eigenvalues is plotted instead. Both are shown in Fig. 3.12 for a simulated dataset of rank four and with various amounts of noise added. It is seen that the eigenvalues level off after four components, but the details are difficult to see in the raw eigenvalues unless zoomed in. It is also seen, that the distinction between 'real' and noise eigenvalues are difficult to discern at high noise levels.

For real data, the plots may even be more difficult to use as also exemplified in the original publication of Cattell as well as in many others[11-13]. Cattell himself admitted

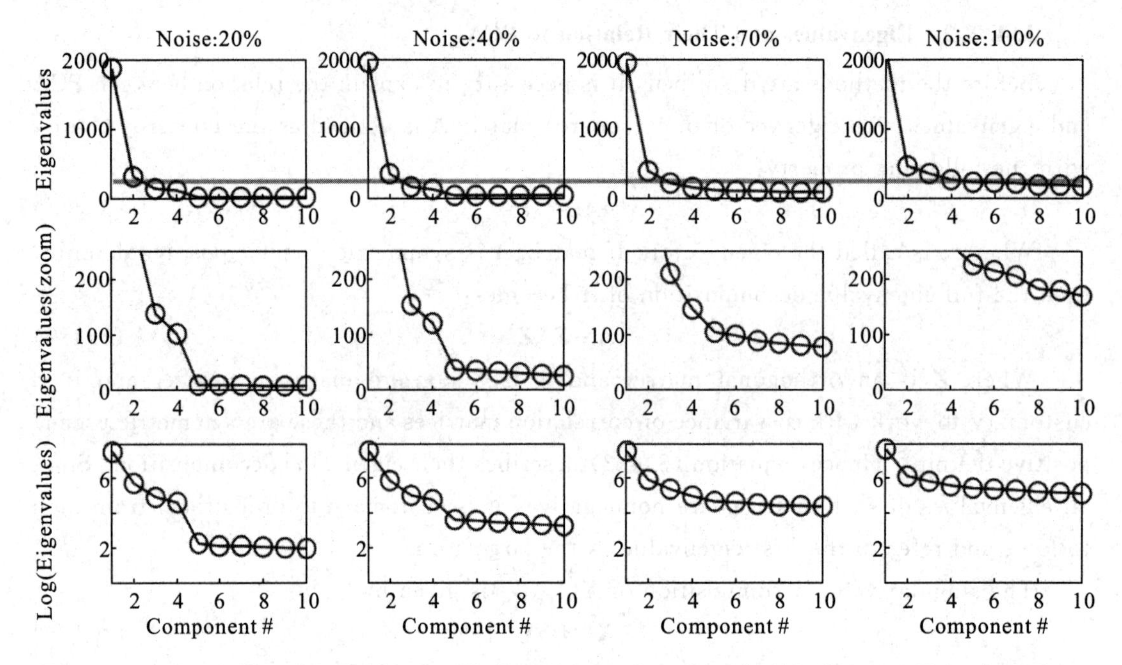

Fig. 3. 12 Scree plots for simulated rank four data with various levels of noise. Top plots show eigenvalues. Middle plots show the same but zoomed in on the y-axis to the line indicated in the top plot. Lower plots show the logarithm of the eigenvalues

that: "Even a test as simple as this requires the acquisition of some art in administering it." This, in fact, is not particular to the scree test but goes for all methods for selecting the number of components.

For the wine data, it is not easy to firmly assess the number of components based on the scree test (Fig. 3. 13). One may argue that seven or maybe nine components seem feasible, but this would imply incorporating components that explain very little variation. A more obvious choice would probably be to assess three components as suitable based on the scree plot and then be aware that further components may also contain useful information.

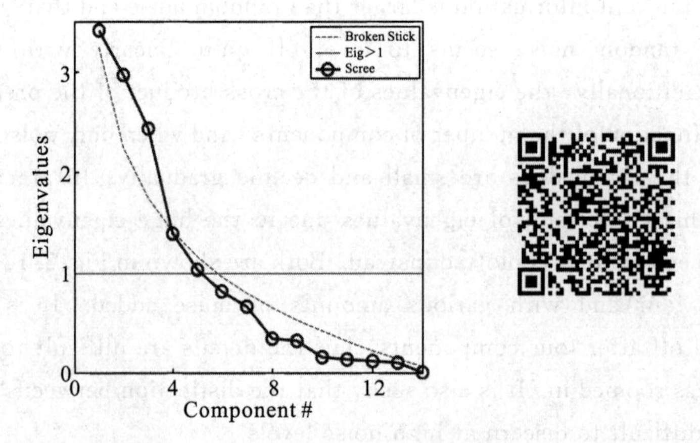

Fig. 3. 13 Scree plot for the autoscaled wine data. The decision lines for having eigenvalues larger than one and the broken stick is also shown

3.4.2.4 Eigenvalue below One

If the data is autoscaled, each variable has a variance of one. If all variables are orthogonal to each other, then every component in a PCA model would have an eigenvalue of one since the preprocessed cross-product matrix (the correlation matrix) is identity. It is then fair to say, that if a component has an eigenvalue larger than one, it explains variation of more than one variable. This has led to the rule of selecting all components with eigenvalues exceeding one (see the full line in Fig. 3.13). It is sometimes also referred to as the Kaisers 'rule or Kaiser-Guttmans' rule and many additional arguments have been provided for this method[14-16]. While it remains a very ad hoc approach, it is nevertheless a useful rule-of-thumb to get an idea about the complexity of a dataset. For the wine data (Fig. 3.13), the rule suggests that around four or five components are reasonable. Note, that for very precise data, it is perfectly possible that even components with eigenvalues far below one can be real and significant. Real phenomena can be small in variation, yet accurate.

3.4.2.5 Broken Stick

A more realistic cut off for the eigenvalues is obtained with the so called broken stick rule[16]. A line is added to the scree plot that shows the eigenvalues that would be expected for random data (the dotted line in Fig. 3.13). This line is calculated assuming that random data will follow a so-called broken stick distribution. The broken stick distribution hypothesizes how random variation will partition and uses the analogy of how the lengths of pieces of a stick will be distributed when broken at random places into J pieces[18]. It can be shown that for auto-scaled data, this theoretical distribution can be calculated as

$$b_r = \sum_{j=r}^{J} \frac{1}{j} \tag{3.4.5}$$

As seen in Fig. 3.13, the broken stick would seem to indicate that three to four components are reasonable.

3.4.2.6 High Fraction of Variation Explained

If the data measured has e.g. one percent noise, it is expected that PCA will describe all the variation down to around one percent. Hence, if a two-component model describes only 50% of the variation and is otherwise sound, it is probable that more components are needed. On the other hand, if the data are very noisy coming e.g. from process monitoring or consumer preference mapping and has an expected noise fraction of maybe 40%, then an otherwise sound model fitting 90% of the variation would imply overfitting and fewer components should be used. Having knowledge on the quality of the data can help in assessing the number of components. In Fig. 3.14, the variation explained is shown. The plot is equivalent to the eigenvalue plot except it is cumulative and on a different scale. For the wine data, the uncertainty is different for each variable, and varies from approximately 5 and even up to 50% relative to the variation in the data. This is quite variable and makes it difficult to estimate how much variation should be explained, but most certainly less than 50% would mean that all is not explained and explaining more than, say 90%－95% of the variation would be meaningless and just modelling of noise. Therefore, based on variation explained, it is likely that there is more than two but less than, say, seven components.

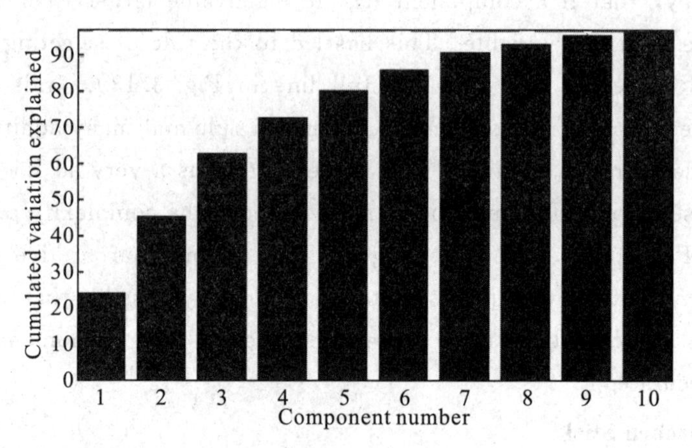

Fig. 3. 14 Cumulated percentage variation explained

3. 4. 2. 7 Valid Interpretation

As indicated by the results, the different rules above seldom agree. This is not as big a problem as it might seem. Quite often, the only thing needed is to know the neighbourhood of how many components are needed. Using the above methods "informally" and critically, will often provide that answer. Furthermore, one of the most important strategies for selecting the number of components is to supplement such methods with interpretations of the model. For the current data, it may be questioned whether e. g. three or four components should be used.

In Fig. 3. 15, it is shown, that there is distinct structure in the scores of component

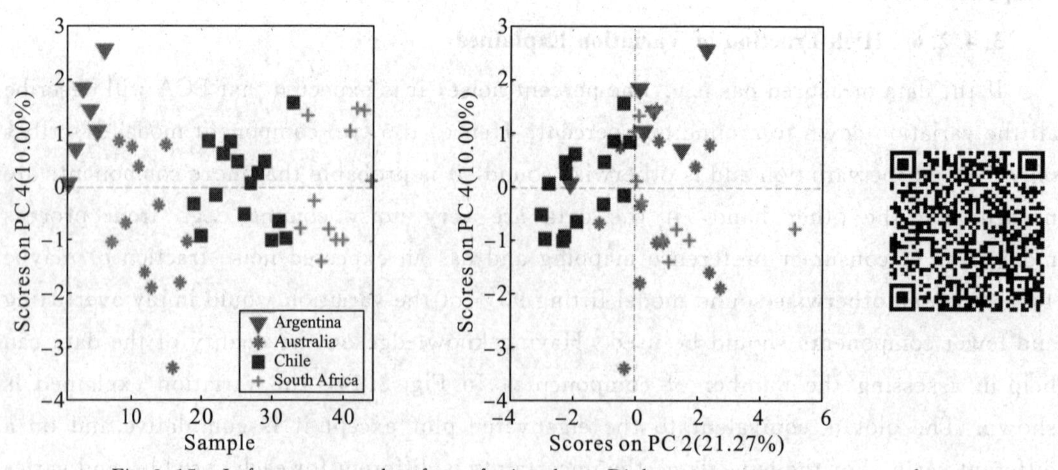

Fig. 3. 15 Left: score number four of wine data; Right: score two versus score four

four. For example, the wines from Argentina all have positive scores. Such a structure or grouping will not happen accidentally unless unfortunate confounding has occurred. Hence, as long as Argentinian wines were not measured separately on a different system or something similar, the mere fact that component four (either scores or loadings) shows distinct behaviour is an argument in favour of including that component. This holds regardless of what other measures might indicate.

The loadings may also provide similar validation by highlighting correlations expected from a priori knowledge. In the case of continuous data such as time series or spectral data, it is also instructive to look at the shape of the residuals. An example is provided in Fig. 3. 16. A dataset consisting of visual and nearinfrared spectra of 40 beer samples is shown in grey. After one component, the residuals are still fairly big and quite structured from a spectral point of view. After six components, there is very little information left indicating that most of the systematic variation has been modeled. Note from the title of the plot, that 95% of the variation explained is quite low for this dataset whereas that would be critically high for the wine data as discussed above.

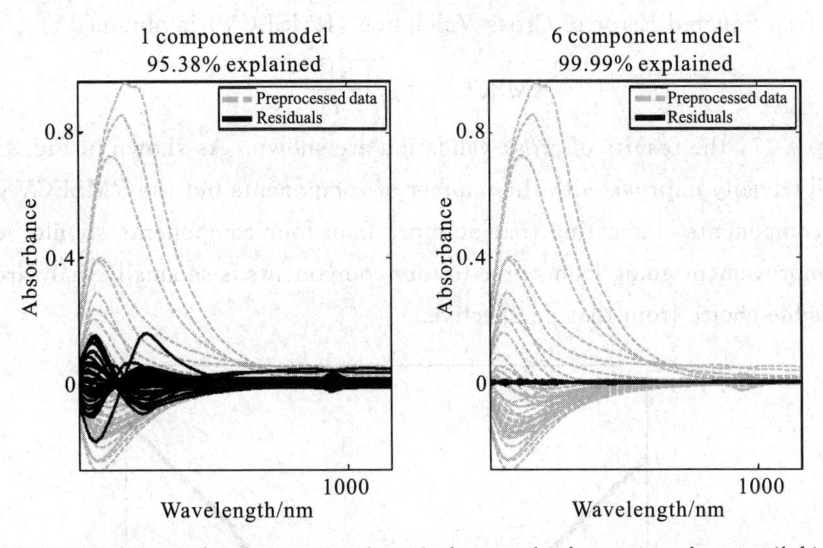

Fig. 3. 16 Example of spectral data (grey) and residual spectral infor-mation after one (left) and six (right) components

3. 4. 2. 8 Cross-validation

In certain cases, it is necessary to establish the appropriate number of components more firmly than in the exploratory or casual use of PCA. For example, a PCA model may be needed to verify if the data of a new patient indicate that this patient is similar to diseased persons. This may be accomplished by checking if the sample is an outlier when projected into a PCA model (see next section on outliers). Because the outlier diagnostics depend on the number of components chosen, it is necessary to establish the number of components before the model can be used for its purpose. There are several ways do to this including the above-mentioned methods. Oftentimes, though, they are considered too ad hoc and other approaches are used. One of the more popular approaches is cross-validation. S. Wold was the first to introduce cross-validation of PCA models[19] and several slightly different approaches have been developed subsequently.

The idea in cross-validation is to leave out part of the data and then estimate the left-out part. If this is done wisely, the prediction of the left-out part is independent of the actual left-out part. Hence, overfitting leading to too optimistic models is not possible. Conceptually, a single element (typically more than one element) of the data matrix is left out. A PCA model handling missing data, can then be fitted to the dataset and based on this PCA model, an estimate of the left out element can be obtained. Hence, a set of residuals is obtained where there are no problems with overfitting. Taking the sum of squares of these yields the so-called Predicted REsidual Sums of Squares (PRESS)

$$\text{PRESS}_r = \sum_{i=1}^{I}\sum_{j=1}^{J}(x_{ij}{}^{(r)})^2 \tag{3.4.6}$$

where $x_{ij}{}^{(r)}$ is the residual of sample i and variable j after r components. From the PRESS the Root Mean Squared Error of Cross-Validation (RMSECV) is obtained as

$$\text{RMSECV}_r = \sqrt{\frac{\text{PRESS}_r}{IJ}} \tag{3.4.7}$$

In Fig. 3. 17, the results of cross-validation are shown. As shown in Fig. 3. 15 the fit to data will trivially improve with the number of components but the RMSECV gets worse after four components, indicating that no more than four components should be used. In fact, the improvement going from three to four components is so small, that three is likely a more feasible choice from that perspective.

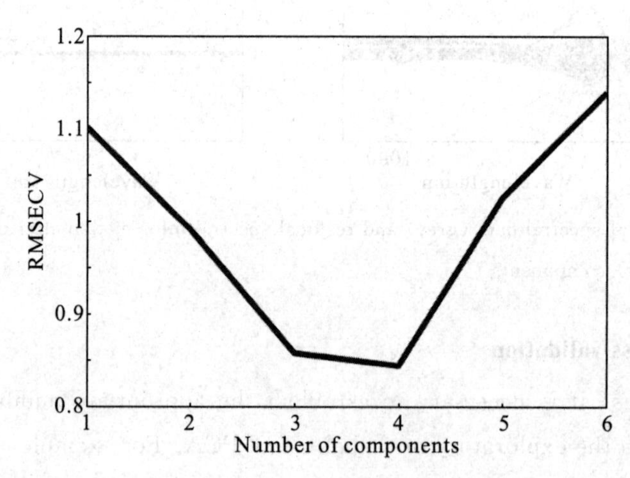

Fig. 3. 17 A plot of RMSECV for PCA models with different number of components

The cross-validated error, RMSECV, can be compared to the fitted error, the Root Mean Squared Error of Calibration, RMSEC. In order for the two to be comparable though, the fitted residuals must be corrected for the degrees of freedom consumed by the model.

3.4.3 When Using PCA for Other Purposes

It is quite common to use PCA as a preprocessing step in order to get a nicely compact representation of a dataset. Instead of the original many (J) variables, the dataset can be expressed in terms of the few (R) principal components. These components can then in turn be used for many different purposes (Fig. 3. 18).

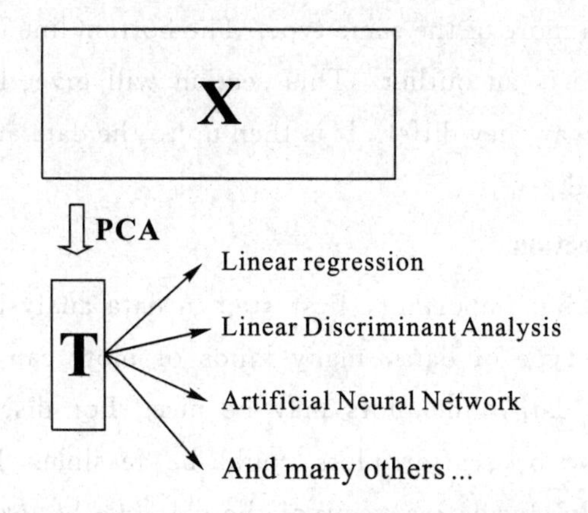

Fig. 3. 18 Using the scores of PCA for further modelling

It is common practice to use, for example, cross-validation for determining the number of components and then use that number of components in further modelling. For example, the scores may be used for building a classification model using linear discriminant analysis. While this approach to selecting components is both feasible and reasonable there is a risk that components that could help improve classification would be left out. For example, cross-validation may indicate that five components are valid, but it turns out that component seven can reliably improve classification. In order to be certain that useful information is retained in the PCA model, it is generally advised to validate the number of components in terms of the actual goal. Instead of validating the number of components that best describe X in some sense (PCA cross-validation), it will often make more sense to use the number of components that provides the best classification results if PCA is used in conjunction with discriminant analysis.

3.4.4 Detecting Outliers

Outliers are samples that are somehow disturbing or unusual. Often, outliers are downright wrong samples. For example, in determining the height of persons, five samples are obtained ([1. 78, 1. 92, 1. 83, 1. 67, 1. 87]). The values are in meters but accidentally, the fourth sample has been measured in centimeters. If the sample is not either corrected or removed, the subsequent analysis is going to be detrimentally disturbed by this outlier. Outlier detection is about identifying and handling such samples. An alternative or supplement to outlier handling is the use of robust methods, which will however, not be treated in detail here.

This section is mainly going to focus on identifying outliers, but understanding the outliers is really the critical aspect. Often outliers are mistakenly taken to mean wrong samples and nothing could be more wrong! Outliers can be absolutely right, but e. g. just badly represented. In such a case, the solution is not to remove the outlier, but to supplement the data with more of the same type. The bottom line is that it is imperative to understand why a sample is an outlier. This section will give the tools to identify the samples and see in what way they differ. It is then up to the data analyst to decide how the outliers should be handled.

3. 4. 4. 1 Data Inspection

An often forgotten, but important, first step in data analysis is to inspect the raw data. Depending on the type of data, many kinds of plots can be relevant as already mentioned. For spectral data, line plots may be nice. For discrete data, histograms, normal probability plots, or scatter plots could be feasible. In short, any kind of visualization that will help elucidate aspects of the data can be useful. Several such plots have already been shown throughout this paper. It is also important, and frequently forgotten, to look at the pre-processed data. While the raw data are important, they actually never enter the modeling. It is the preprocessed data that will be modeled and there can be big differences in the interpretations of the raw and the preprocessed data.

3. 4. 4. 2 Score Plots

While raw and preprocessed data should always be investigated, some types of outliers will be difficult to identify from there. The PCA model itself can provide further information. There are two places where outlying behavior will show up most evidently: in the scores and in the residuals. It is appropriate to go through all selected scores and look for samples that have strange behaviour. Often, it is only component one and two that are investigated but it is necessary to look at all the relevant components.

As for the data, it is a good idea to plot the scores in many ways, using different combinations of scatter plots, line plots, histograms, etc. Also, it is often useful to go through the same plot but coloured by all the various types of additional information available. This could be any kind of information such as temperature, storage time of sample, operator or any other kind of either qualitative or quantitative information available. For the wine data model, it is seen in Fig. 3. 19 that one sample is behaving differently from the others in score plot one versus two (upper left corner).

Fig. 3. 19

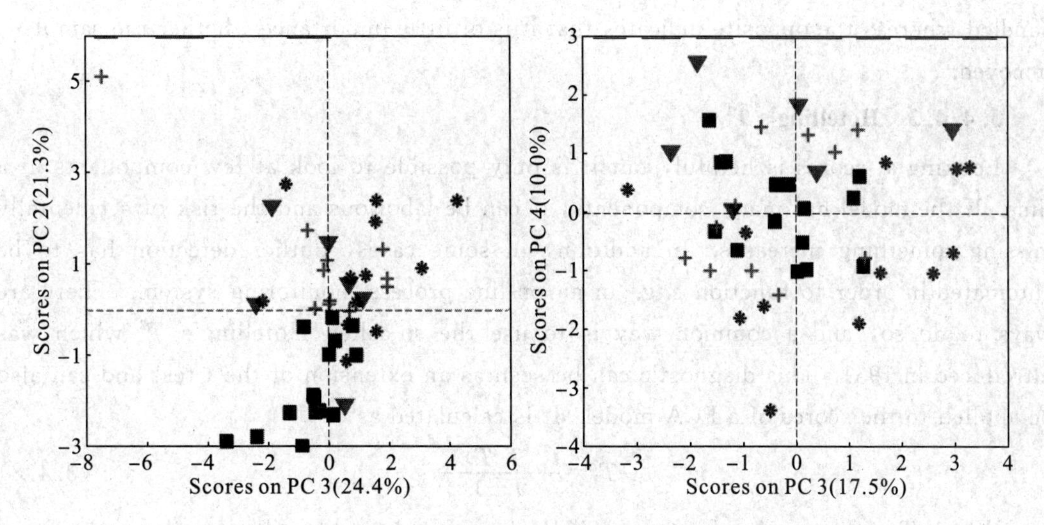

Fig. 3. 19 Score plot of a four component PCA model of the wine data

Looking at the loading plot (Fig. 3. 20) indicates that the sample must be (relatively) high in volatile and lactic acid and low in malic acid. This should then be verified in the raw data. After removing this sample, the model is rebuilt and reevaluated. No more extreme samples are observed in the scores.

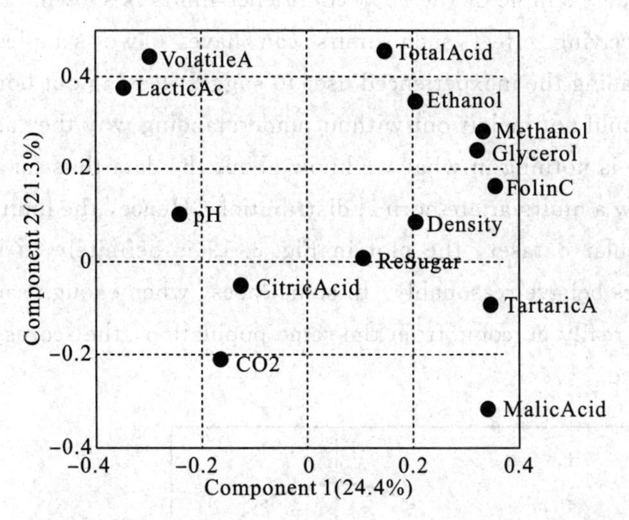

Fig. 3. 20 Scatter plot of loading 1 versus loading 2

Before deciding on what to do with an outlier, it is necessary to look at how important the component is. Imagine a sample that is doing an "excellent job" in the first seven components, but in the eighth has an outlying behaviour. If that eighth component is very small in terms of variation explained and not the most important for the overall use of the model; then it is probably not urgent to remove such a sample.

Whenever in doubt as to whether to remove an outlier or not, it is often instructive to compare the models before and after removal. If the interpretation or intended use changes dramatically, it indicates that the sample has an extreme behaviour that needs to be handled whereas the opposite indicates that it is of little importance whether the sample is removed.

3.4.4.3 Hotelling's T^2

Looking at scores is helpful, but it is only possible to look at few components at a time. If the model has many components, it can be laborious and the risk of accidentally missing something increases. In addition, in some cases, outlier detection has to be automated in order to function e. g. in an on-line process monitoring system. There are ways to do so, and a common way is to use the so-called Hotelling's T^2 which was introduced in 1931. This diagnostic can be seen as an extension of the t-test and can also be applied to the scores of a PCA model. It is calculated as

$$T_1^2 = \frac{\boldsymbol{t}^{\mathrm{T}}(\boldsymbol{T}^{\mathrm{T}}\boldsymbol{T})^{-1}}{I-1} \tag{3.4.8}$$

Where $\boldsymbol{T}$ is the matrix of scores ($I \times R$) from all the calibration samples and $\boldsymbol{t}_i$ is an $R \times 1$ vector holding the R scores of the ith sample. Assuming that the scores are normally distributed, then confidence limits for T_1^2 can be assigned as

$$T_{i(I,R)}^2 = \frac{R(I-1)}{I-R} F_{R,I-R,a} \tag{3.4.9}$$

In Fig. 3.21, an example of the 95% confidence limits is shown. This plot illustrates the somewhat deceiving effect such limits can have. Two samples are outside the confidence limit leading the inexperienced user to suggest leaving out both. However, first of all, samples should not be left out without understanding why they are wrong and more importantly, there is nothing in what we know about the data thus far, that suggests the scores would follow a multivariate normal distribution. Hence, the limit is rather arbitrary and for this particular dataset, the plot in Fig. 3.19 is definitely to be preferred when assessing if samples behave reasonably. In some cases, when enough samples are available and those samples really do come from the same population, the scores are approximately normally distributed.

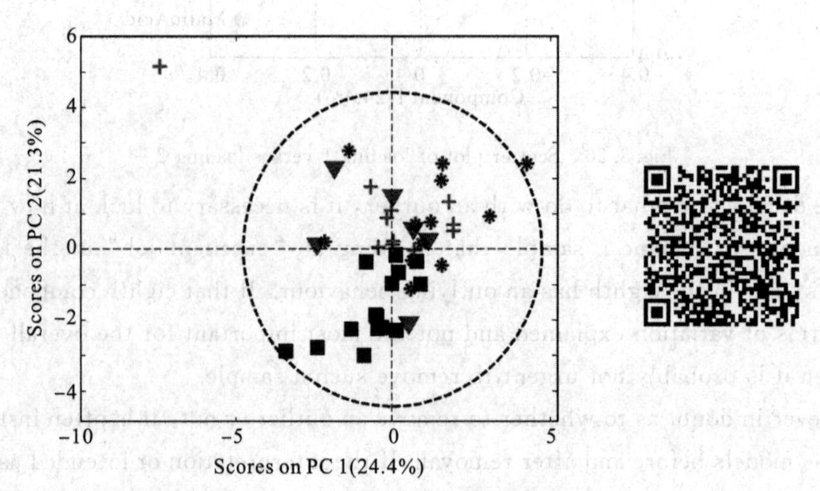

Fig. 3.21　PCA score plot similar to Fig. 3.19 (left) but now with a 95% confidence limit shown

The limits provided by Hotelling's T^2 can be quite misleading for grouped data. As an example, Fig. 3. 22 shows the score plot of a dataset, where the samples fall in four distinct groups (based on the geological background). The sample in the middle called outlier? is by no means extreme with respect to Hotelling's T^2 even though the sample is relatively far from all other samples.

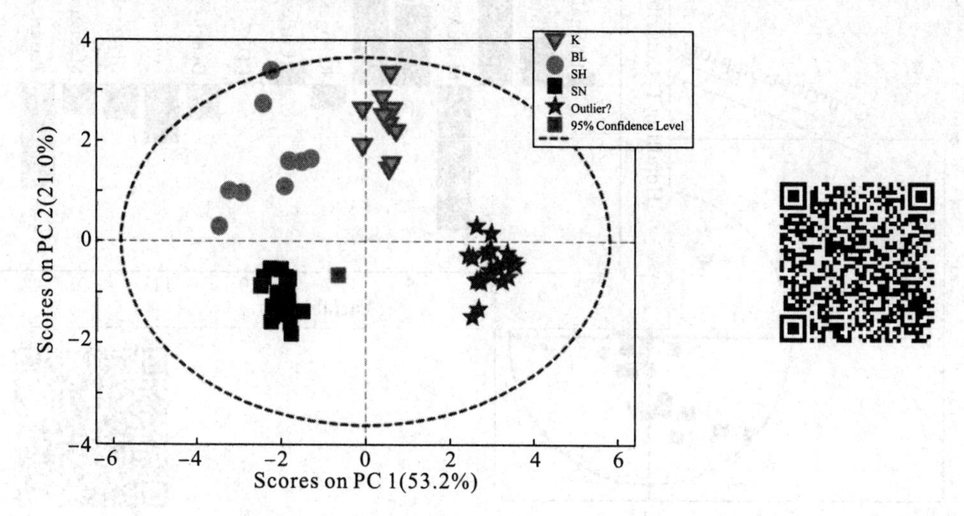

Fig. 3. 22 PCA scores plot (1 vs. 2) for a dataset consisting of ten concentrations of trace elements in obsidian samples from four specific quarries—data from a study by Kowalski

3. 4. 4. 4 Score Contribution Plots

When a sample has been detected as being an outlier, it is often interesting to try to investigate the reason. Extreme scores indicate that the sample has high levels of whatever, the specific component reflects in its corresponding loading vector. Sometimes, it is difficult to verify directly what is going on and the so-called contribution plot can help. There are several different implementations of contribution plots but one common version was originally developed by Nomikos. The contribution for a given sample indicates what variables caused that sample to get an extreme set of scores. For a given set of components (e. g. component one and two in Fig. 3. 23), this contribution can be calculated as

$$c_J^D = \sum_{r=1}^{R} \frac{t_r^{\text{new}} x_j^{\text{new}} p_{jr}}{\dfrac{\boldsymbol{t}_r^{\mathrm{T}} \boldsymbol{t}_r}{I-1}} \tag{3.4.10}$$

The vector $\boldsymbol{t}_r$ is rth score vector from the calibration model, I the number of samples in the calibration set and t_j^{new} is the score of the sample in question. It can come from the calibration set or be a new sample. x_j^{new} is the data of the sample in question for variable j and p_{xj} is the corresponding loading element. In this case, R components are considered, but fewer components can also be considered by adjusting the summation in equation (3. 4. 10).

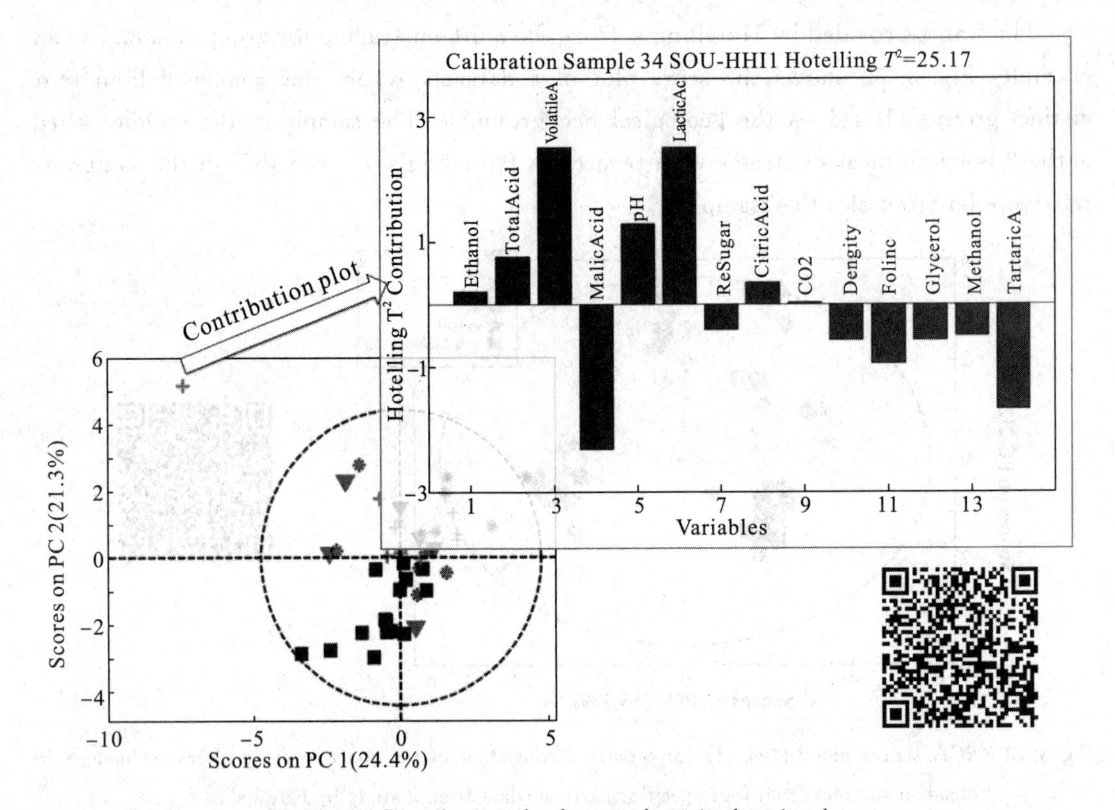

Fig. 3. 23 Contribution plot for sample 34 in the wine data

The contribution plot indicates what variables make the selected sample have an extreme Hotelling's T^2 and in Fig. 3. 23, the most influential variables are also the ones that that are visible in the raw data (not shown). Equation (3. 4. 10) explains the simplest case of contribution plots with orthogonal $\boldsymbol{P}$ matrices. Generalized contributions are available for non-orthogonal cases. Note that if x_j^{new} is a part of the calibration set, it influences the model. A more objective measure of whether x_j^{new} fits the model can be obtained by removing it from the data and then afterwards projecting it onto the model thereby obtaining more objective scores and residuals.

3. 4. 4. 5 Lonely Wolfs

Imagine a situation where the samples are constituted by distinct groups rather than one distribution as also exemplified in Fig. 3. 22. Hotelling's T^2 is not the most obvious choice for detecting samples that are unusually positioned but not far from the center. A way to detect such samples, is to measure the distance of the sample to the nearest neighbor. This can also be generalized e. g. to the average distance to the k nearest neighbors and various distance measures can be used if so desired.

In Fig. 3. 24, it is seen that colouring the scores by the distance to the nearest neighbour, highlights that there are, in fact, several samples that are not very close to other samples. When the samples are no longer coloured by class as shown in Fig. 3. 22, it is much less obvious that the green "K" class is indeed a well-defined class.

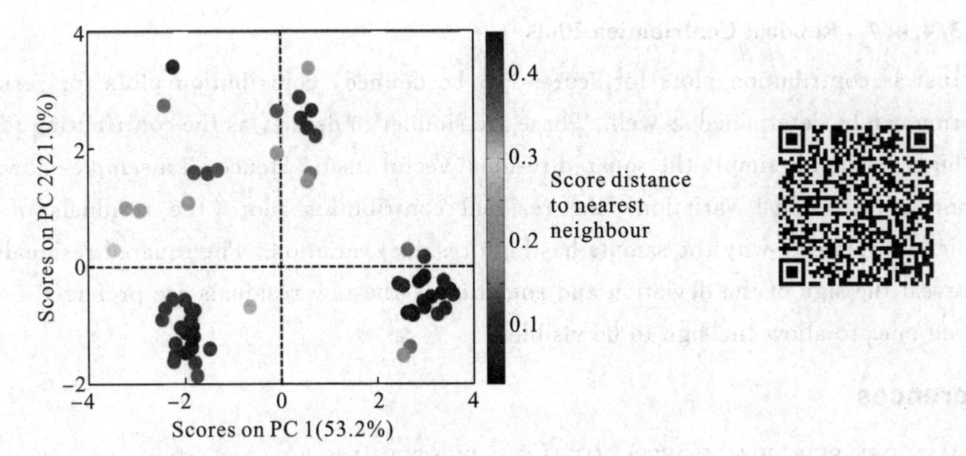

Fig. 3. 24 Score plot of Fig. 3. 22. Samples are coloured according to the distance of the sample to the nearest neighbour

3. 4. 4. 6 Residuals

The use of residuals has already been described in detail. For outlier detection, it is common to use the sum squared residuals, often called the Q-statistics, of each sample to look for samples that are not well-described by the PCA model. When Q is plotted against T^2, it is often referred to as an influence plot. Note, that both residuals and T^2 will change with the number of components, so if the number of components are not firmly defined, it may be necessary to go back and forth a bit between different numbers of components .

In the influence plot in Fig. 3. 25, it is clear that one sample stands out with a high Hotelling's T^2 in the PCA model and no samples have extraordinarily large residuals. It will hence, be reasonable to check the T^2 contribution plot of that sample, to see if an explanation for the extreme behavior can be obtained. The two blue lines are 95% confidence levels. Such lines are often given in software but should not normally be the focus of attention as also described above for score plots.

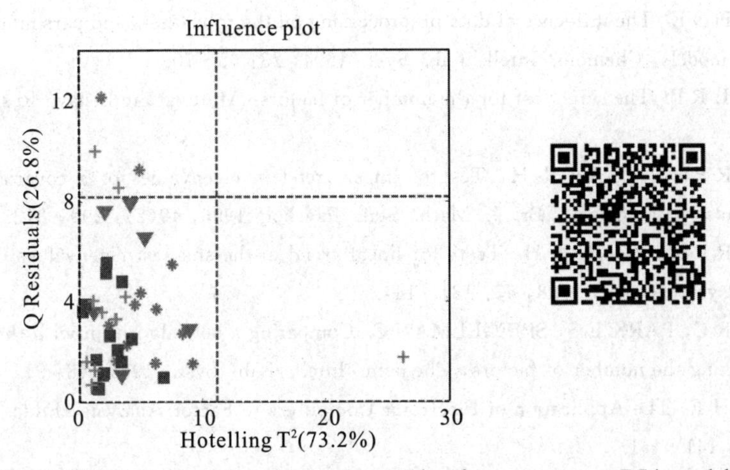

Fig. 3. 25 Influence plot of wine data with a four component PCA model

3.4.4.7 Residual Contribution Plots

Just as contribution plots for scores can be defined, contribution plots for residual variation can be determined as well. These are simpler to define, as the contributing factor to a high residual is simply the squared residual vector itself. Hence, if a sample shows an extraordinary residual variation, the residual contribution plot (the residuals of the sample) can indicate why the sample has high residual variation. The squared residuals do not reveal the sign of the deviation and sometimes, the raw residuals are preferred to the squared ones to allow the sign to be visible.

References

[1] VAN DEN BERG R A, HOEFSLOOT H C J, WESTERHUIS J A, et al. Centering, scaling, and transformations: improving the biological information content of metabolomics data, BMC Genomics, 2006, 7:142.

[2] BRO R, SMILDE A K. Centering and scaling in component analysis, J. Chemom., 2003, 17(1): 16 - 33.

[3] AFSETH N K, SEGTNAN V H, WOLD J P. Raman spectra of biological samples: a study of preprocessing methods, Appl. Spectrosc. 2006, 60(12): 1358 - 1367.

[4] BROWN C D, VEGA-MONTOTO L, WENTZELL P D. Derivative preprocessing and optimal corrections for baseline drift in multivariate calibration, Appl. Spectrosc. , 2000, 54(7): 1055 - 1068.

[5] DEMING S N, PALASOTA J A, NOCERINO J M. The geometry of multivariate object preprocessing, J. Chemom. 1993, 7: 393 - 425.

[6] MARTENS H, STARK E. Extended multiplicative signal correction and spectral interference subtraction: new preprocessing methods for near infrared spectroscopy. J. Pharm. Biomed. Anal. , 1991, 9(8): 625 - 635.

[7] PARDO M, NIEDERJAUFNER G, BENUSSI G, et al. Data preprocessing enhances the classification of different brands of Espresso coffee with an electronic nose, Sens. Actuators B, 2000, 69(3): 397 - 403.

[8] BRO R. Multivariate calibration-What is in chemometrics for the analytical chemist? Anal. Chim. Acta, 2003, 500(1 - 2): 185 - 194.

[9] DE NOORD O E. The influence of data preprocessing on the robustness and parsimony of multivariate calibration models, Chemom. Intell. Lab. Syst. 1994, 23: 65 - 70.

[10] CATTELL R B. The scree test for the number of factors. Multivariate Behav. Res. 1966, 1: 245 - 276.

[11] BENTLER P M, YUAN K H. Test of linear trend in eigenvalues of a covariance matrix with application to data analysis. Br. J. Math. Stat. Psychol. 1996, 49(2): 299 - 312.

[12] BENTLER P M, YUAN K H. Tests for linear trend in the smallest eigenvalues of the correlation matrix. Psychometrika, 1998, 63: 131 - 144.

[13] HENRY R C, PARK E S, SPIEGELMAN C. Comparing a new algorithm with the classic methods for estimating the number of factors, Chemom. Intell. Lab. Syst., 1999, 48: 91 - 97.

[14] KAISER H F. The Application of Electronic Computers to Factor Analysis, Educ. Psychol. Meas. 1960, 20, 141 - 151.

[15] CLIff N. The Eigenvalues-Greater-Than-One Rule and the Reliability of Components. Psychol.

Bull. 1988, 103(2): 276 - 279.

[16] GUTTMAN L. Some necessary conditions for common-factor analysis, Psychometrika, 1954, 19 (2): 149 - 161.

[17] FRONTIER S. Etude de la decroissance des valeurs propers dans une analyze en composantes principales: comparison avec le modele de baton brise. J. Exp. Mar. Biol. Ecol. 1976, 25: 67 - 75.

[18] MACARTHUR R H. On the relative abundance of bird species, Proc. Natl. Acad. Sci. U. S. A. 1957, 43(3): 293 - 295.

[19] WOLD S. Cross-validatory estimation of the number of components in factor and principal components models. Technometrics, 1978, 20: 397 - 405.

3.5 Sklearn PCA

This section introduces the sklearn PCA code in detail, so that readers can understand how to construct a python data analysis program.

3.5.1 Source Code

Here we adopted the sklearn. decomposition. PCA, its version is 0. 18. 0.

```
class PCA(_BasePCA):
    Parameters
        ----------
    n_components : int, float, None or string
            Number of components to keep.
            if n_components is not set all components are kept: :
                n_components == min(n_samples, n_features)
            If "n_components == 'mle'" and "svd_solver == 'full'", Minka\'s
            MLE is used to guess the dimension. Use of "n_components == 'mle'"
            will interpret "svd_solver == 'auto'" as "svd_solver == 'full'".
            If "0 < n_components < 1" and "svd_solver == 'full'", select the
            number of components such that the amount of variance that needs to be
            explained is greater than the percentage specified by n_components.
            If "svd_solver == 'arpack'", the number of components must be
            strictly less than the minimum of n_features and n_samples.
        svd_solver : string {'auto', 'full', 'arpack', 'randomized'}
            auto :
                the solver is selected by a default policy based on 'X. shape' and
                'n_components': if the input data is larger than 500x500 and the
                number of components to extract is lower than 80% of the smallest
                dimension of the data, then the more efficient 'randomized'
                method is enabled. Otherwise the exact full SVD is computed and
                optionally truncated afterwards.
            full :
```

```
            run exact full SVD calling the standard LAPACK solver via
            'scipy. linalg. svd' and select the components by postprocessing
        arpack :
            run SVD truncated to n_components calling ARPACK solver via
            'scipy. sparse. linalg. svds'. It requires strictly
            0 < n_components < min(X. shape)
        randomized :
            run randomized SVD by the method of Halko et al.
        .. versionadded: : 0.18.0
Attributes
    ----------
    components_ : array, shape (n_components, n_features)
            Principal axes in feature space, representing the directions of
            maximum variance in the data. The components are sorted by
        "explained_variance_".
    explained_variance_ratio_ : array, shape (n_components, )
        Percentage of variance explained by each of the selected components.
        If "n_components" is not set then all components are stored and the
        sum of the ratios is equal to 1.0.
    singular_values_ : array, shape (n_components, )
        The singular values corresponding to each of the selected components.
        The singular values are equal to the 2-norms of the "n_components"
        variables in the lower-dimensional space.
    mean_ : array, shape (n_features, )
        Per-feature empirical mean, estimated from the training set.
        Equal to 'X. mean(axis=0)'.
    n_components_ : int
        The estimated number of components. When n_components is set
        to 'mle' or a number between 0 and 1 (with svd_solver == 'full') this
        number is estimated from input data. Otherwise it equals the parameter
        n_components, or the lesser value of n_features and n_samples
        if n_components is None.

def __init__(self, n_components=None, copy=True, whiten=False, svd_solver='auto', tol=
0.0, iterated_power='auto', random_state=None):
    self. n_components = n_components
    self. copy = copy
    self. whiten = whiten
    self. svd_solver = svd_solver
    self. tol = tol
    self. iterated_power = iterated_power
    self. random_state = random_state
```

```
def fit(self, X, y=None):
self._fit(X)
return self

def fit_transform(self, X, y=None):
        U, S, V =self._fit(X)
        U = U[:, :self.n_components_]

if self.whiten:
# X_new = X * V / S * sqrt(n_samples) = U * sqrt(n_samples)
            U *= sqrt(X.shape[0])
else:
# X_new = X * V = U * S * V^T * V = U * S
            U *= S[:self.n_components_]

#           return U
return U, V[:self.n_components_].T

def _fit(self, X):
# Raise an error for sparse input.
# This is more informative than the generic one raised by check_array.
if issparse(X):
raise TypeError('PCA does not support sparse input. See '
'TruncatedSVD for a possible alternative.')

        X = check_array(X, dtype=[np.float64], ensure_2d=True,
                               copy=self.copy)

# Handle n_components==None
if self.n_components is None:
            n_components = X.shape[1]
else:
            n_components =self.n_components

# Handle svd_solver
        svd_solver =self.svd_solver
if svd_solver == 'auto':
# Small problem, just call full PCA
if max(X.shape) <= 500:
              svd_solver ='full'
elif n_components >= 1 and n_components < .8 * min(X.shape):
              svd_solver ='randomized'
# This is also the case of n_components in (0, 1)
```

```
else:
                svd_solver ='full'

# Call different fits for either full or truncated SVD
if svd_solver == 'full':
return self._fit_full(X, n_components)
elif svd_solver in ['arpack', 'randomized']:
return self._fit_truncated(X, n_components, svd_solver)

def _fit_full(self, X, n_components):
"""Fit the model by computing full SVD on X"""
        n_samples, n_features = X.shape

if n_components == 'mle':
if n_samples < n_features:
raise ValueError("n_components='mle' is only supported "
"if n_samples >= n_features")
elif not 0 <= n_components <= n_features:
raise ValueError("n_components=%r must be between 0 and "
"n_features=%r with svd_solver='full'"
                              % (n_components, n_features))

# Center data
self.mean_ = np.mean(X, axis=0)
        X -=self.mean_

        U, S, V = linalg.svd(X, full_matrices=False)
# flip eigenvectors' sign to enforce deterministic output
        U, V = svd_flip(U, V)

        components_ = V

# Get variance explained by singular values
        explained_variance_ = (S ** 2) / n_samples
        total_var = explained_variance_.sum()
        explained_variance_ratio_ = explained_variance_ / total_var

# Postprocess the number of components required
if n_components == 'mle':
            n_components = \
                _infer_dimension_(explained_variance_, n_samples, n_features)
elif 0 < n_components < 1.0:
# number of components for which the cumulated explained
```

```
# variance percentage is superior to the desired threshold
            ratio_cumsum = explained_variance_ratio_.cumsum()
            n_components = np.searchsorted(ratio_cumsum, n_components) +1

# Compute noise covariance using Probabilistic PCA model
# The sigma2 maximum likelihood (cf. eq. 12.46)
if n_components < min(n_features, n_samples):
self.noise_variance_ = explained_variance_[n_components:].mean()
else:
self.noise_variance_ = 0.

self.n_samples_, self.n_features_ = n_samples, n_features
self.components_ = components_[:n_components]
self.n_components_ = n_components
self.explained_variance_ = explained_variance_[:n_components]
self.explained_variance_ratio_ = \
            explained_variance_ratio_[:n_components]

return U, S, V


def _fit_truncated(self, X, n_components, svd_solver):
        n_samples, n_features = X.shape

if isinstance(n_components, six.string_types):
raise ValueError("n_components=%r cannot be a string "
"with svd_solver='%s'"
                                 % (n_components, svd_solver))
elif not 1 <= n_components <= n_features:
raise ValueError("n_components=%r must be between 1 and "
"n_features=%r with svd_solver='%s'"
                                 % (n_components, n_features, svd_solver))
elif svd_solver == 'arpack' and n_components == n_features:
raise ValueError("n_components=%r must be stricly less than "
"n_features=%r with svd_solver='%s'"
                                 % (n_components, n_features, svd_solver))

        random_state = check_random_state(self.random_state)

# Center data
self.mean_ = np.mean(X, axis=0)
        X -=self.mean_

if svd_solver == 'arpack':
```

```
# random init solution, as ARPACK does it internally
            v0 = random_state.uniform(-1, 1, size=min(X.shape))
            U, S, V = svds(X, k=n_components, tol=self.tol, v0=v0)
# svds doesn't abide by scipy.linalg.svd/randomized_svd
# conventions, so reverse its outputs.
            S = S[::-1]
# flip eigenvectors' sign to enforce deterministic output
            U, V = svd_flip(U[:, ::-1], V[::-1])

elif svd_solver == 'randomized':
# sign flipping is done inside
            U, S, V = randomized_svd(X, n_components=n_components,
                                     n_iter=self.iterated_power,
                                     flip_sign=True,
                                     random_state=random_state)

self.n_samples_, self.n_features_ = n_samples, n_features
self.components_ = V
self.n_components_ = n_components

# Get variance explained by singular values
self.explained_variance_ = (S ** 2) / n_samples
        total_var = np.var(X, axis=0)
self.explained_variance_ratio_ = \
self.explained_variance_ / total_var.sum()
if self.n_components_ < n_features:
self.noise_variance_ = (total_var.sum() -
self.explained_variance_.sum())
else:
self.noise_variance_ = 0.

return U, S, V

def score_samples(self, X):
        check_is_fitted(self, 'mean_')

        X = check_array(X)
        Xr = X-self.mean_
        n_features = X.shape[1]
        log_like = np.zeros(X.shape[0])
        precision =self.get_precision()
        log_like = -.5 * (Xr * (np.dot(Xr, precision))).sum(axis=1)
        log_like -= .5 * (n_features * log(2. * np.pi) -
```

```
                fast_logdet(precision))
        return log_like

    def score(self, X, y=None):
        return np.mean(self.score_samples(X))
```

3.5.2 Examples

Linear dimensionality reduction using Singular Value Decomposition of the data to project it to a lower dimensional space.

It uses the LAPACK implementation of the full SVD or a randomized truncated SVD by the method of Halko et al. 2009, depending on the shape of the input data and the number of components to extract.

It can also use the scipy.sparse.linalg ARPACK implementation of the truncated SVD.

```
>>>import numpy as np
>>>from sklearn.decomposition import PCA
>>> X = np.array([[-1, -1], [-2, -1], [-3, -2], [1, 1], [2, 1], [3, 2]])
>>> pca = PCA(n_components=2)
>>> pca.fit(X)
PCA(copy=True, iterated_power='auto', n_components=2, random_state=None, svd_solver
='auto', tol=0.0, whiten=False)
>>>print(pca.explained_variance_ratio_)
[0.9924… 0.0075…]
>>>print(pca.singular_values_)
[6.30061… 0.54980…]
```

Set n_components=2 for the PCA model, then fit the model with X. Output the percentage of variance explained by each of the selected components and singular values.

```
>>> pca = PCA(n_components=2, svd_solver='full')
>>> pca.fit(X)
PCA(copy=True, iterated_power='auto', n_components=2, random_state=None, svd_solver
='full', tol=0.0, whiten=False)
>>>print(pca.explained_variance_ratio_)
[0.9924… 0.00755…]
>>>print(pca.singular_values_)
[6.30061… 0.54980…]
```

Set n_components=2 and svd_solver='full' for the PCA model, then fit the model with X. Output the percentage of variance explained by each of the selected components and singular values.

```
>>> pca = PCA(n_components=1, svd_solver='arpack')
>>> pca.fit(X)
PCA(copy=True, iterated_power='auto', n_components=1, random_state=None, svd_solver
='arpack', tol=0.0, whiten=False)
>>>print(pca.explained_variance_ratio_)
[0.99244…]
>>>print(pca.singular_values_)
[6.30061…]
```

Set n_components = 1 and svd_solver = 'arpack' for the PCA model, then fit the model with X. Output the percentage of variance explained by each of the selected components and singular values.

3.6 Principal Component Regression

In statistics, principal component regression (PCR) is a regression analysis technique that is based on principal component analysis (PCA). Typically, it considers regressing the outcome (also known as the response or the dependent variable) on a set of covariates (also known as predictors, or explanatory variables, or independent variables) based on a standard linear regression model, but uses PCA for estimating the unknown regression coefficients in the model.

In PCR, instead of regressing the dependent variable on the explanatory variables directly, the principal components of the explanatory variables are used as regressors. One typically uses only a subset of all the principal components for regression, thus making PCR some kind of a regularized procedure. Often the principal components with higher variances (the ones based on eigenvectors corresponding to the higher eigenvalues of the sample variance-covariance matrix of the explanatory variables) are selected as regressors. However, for the purpose of predicting the outcome, the principal components with low variances may also be important, in some cases even more important.

One major use of PCR lies in overcoming the multicollinearity problem which arises when two or more of the explanatory variables are close to being collinear. PCR can aptly deal with such situations by excluding some of the low-variance principal components in the regression step. In addition, by usually regressing on only a subset of all the principal components, PCR can result in dimension reduction through substantially lowering the effective number of parameters characterizing the underlying model. This can be particularly useful in settings with high-dimensional covariates. Also, through appropriate selection of the principal components to be used for regression, PCR can lead to efficient prediction of the outcome based on the assumed model.

3.6.1 Source Code

```
#-*-coding: utf-8-*-

from Cross_Validation import Cross_Validation
import numpy as np
from sklearn.cross_validation import train_test_split
# from sklearn.decomposition import PCA
from principal_component_analysis.pca import PCA
from scipy.io.matlab.mio import loadmat

class PCR():
    def __init__(self, n_folds=10, max_components=5):
    self.n_folds = n_folds
    self.max_components = max_components

    def fit(self, X, y, comp_best):

    self.x_mean = np.mean(X, axis=0)
    self.y_mean = np.mean(y, axis=0)

            pca = PCA(n_components=comp_best)
            T, P = pca.fit_transform(np.subtract(X, self.x_mean))

            y_center = np.subtract(y, self.y_mean)
            coefs_B = np.linalg.lstsq(T, y_center)[0]

    return coefs_B, P

def cv_choose_param(self, X, y):

        pcr_cv = Cross_Validation(X, y, self.n_folds, self.max_components)
        y_allPredict, y_measure = pcr_cv.predict_cv()
        RMSECV, min_RMSECV, comp_best = pcr_cv.mse_cv(y_allPredict, y_measure)

return RMSECV, min_RMSECV, comp_best

def predict(self, X, y, coefs_B, P):

        T = np.dot(np.subtract(X, self.x_mean), P)

        y_pre = np.dot(T, coefs_B)
        y_predict = np.add(y_pre, self.y_mean)
```

```
            press = np.square(np.subtract(y, y_predict))
            all_press = np.sum(press, axis=0)
            RMSEP = np.sqrt(all_press / X.shape[0])

    return y_predict, RMSEP


if __name__ == '__main__':

    D = loadmat(\NIRCorn.mat')
    y = D['cornprop'][:, [0]]
    x_m5spec=D['m5spec']['data'][0][0]
    x_mp5spec=D['mp5spec']['data'][0][0]
    x_mp6spec=D['mp6spec']['data'][0][0]
    x = x_mp5spec#x_mp6spec

    x_cal, x_test, y_cal, y_test = train_test_split(x, y, test_size=0.2, random_state=0)

    pcr = PCR(n_folds=10, max_components=15)
    RMSECV, min_RMSECV, comp_best = pcr.cv_choose_param(x_cal, y_cal)
    coefs_cal, P_m_cal = pcr.fit(x_cal, y_cal, comp_best)
    y_cal_predict, RMSEC = pcr.predict(x_cal, y_cal, coefs_cal, P_m_cal)
    y_test_predict, RMSEP = pcr.predict(x_test, y_test, coefs_cal, P_m_cal)
print RMSEC, RMSEP, min_RMSECV, comp_best
```

Cross_Validation.py

```
#-*-coding: utf-8-*-

import numpy as np
from sklearn import cross_validation
# from sklearn.decomposition import PCA
from principal_component_analysis.pca import PCA

class Cross_Validation():
def __init__(self, x, y, n_folds, max_components):
self.x = x
self.y = y
self.n_folds = n_folds
self.max_components = max_components
self.n = x.shape[0]

def cv(self):
        kf = cross_validation.KFold(self.n, self.n_folds)
        x_train = []
        x_test = []
```

```
            y_train = []
            y_test = []
        for train_index, test_index in kf:
            xtr, xte =self.x[train_index], self.x[test_index]
            ytr, yte =self.y[train_index], self.y[test_index]
            x_train.append(xtr)
            x_test.append(xte)
            y_train.append(ytr)
            y_test.append(yte)

    return x_train, x_test, y_train, y_test

    def predict_cv(self):
        x_train, x_test, y_train, y_test =self.cv()
        y_allPredict = np.ones((1, self.max_components))
        B = []

    for i in range(self.n_folds):

            y_predict = np.zeros((y_test[i].shape[0], self.max_components))
            x_train_mean = np.mean(x_train[i], axis=0)
            y_train_mean = np.mean(y_train[i], axis=0)
            x_testCenter = np.subtract(x_test[i], x_train_mean)

    for j in range(self.max_components):
                pca = PCA(n_components=j +1)
                T, P = pca.fit_transform(x_train[i])
                B =self.fit(T, y_train[i])

                y_pre = np.dot(np.dot(x_testCenter, P), B)
                y_pre = y_pre + y_train_mean
                y_predict[:, j] = y_pre.ravel()

            y_allPredict = np.vstack((y_allPredict, y_predict))
        y_allPredict = y_allPredict[1:]

    return y_allPredict, self.y

    def fit(self, T, Y):   # 求原系数
        Y_mean = np.mean(Y, axis=0)
        y_cen = np.subtract(Y, Y_mean)
        coefs = np.linalg.lstsq(T, y_cen)[0]

    return coefs

    def mse_cv(self, y_allPredict, y_measure):
```

```
        press = np.square(np.subtract(y_allPredict, y_measure))
        press_all = np.sum(press, axis=0)
        RMSECV = np.sqrt(press_all /self.n)
        min_RMSECV = min(RMSECV)
        comp_array = RMSECV.argsort()
        comp_best = comp_array[0] + 1

    return RMSECV, min_RMSECV, comp_best
```

3.6.2 K-Fold Cross-Validation

When evaluating different settings ("hyperparameters") for estimators, such as the C setting that must be manually set for an SVM, there is still a risk of overfitting on the test set because the parameters can be tweaked until the estimator performs optimally. This way, knowledge about the test set can "leak" into the model and evaluation metrics no longer report on generalization performance. To solve this problem, yet another part of the dataset can be held out as a so-called "validation set": training proceeds on the training set, after which evaluation is done on the validation set, and when the experiment seems to be successful, final evaluation can be done on the test set.

However, by partitioning the available data into three sets, we drastically reduce the number of samples which can be used for learning the model, and the results can depend on a particular random choice for the pair of (train, validation) sets.

In fig. 3.26, a solution to this problem is a procedure called cross-validation (CV for short). A test set should still be held out for final evaluation, but the validation set is no longer needed when doing CV. In the basic approach, called K-Fold CV, the training set is split into k smaller sets (other approachesare described below, but generally follow the same principles). The following procedure is followed for each of the k "folds":

- A model is trained using of the folds as training data;
- the resulting model is validated on the remaining part of the data (i. e., it is used as a test set to compute a performance measure such as accuracy).

The performance measure reported by K-Fold cross-validation is then the average of the values computed in the loop. This approach can be computationally expensive, but does not waste too much data (as is the case when fixing an arbitrary validation set), which is a major advantage in problems such as inverse inference where the number of samples is very small.

3.6.3 Examples

3.6.3.1 Dataset Description

1. Corn Dataset

The corn dataset which contains 80 samples was measured on three NIR spectrometers (m5, mp5, and mp6). Each sample consists of four components: moisture, oil, protein, and

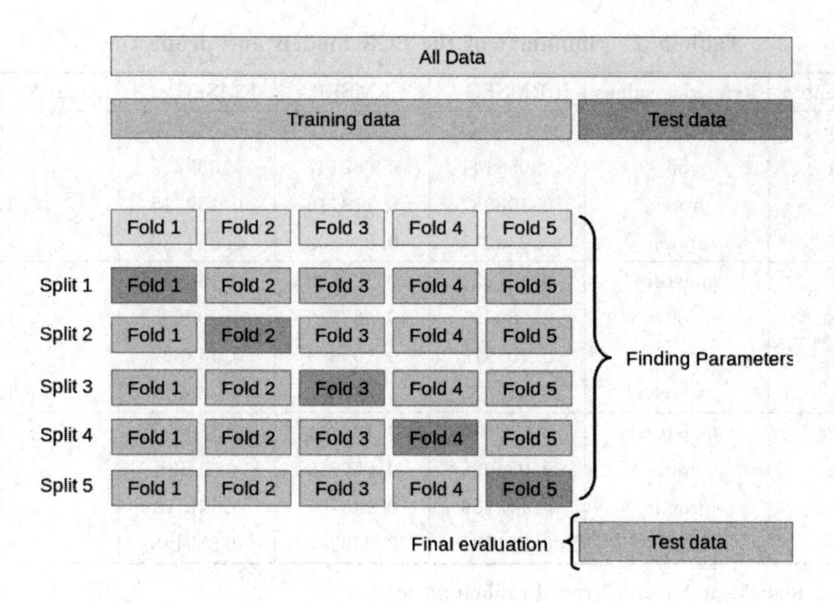

Fig. 3. 26 5-fold cross-validation

starch. The wavelength range is 1100-2400 nm with interval 2 nm (700 channels). The spectrum measured in m5spec is used as the master spectrum, and the spectrum measured by mp6spec is used as the secondary spectrum. The dataset was divided into a calibration set of 64 samples and a test set of 16 samples based on Kennard-Stone (KS) algorithm.

2. Wheat Dataset

The wheat dataset was used as the shootout data for the International Diffuse Conference 2016, and the protein content was chosen as the property. 248 samples of the wheat dataset from three different NIR instrument manufacturers (B1, B2 and B3) will be analyzed. According to KS algorithm, 198 samples were chosen as the calibration set and the remainder of samples formed the test set. The wavelength range is 570-1100 nm with an interval of 0. 5 nm.

3. Pharmaceutical Tablet Dataset

The third dataset comes from the IDRC shootout 2002, which contains 655 pharmaceutical tablets measured on two spectrometers, with the range from 600 to 1898 nm, and the interval is 2 nm. We can obtain it from http: //www. eigenvector. com/data/tablets/index. html. There are three reference values associated with this dataset, but we are only interested in weight content for each sample.

3. 6. 3. 2 Results of the Different Datasets' PCR Models

1. Corn Dataset

In this section the corn dataset was used for experiments. Table 3. 2 shows the training error, cross-validation error, prediction error, and principal component number of the PCR model for moisture, oil, protein, and starch content directly using the corn dataset.

Table 3.2 Summary of the PCR models and properties

Instrument	Reference values	RMSEC	RMSEP	$RMSECV_{min}$	Latent Variable (LV)
m5spec	moisture	0.015763	0.024712	0.021779	14
m5spec	oil	0.050141	0.062137	0.069252	15
m5spec	protein	0.108987	0.119118	0.139915	12
m5spec	starch	0.238211	0.235303	0.321736	12
mp5spec	moisture	0.107022	0.126895	0.138539	12
mp5spec	oil	0.080531	0.068145	0.100688	10
mp5spec	protein	0.107912	0.165999	0.142857	13
mp5spec	starch	0.262573	0.337348	0.369614	15
mp6spec	moisture	0.11744	0.103741	0.158303	15
mp6spec	oil	0.084096	0.066929	0.10598	10
mp6spec	protein	0.112759	0.146784	0.143704	12
mp6spec	starch	0.261581	0.359245	0.359148	14

RMSEC: Root Mean Square Error of calibration set

RMSEP: Root Mean Square Error of test set

$RMSECV_{min}$: Minimum Root Mean Square Error of Cross-Validation

LV: The optimal number of latent variables is selected only when the lowest RMSECV

In this paper, the principal component of the PCR algorithm is selected by the 10-fold cross-validation method. The pictures of RMSECV are given in Fig. 3.27, Fig. 3.29, Fig. 3.31. The comparison between measured values and predicted values is also presented in Fig. 3.28, Fig. 3.30, Fig. 3.32.

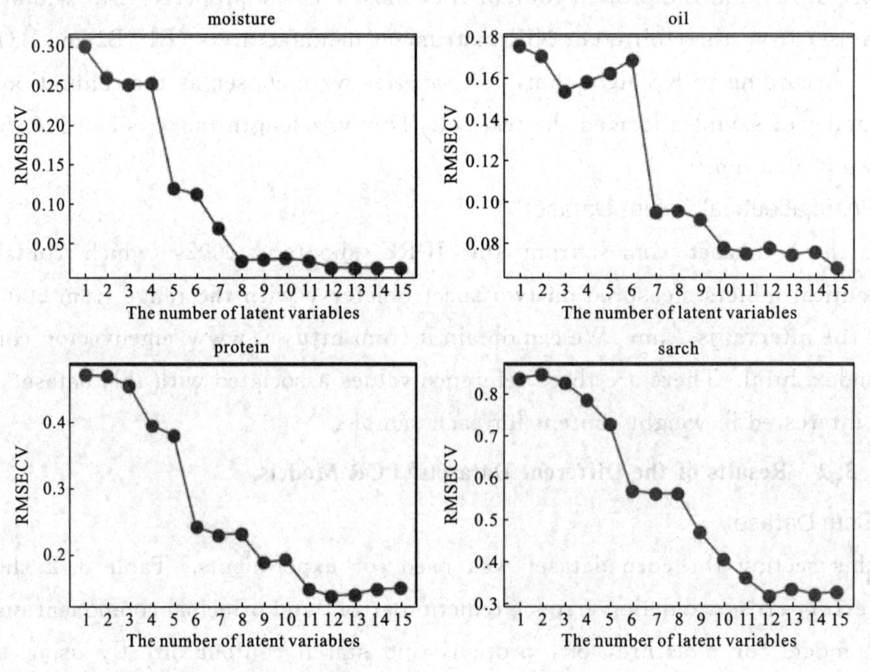

Fig. 3.27 The principal component number selection process of PCR model about the m5spec instrument

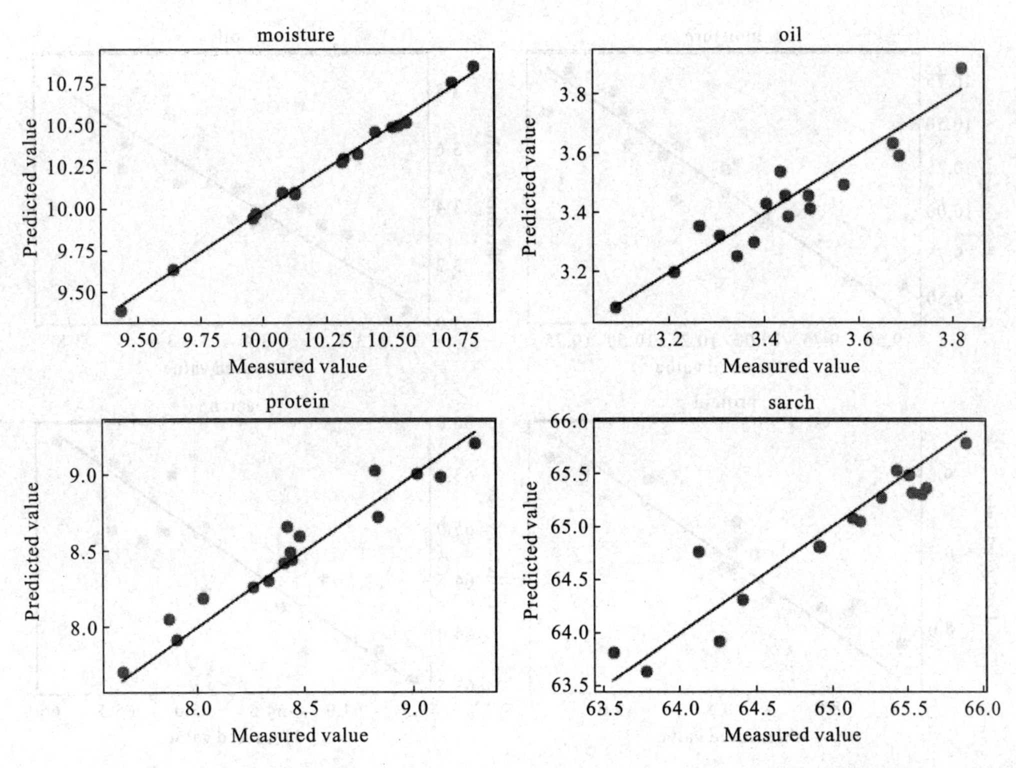

Fig. 3. 28 Measured values versus predicted values of the m5 spec instrument as determined by PCR

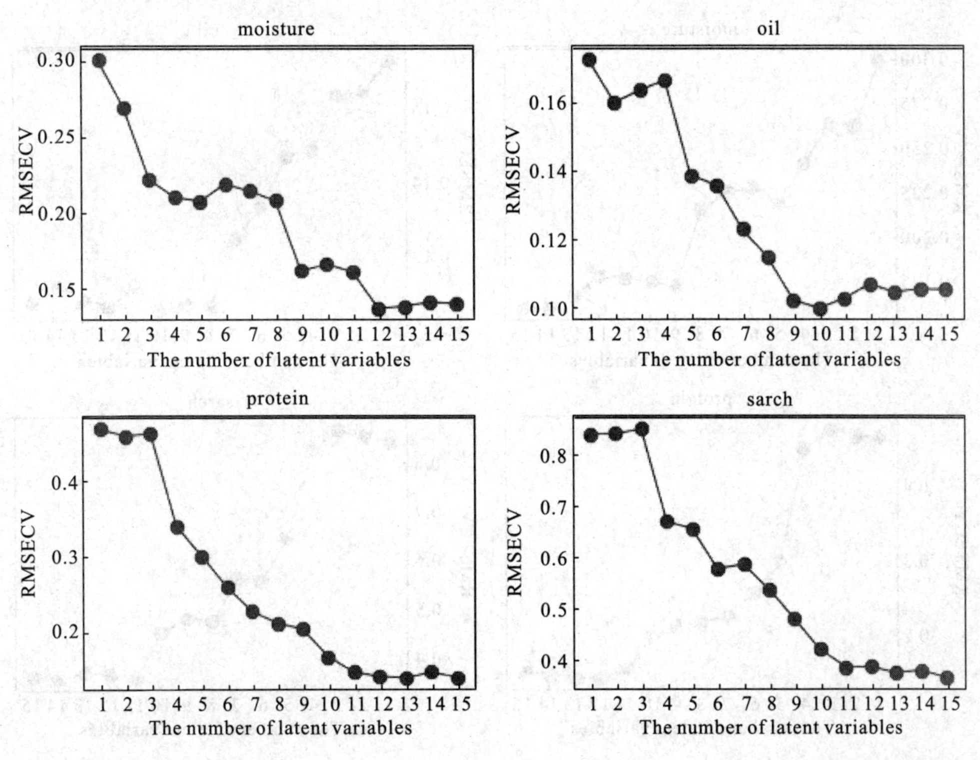

Fig. 3. 29 The principal component number selection process of PCR model about the mp5 spec instrument

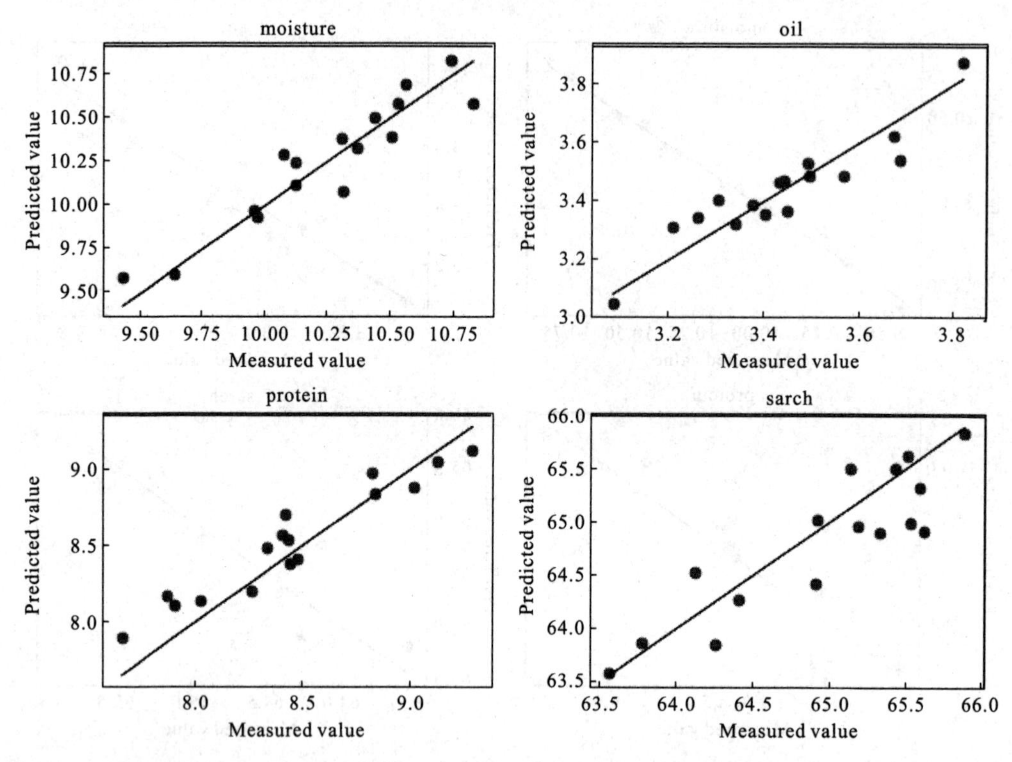

Fig. 3.30　Measured values versus predicted values of the mp5 spec instrument as determined by PCR

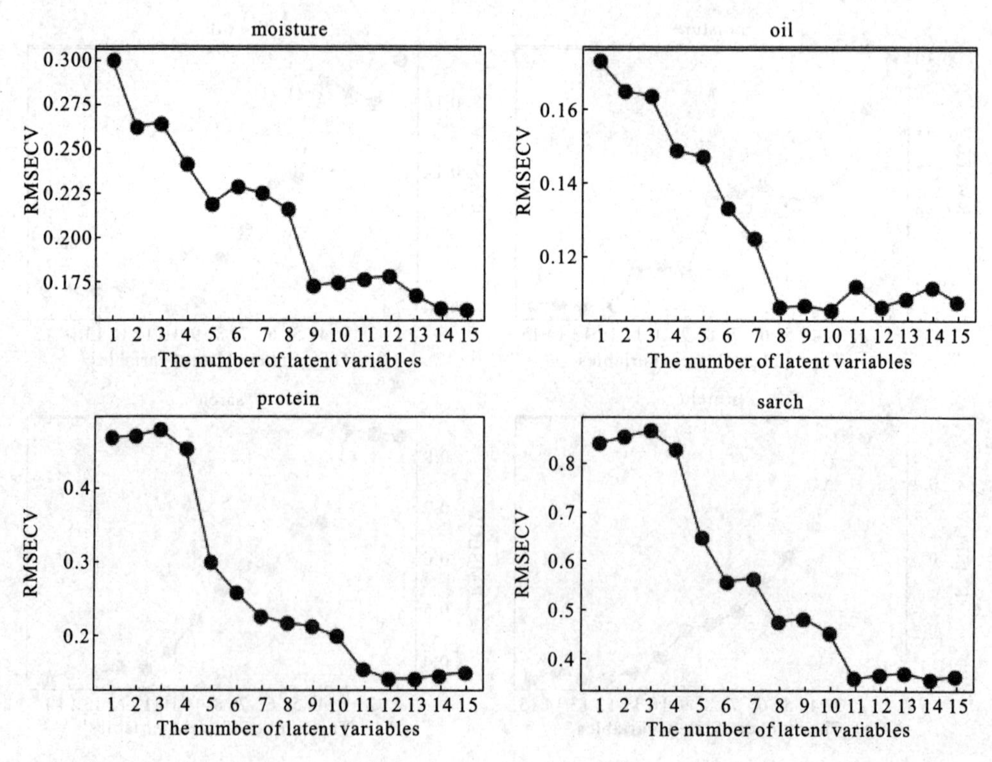

Fig. 3.31　The principal component number selection process of PCR model about the mp6 spec instrument

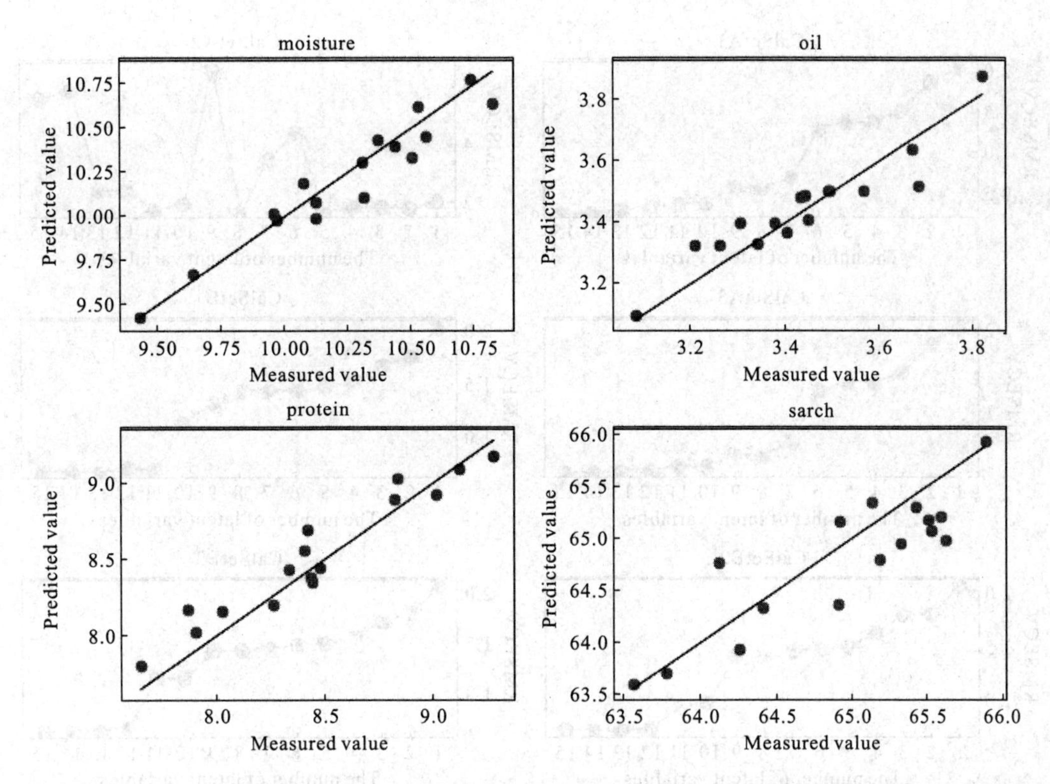

Fig. 3. 32 Measured values versus predicted values of the mp6spec instrument as determined by PCR

2. Wheat Dataset

In this section the wheat dataset was used for experiments Fig. 3. 33, Fig. 3. 34. Table 3. 3 shows the training error, cross-validation error, prediction error, and principal component number of the PCR model for protein content directly using the wheat dataset.

Table 3. 3 Summary of the PCR models and properties

Instrument	Reference values	RMSEC	RMSEP	$RMSECV_{min}$	Latent Variable (LV)
B1	protein	0. 229176	0. 309061	0. 260882	15
B2	protein	1. 377594	1. 352138	1. 673944	5
B3	protein	0. 212653	0. 276625	0. 253315	15
A1	protein	0. 618719	0. 498393	0. 693591	13
A2	protein	0. 636495	0. 400822	0. 710181	13
A3	protein	0. 584967	0. 531832	0. 637815	14

RMSEC: Root Mean Square Error of calibration set

RMSEP: Root Mean Square Error of test set

$RMSECV_{min}$: Minimum Root Mean Square Error of Cross-Validation

LV: The optimal number of latent variables is selected only when the lowest RMSECV

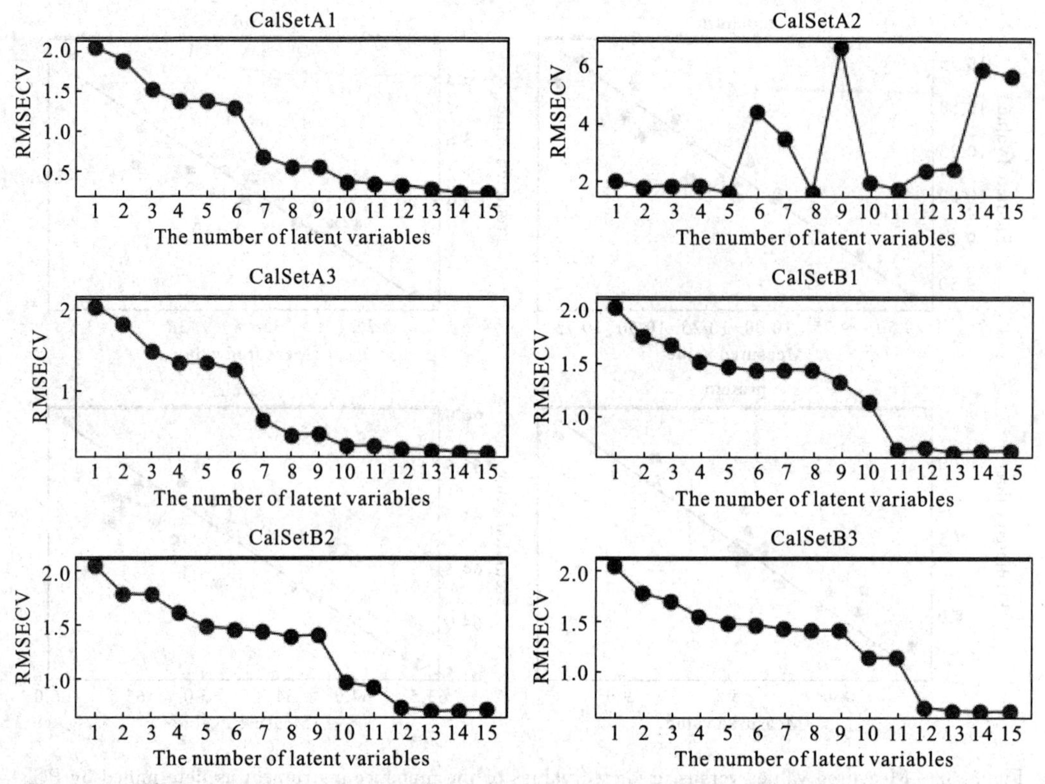

Fig. 3. 33 The principal component number selection process of PCR model about the wheat dataset

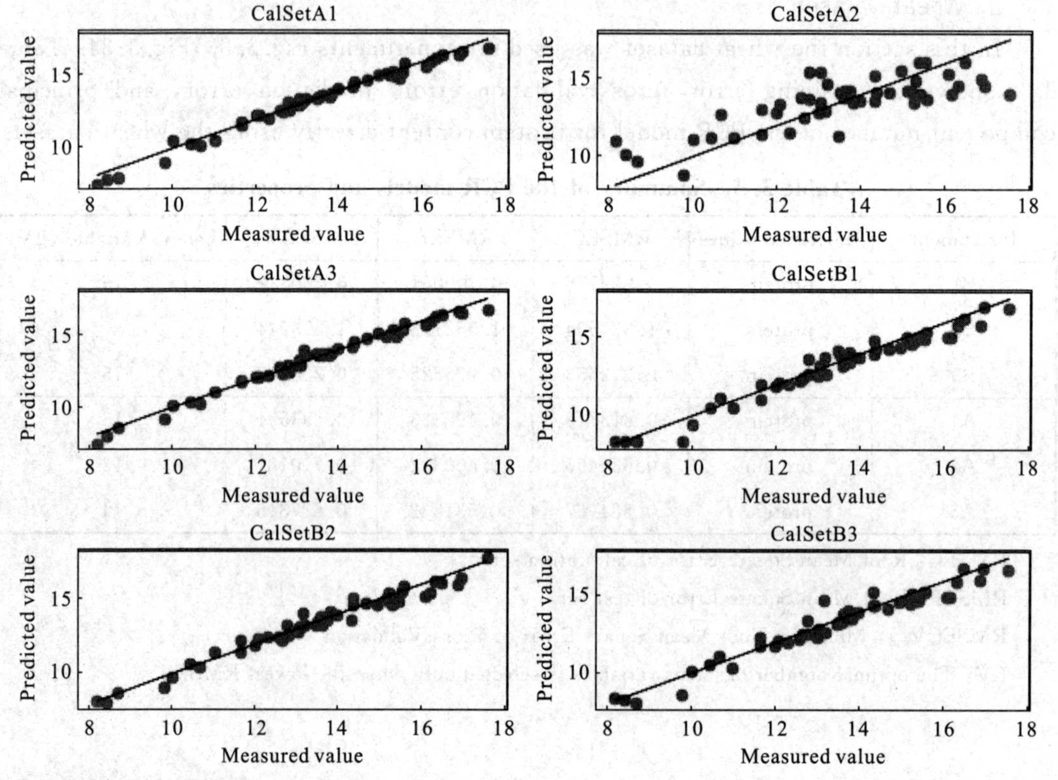

Fig. 3. 34 Measured values versus predicted values of the wheat dataset as determined by PCR

3. Pharmaceutical Tablet Dataset

In this section the pharmaceutical tablet dataset was used for experiments. Table 3. 4 shows the training error, cross-validation error, prediction error, and principal component number of the PCR model for weight, hardness and assay content directly using the pharmaceutical tablet dataset. Fig. 3. 35 gives the PC number selection process.

Table 3. 4 Summary of the PCR models and properties

Instrument	Reference values	RMSEC	RMSEP	$RMSECV_{min}$	Latent Variable (LV)
spectrometer1	weight	3. 38803	4. 22138	3. 762873	9
spectrometer1	hardness	1. 316597	0. 677325	1. 368493	3
spectrometer1	assay	4. 312213	5. 867524	4. 848849	13
spectrometer2	weight	2. 953375	4. 14708	3. 53662	14
spectrometer2	hardness	1. 359982	0. 630843	1. 403434	2
spectrometer2	assay	4. 596838	5. 95695	4. 917867	7

RMSEC: Root Mean Square Error of calibration set
RMSEP: Root Mean Square Error of test set
$RMSECV_{min}$: Minimum Root Mean Square Error of Cross-Validation
LV: The optimal number of latent variables is selected only when the lowest RMSECV

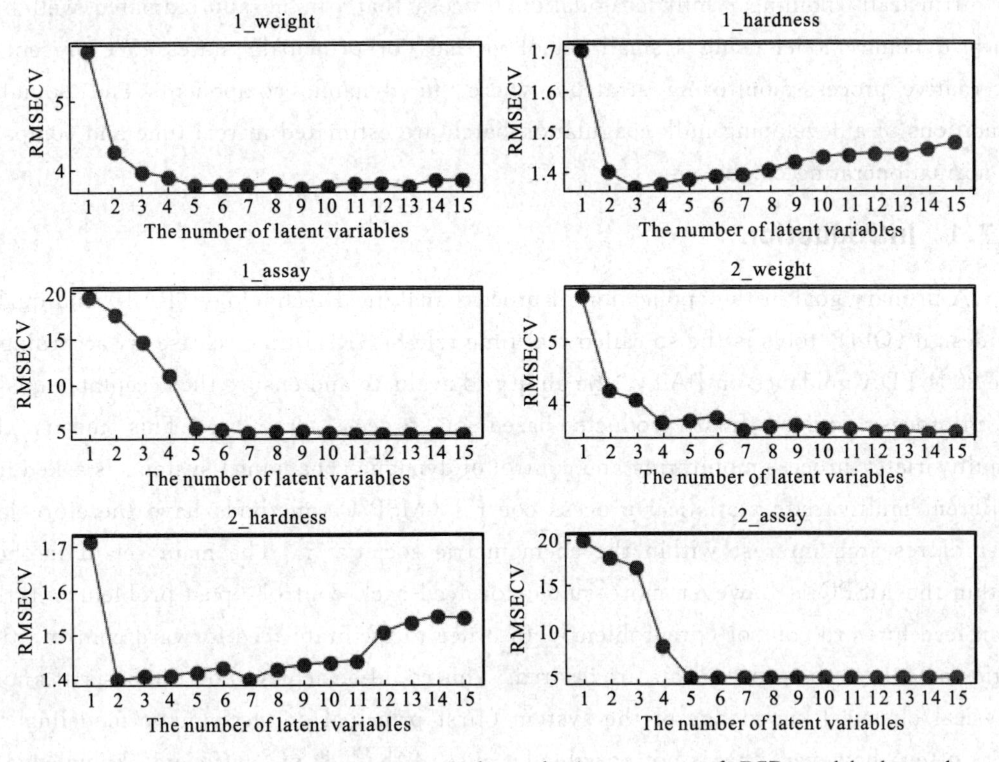

Fig. 3. 35 The principal component number selection process of PCR model about the pharmaceutical tablet dataset

3.7 Subspace Methods for Dynamic Model Estimation in PAT Applications

One primary goal in the application of process analytical technology tools is improved process monitoring and control. A second is to obtain a better understanding of how a normal process behaves (i. e. the normal dynamics). In order to perform feed-forward control, time series models of the process data are required. Such models could be developed on the basis of known physical/chemical knowledge of the system (i. e. first principal or mechanistic modeling). However, very often, this is not possible because of the lack of sufficient information. This leads to the need of system identification (SI). One class of models within SI is the state space models, linear models that relate the input of the system at time k to the output at time k via estimation of the so-called system states. State space models may be fitted using what is known as the subspace methods. Subspace methods are based on the projection of data on subspaces identified by, for example, the singular value decomposition of time-shifted data during a training phase. This paper introduces state space models, illustrates how subspace methods are closely related to known chemometric tools, and how they can be applied in, for example, model-based feed-forward process monitoring and control. The concepts are illustrated using a data set from an intrinsically nonlinear milk coagulation process that can be approximated well by a linear dynamic model using a small set of virtual (or principal) states. We present an alternative process-monitoring strategy where the dynamic components and boundary conditions of a developing milk coagulation batch are estimated in real-time and compared to normal operating conditions.

3.7.1 Introduction

A primary goal in the application of process analytical technology (PAT) and quality by design (QbD) tools is the so-called real-time release. Real-time release is, according to the 2004 FDA guidance on PAT, "the ability to evaluate and ensure the acceptable quality of in-process and/or final products based on process data"[1]. This means that (multivariate) process monitoring and control of dynamic (changing) systems is asked for. Different multivariate statistical process control (MSP[6]C) methods have therefore long been of research interest within the chemometric society[2-5]. The main set of methods within the MSPC is however more suited for feed-back control (post-problem), rather than feed-forward control (preproblem). In order to facilitate feed-forward control, time series models of the process data are desired. This could be achieved on the basis of known physical/chemical knowledge of the system (first principal or mechanistic modeling[6]). Very often, however, this is not possible because of the lack of (sufficient) knowledge on the system, for example in such complex processes as food production, leading to the need of system identification (SI). One class of models within SI is the state space models. They are linear, time-invariant models that relate the in-put to the system at time k to the output at time k via estimation of the system states. These states try to capture or model the dynamic behavior or development of a system without having a direct physical meaning, much like the concept principal component or latent variable in, for example, principal component analysis (PCA). State space models may either be fitted using iterative predictor error algorithms or by using the so-called subspace methods that are based on the projection of data on subspaces identified by, for example, singular value decomposition. The aim of this paper is to discuss state space models, illustrate that subspace methods are closely related to known chemometric tools, and show how they can be applied in feed-forward process monitoring and control.

3.7.2 Theory

Discrete time state space models can be written via vector/matrix products as in equation (3.7.1) and (3.7.2). They are linear models that link the input to the system at time k (u_k), to the output at time k (y_k), via the system state vector $\boldsymbol{x}_k$ (size $n \times 1$; for ease of notation, we will assume univarite inputs and outputs here, but expansion is straight forward):

$$\boldsymbol{x}_{k+1} = \boldsymbol{A}\boldsymbol{x}_k + \boldsymbol{B}u_k + \boldsymbol{w}_k \tag{3.7.1}$$

$$y_k = \boldsymbol{C}\boldsymbol{x}_k + \boldsymbol{D}u_k + v_k \tag{3.7.2}$$

The $\boldsymbol{A}$ matrix (size $n \times n$) is called the system matrix and describes the system dynamics (i.e. how the system states in vector $\boldsymbol{x}_k$ evolve from one time step to the next $\boldsymbol{x}_{k+1}$). The order (or rank) of this matrix determines how many distinct components or states are identified in the system and their connectivity (determined by the entries in $\boldsymbol{A}$). As reported, the states do not necessarily coincide with the physical phenomena in the system (e.g. biomass in a bioreactor) but can be seen as a latent representation of the dynamic behavior. $\boldsymbol{B}$ is called the input matrix that relates the control input at the current time step to the system states one step ahead. It shows how an input to a system at time k (e.g. feed to a bioreactor) would influence the state of the system at time $k+1$ (e.g. biomass growth conditions in the reactor). $\boldsymbol{C}$ is the output matrix that describes how the system states are reflected in the measurable system output (which, typically, is a physically identifiable entity). It is the link between the principal model at time k and the physical world at time k (e.g. between biomass and growth conditions and actual cell count). $\boldsymbol{D}$ is called the (direct) feed-through term (e.g. how the feed of the reactor is seen instantaneously in the measured response, hence, not how feed changes the system); this term is often not included in the modeling. The careful reader will notice that for the univariate case $\boldsymbol{B}$, $\boldsymbol{C}$, and $\boldsymbol{D}$ in equation (3.7.1) and (3.7.2) should officially be lowercase vectors. However, to stay with common notation, we will keep using matrix capitals instead. $\boldsymbol{w}_k$ and v_k are noise sequences representing model inaccuracy and measurement uncertainty, respectively. Equation (3.7.1) is often referred to as the system equation (reflecting that it describes how the system evolves over time), whereas equation (3.7.2) is called the measurement equation (indicating that it describes how the measured output is related to the state of the system). We will only use discrete time state space models in our study where the effective time between two observations (delta-time, $k+1$ minus k) is the clock time, assumed equidistant and decided by the measurement instrumentation. This could, for example, be the measurement frequency of a spectroscopic determination (or, more accurately, the inverse of the sampling frequency, being the time between two measurements, becoming available).

In order to employ equation (3.7.1) and (3.7.2) in a time series-based process-monitoring scheme, the system matrices must be know or estimated. As stated previously, this could theoretically be achieved on the basis of known physical/chemical knowledge of the system, but very often, this is not possible because of the lack of (sufficient) knowledge on the system. This, leads to the need of SI, with one class of algorithms within SI being sub-space identification. The reader is referred to the Appendix and van Overschee and De Moor for more details on how estimation of the system matrices is performed for the subspace algorithm that is used in this paper. Because several studies have found canonical variates analysis-based (CVA) algorithms outperform others (see subsequent sections for details), the state space model is fitted using this method (see Appendix). The systems studied in our research are pure batch processes with no external inputs, resulting in a so-called stochastic time series (e.g. beer production in a bioreactor is often run as pure batch with no active input). The computations/estimations remain the same, where all input-related parts are canceled by zero-entries. The applied algorithm produces state space models in a forward innovation form, meaning that an optimal least squares gain ($\boldsymbol{K}$) is used as driving term in the prediction (to substitute the deterministic input $\boldsymbol{B}_{u_k}$ Buk, plus purely stochastic part $\boldsymbol{w}_k$ in equation (3.7.1)). The state space model for the (reduced) stochastic case is, thus, to be reformulated as (see Appendix):

$$\boldsymbol{x}_{k+1}=\boldsymbol{A}\boldsymbol{x}_k+\boldsymbol{K}_{e_k} \tag{3.7.3}$$

$$y_k=\boldsymbol{C}\boldsymbol{x}_k+e_k \tag{3.7.4}$$

Here e_k, the innovation at time step k, is equal to the difference between the observed output and the output predicted by the model (the prediction error at time k). Notice that despite their apparent simplicity, stochastic systems can still approximate complex, nonlinear phenomena because of the interaction between the states in $\boldsymbol{x}_k$ via the entries in system matrix $\boldsymbol{A}$, plus the initial conditions for the system (e.g. in the bioreactor for beer production, for example, the biomass growth and amount of sugar available for conversion into alcohol might lead to a complex development over time on the basis of internal feedbacks, interacting with the raw material properties and quantities at the batch start/charge plus the yeast's biological efficiency).

3.7.3 State Space Models in Chemometrics

The idea of applying state space models in chemometrics is not new, albeit not widely spread. A series of papers on state space modeling were published in the chemometric literature in the late 1990s and early 2000s, but the research area has received less attention during the last 10 years. Here, we will give an (nonexhaustive) overview. A chemometric paper on state space models was published in 1997 by Negiz and Cinar. It was shown that partial least squares (PLS) can be used to fit state space equations, but it was, at the same time, shown that modifications of the PLS algorithm were necessary to give useful results. A method, based on CVA, proved to give the best outcome. Hartnett and coworkers published two different papers in the end of the 1990s. In the first of the two papers, genetic algorithms are used in combination with principal components regression (PCR), to do dynamic inferential estimation of process variables[9]. The measurement equation of an underlying state space model is used, but the focus is not on the state space model itself. This is performed later in the paper[10], where how a nonlinear multivariable production plant can be modeled using a combination of PCA and state space modeling is shown. The idea is to perform PCA on the process outputs; the scores obtained are then used as states, and the loadings used as the C matrix (no input is used in the measurement equation in this paper[10]). The system equation is subsequently identified by concatenating state matrices and input-matrices and by regressing the concatenated matrix on future states by means of PCR. The PCA-based state space model was compared with an analytical state space model. Both had good performance in approximating the nonlinear system. The authors note that no prior decision on model order needs to be taken when using this PCA-based approach. This is correct, but a decision on the number of principal components in both the PCA and the PCR step needs to be made. Ergon[11] uses state space equations, PCR and PLS, to derive relations that can be used to predict one output variable from another. An example of state space modeling is also given, but this is via the predictor error method (PEM). Dynamic system PCR and PLS solutions for output predictions are also presented, again, based on a PEM state space model. In a later paper by Ergon and Halstensen[12], these results are elaborated for a system with low-sampling-rate reference measurements—a combination of PCA and PEM is utilized to produce predictions of the reference measurements with a better performance as compared with the PLS. Shi and MacGregor give a complete review[13] of different subspace methods and compare them to different latent variable techniques (PCA, PLS, and PCR). They come to two overall conclusions: (i) for process monitoring ("Is my process on track?"), latent variable methods are to be preferred, but for process identification ("What are the process dynamics?" or "Where is my process heading?"), dedicated sub-space identification methods are preferential; and (ii) CVA and the N4SID algorithm[7] have the best performances of the subspace methods they tested. In the most recent paper, applying state space models in the chemometric literature, Pan et al.[14] showed, contrary to the first conclusion by Shi and MacGregor[13], that better monitoring performances could be archived if a state space/subspace method was applied compared to PCA-based monitoring. The authors use PCA to reduce the dimensionality of the output, followed by fitting state space models by means of N4SID. A Kalman filter is subsequently used. A large part of the advantage from this model is, according to the authors, a result of the implementation of the Kalman filter.

3.7.4 Milk Coagulation Monitoring

In this paper, it is shown how the dynamics of the coagulation of milk can be observed and modeled by combining near infrared (NIR) spectroscopy, PCA, and subspace-based state space estimation. The example, milk coagulation for cheese production, is a purely stochastic time series with no input or control, and is therefore modeled according to equation (3.7.3) and (3.7.4). Twelve batches of coagulating milk were monitored by NIR spectroscopy. The data was first published by Lyndgaard et al[15], and for further details on the procedures and measurements, including the batch numbering, we refer to this publication. In that paper, it was shown that scores from a PCA decomposition of the NIR data could be modeled by mechanistic models[15]. In this manuscript, it is shown that it also is possible to model the data via SI without prior assumptions on the process dynamics. Throughout this paper, we will follow the standard notation used in state space literature. x_k is therefore the process states at time point k, and y_k the system output at time point k in our case, it is the first PCA score vector from the decomposition of the NIR spectra recorded during the batch process. Fig. 3.36(a) shows the NIR spectra of a representative normal operating conditions (NOC) batch color-coded by runtime, whereas Fig. 3.36(b) shows the first PCA score of all 12 batches used in this research[15].

Fig. 3.36

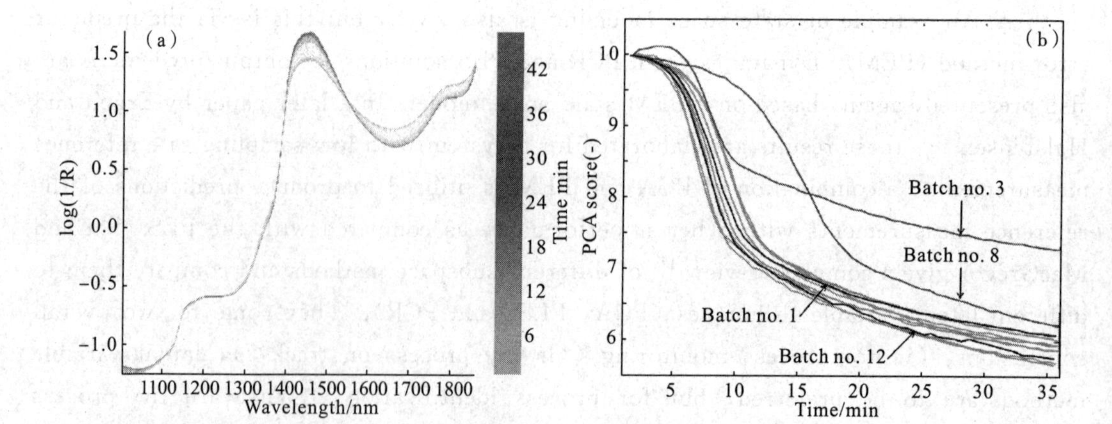

Fig. 3.36 (a) Near infrared (NIR) reflection spectra of coagulating milk in one batch. (b) PCA scores over time from the NIR spectra of 12 different coagulating milk batches

It is seen that the main effect over time is a narrowing and an increase of the water band around 1400 - 1500 nm which can be attributed to gel formation and hardening[16]. This is a general trend for all twelve batches. In order not to mix symbols, this PCA step is written as $\boldsymbol{Z}=\boldsymbol{y}\boldsymbol{p}^{\mathrm{T}}+\boldsymbol{E}$ with $\boldsymbol{Z}$, $\boldsymbol{y}$, $\boldsymbol{p}$, and $\boldsymbol{E}$, respectively, being a set of NIR spectra sorted as a function of time, the first PCA score vector (equaling the output of our state space model), the corresponding PCA loading vector, and the residuals. The data flow for the state space-based process monitoring is shown in Fig. 3. 41 in the Section 3. 7. 8.

Eight batch runs will be considered as training NOC batches, four runs as the test set (Batches 1 and 12 as NOC, and Batches 3 and 8 as extremes/non-NOC). The general PCA score trajectory can be described by three different phases: a very short lag-phase plus a decaying sigmoidal curve and an exponential decay, all superimposed on each other with ill-defined boundaries/transition times[15]. Knowing/predicting the development of the last phase (gel hardening) is of great importance for cheese manufacturing because it gives information on the optimal cutting time (the following step in production[15]), and thus, the quality of the end product. It is, in spite of the clear nonlinear tendencies that can be observed, expected that the data can be well approximated by the linear state space models of sufficient rank. It can furthermore be noticed that two batches differ noticeable from the others: Batch 3 has increasing score values during the first 5 min followed by a very short sigmoidal part which results in the highest "end-value" of all the batches. Batch 8 has a low "end-value" and a longer lag-phase. It should be noted that all experimental runs were performed as similar as possible, and outlying behavior is thus caused by unanticipated but natural variation[15].

3.7.5 State Space Based Monitoring

In our milk coagulation investigation, delta-time is the measurement frequency of the NIR spectrometer (which gives a new outcome every 36s, hence, k to $k+1$ takes 36s). Each new NIR data collection is scatter corrected by means of standard normal variate scaling right after collection. The new, expanded spectral matrix $\boldsymbol{Z}$ is centered, and a PCA decomposition is performed. During the calibration phase, it was established that a one-component PCA model using centered data was essentially the same as a two-component model on the noncentered data, with the first principal component being close to the average NIR spectra and the second showing the dynamics of interest in this study. Following the notion of parsimony, a one-component PCA model on the centered data is therefore preferred. In order to avoid numerical problems due to a sign change in the score-values vector during the subsequent state space modeling, all scores are lifted to be positive. This is performed by simply adding the right, same amount to all score values collected thus far to make the first score value equal to 10 ($y_0=10$). This operation is performed after each new PCA decomposition. The PCA score time trajectory is zero-order-hold resampled[6] to double the number of data points, after which, a state space model can be fitted. In this procedure, the number of data points is doubled by repeating a measured data value once at the intermediate time for this true value and its proceeding measured neighbor (giving a "staircase" resampled signal that will not introduce false dynamics in the system). This is performed to achieve a more stable estimate of the Hankel matrices (see Appendix) in the beginning of the batch monitoring where only a few measurement observations are available. The models are fitted recursively, meaning that real-time acquisition of data is simulated by step-wise, including more and more observations. The first state space model for each batch is fitted when the first $k=16$ NIR spectra are collected (corresponding to approximately 9 min into the coagulation process, well pass the initial lag-phase of gel formation, inside the sigmoidal phase for normal batches[15], Fig. 3. 36 (b)). At the next time step, one more NIR spectrum is included in the data set, and a state space model is determined from the newly computed, offset-corrected and resampled PCA scores for $k=17$. In this way, it is possible to make an estimate of the $\boldsymbol{A}$ matrix (which contains the estimated dynamics of the system) and the initial state vector $\boldsymbol{x}_0$ (which represents the estimated initial or boundary conditions of the system) at each time step. All these computation steps take less than 1s on a normal personal computer and can thus be performed in "real-time" for an NIR measurement rate of 36s (see Fig. 3. 41 for the full computational procedure).

The eigenvalues of $\boldsymbol{A}$ reflect the dynamics and stability of the system[6]; stable discrete linear time-invariant systems have eigenvalues within the unit circle (where complex eigenvalues indicate an oscillating system). By comparing the eigenvalues of $\boldsymbol{A}$ for different runs, the development of different batches can be compared. On the basis of a training set of NOC batches, statistical process control (SPC) charts for the system matrix $\boldsymbol{A}$ over time can be constructed and applied to new production runs. The initial state estimate $\boldsymbol{x}_0$ gives information on the boundary conditions of the difference equation in equation (3. 7. 1) or (3. 7. 3). This represents the best estimate for the initial conditions in the batch. For monitoring purposes, we propose 95% confidence intervals (95% CI) for the elements of $\boldsymbol{x}_0$ and λ the eigenvalues 1 of $\boldsymbol{A}$ on the basis of a Student's t-statistic of the values for the NOC training batches. Although it is not guaranteed that normal probabilities are valid for either set of parameters (e. g. because of the mentioned bias in the solution[7], see Appendix), it can serve as a first approximation. The surveillance of the initial conditions in $\boldsymbol{x}_0$ and the system dynamics in $\boldsymbol{A}$, thus, tracks whether the batch evolves according to NOC or not ("Did the batch start at normal conditions, and is it developing as expected?"). The dynamic representation in equation (3. 7. 3) and (3. 7. 4) (or equation (3. 7. 1) and (3. 7. 2) in the nonstochastic case) can further be used to predict progress of the system—by developing, in time, the equations starting from time zero ($\boldsymbol{x}_0$), an estimate of future states and measurement observations can be made.

3.7.6 Results

A critical step in state space modeling is the order or rank selection, the number of states in the system. Different tools can be used for this, and as outlined in the Appendix here, we will use the singular values of the block Hankel matrix (which is built from time-shifted versions of the time series for each batch). As the eigenvalues of the covariance matrix can be used for decision on the number of PCA components, the idea is to inspect the singular values of the block Hankel matrix. Fig. 3. 37 presents the average singular values and the approximate 95% confidence limits on the basis of a student's t-statistic for the Hankel matrix of the eight full NOC batch runs with six block rows. Fig. 3. 40 shows how the first two singular values are very stable, whereas the remaining four have a higher variance/uncertainty. From this plot, the system order is therefore deemed to be two, which also seems reasonable from the observation of the two main phases in the PCA score trajectory (a sigmoidal and an exponential decay, where the lag-phase is too short and is weakly present for our sampling rate of 36s to capture). No significant difference in predictive performance of the model was observed for a rank three system (where one real and two complex eigenvalues were found for system matrix $\mathbf{A}$), whereas a rank one model severely underperformed with a biased prediction (results not shown).

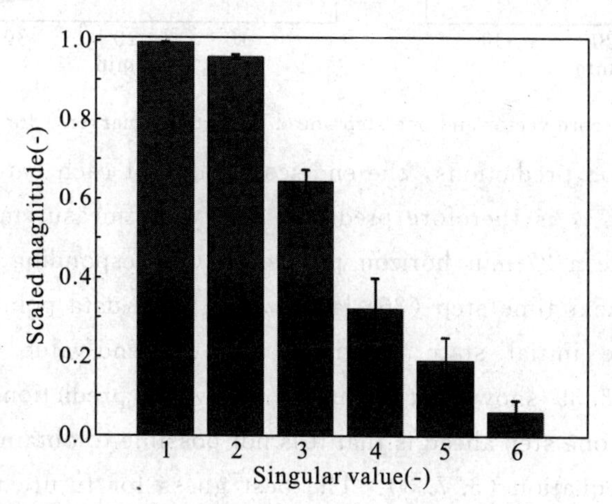

Fig. 3. 37 Scaled singular values of the block Hankel matrix for the normal operating conditions data

Equations (3. 7. 3) and (3. 7. 4) enable the prediction of future system outputs. A natural way of validating state space models is therefore to compare the predicted output to the actual system output. Fig. 3. 38 shows how the models fitted on each individual batch run handle the one-step-ahead prediction for the four test batches. Data is collected for $K=1$ to k; state space modeling is performed on the collected data, and the one-step-ahead prediction is found using equation (3. 7. 3) and (3. 7. 4). Fig. 3. 48 shows that the one-step-ahead predicted score vector is very close to the observed profile for all the batches, including the non-NOC batches. The order two state space models are thus good at capturing the essential dynamics and producing predictions over a short time horizon.

It should, however, be remarked that only testing the one-step-ahead prediction is not very powerful because this may lead to very optimistic prediction errors. In order to test

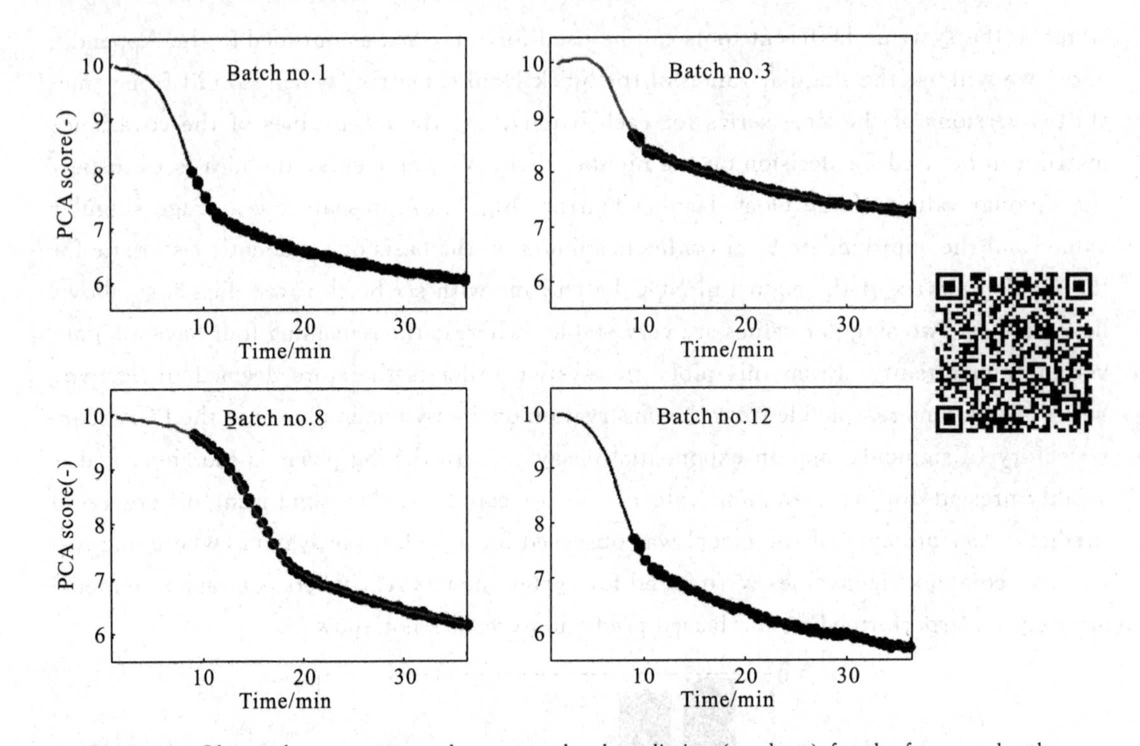

Fig. 3. 38　Observed score vector and one-step-ahead prediction (markers) for the four test batches

the longer time horizon predictions, the end-score value of each batch, taken here as 36 min into coagulation[15], is therefore predicted after each measurement point. The first prediction is therefore a 27-min horizon prediction (corresponding to a 45 steps-ahead prediction). At the next time step (36s later), one more data point is obtained, a new system including the initial state identified and the end-value predicted from this information. Figure 3. 39 shows the error for end-value prediction. A challenge when predicting more than one step ahead is that it is not possible to obtain an innovation (e_{k+n}) for future values in equation (3. 7. 3). The best guess for future time points—used in Fig. 3. 39—is e_k, the last known innovation as a substitute for the remaining time steps.

Several things can be noticed in Fig. 3. 39. As anticipated, the end-value estimate gets better as more and more measurements are available for fitting the model and less extrapolation is required, and already after approximately 12 min, an acceptable estimate of the batch end-value can be obtained for Batches 1, 3, and 12. The initial models have clear difficulties in predicting the end-value of the non-NOC Batch 8; the end-value cannot be predicted with a satisfactory small error until 20min into the batch. This is caused by the longer lag-phase (not present in the NOC set) and delayed response for this batch.

SPC charts for the eigenvalues of $\boldsymbol{A}$ and the elements of $\boldsymbol{x}_0$ are implemented for process monitoring. These charts are shown in Fig. 3. 40 for the four test batches. It can be observed that the NOC batches stay inside or close to the proposed 95% confidence limits in all control charts, whereas the two deviating batches clearly break the limits for several of them. For example, it can be observed that both of the non-NOC batches already break

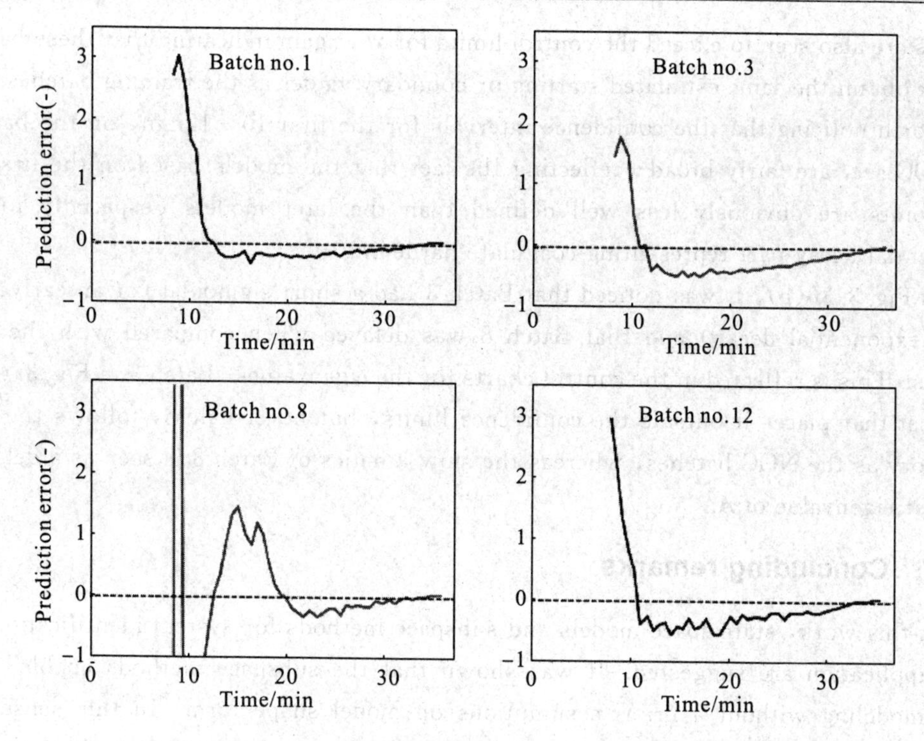

Fig. 3. 39 Prediction error for the end-value prediction of the four test batches

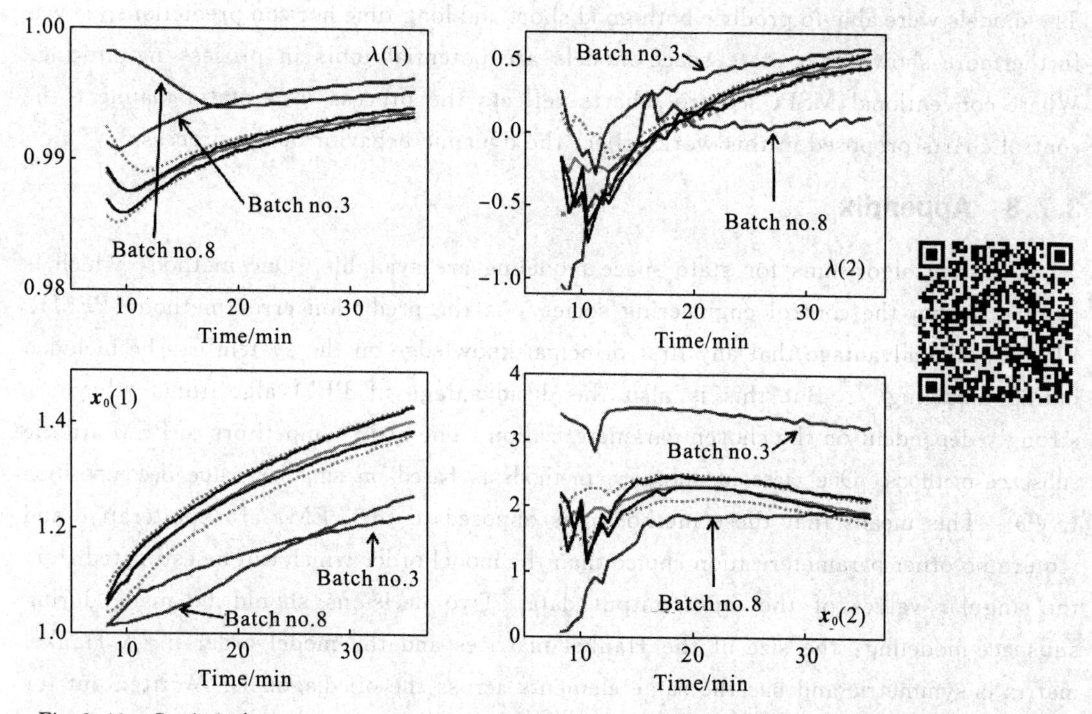

Fig. 3. 40 Statistical process control charts for eigenvalues of the $\boldsymbol{A}$ matrix and the initial state vector $\boldsymbol{x}_0$

the 95% CI in the control chart of the first eigenvalue of $\boldsymbol{A}$ after 9–10min, clearly indicating that these two batches do not follow the NOC dynamics. The two deviating batches are also seen to exceed the control limits for $\boldsymbol{x}_0$, again indicating that these batches did not obtain the same estimated starting or boundary values as the training batches. It is also worth noticing that the confidence intervals for the first 10 - 12min, on the basis of the NOC set, are fairly broad—reflecting the fact that the models based on the first few data points are obviously less well defined than the later models (especially for the exponential decay part representing coagulate hardening)[15].

In Fig. 3.36(b), it was noticed that Batch 3 had a short sigmoidal (or an early onset of the exponential decay) and that Batch 8 was delayed when compared with the NOC batches. This is reflected in the control charts for the eigenvalues. Batch 3 in Fig. 3.40 has an offset that places it outside the confidence limits, but nevertheless, follows the same trajectory as the NOC batches, whereas the slow kinetics of Batch 8 is seen as a delay for the first eigenvalue of $\boldsymbol{A}$.

3.7.7 Concluding remarks

In this work, state space models and subspace methods for system identification in a PAT application are suggested. It was shown that the subspace methods enabled state space modeling without a priori assumptions on model shape/form. In this sense, the subspace methods enabled the modeling to be data-driven rather than hypothesis-driven. The models were able to produce both good short and long time horizon predictions. It was furthermore shown that state space models are potential tools in process monitoring. Where conventional MSPC control charts reflects the process in a static manner, the control charts proposed in this work reflect the dynamic behavior of the process.

3.7.8 Appendix

Different algorithms for state space modeling are available. One method, which is popular within the control engineering society, is the prediction error method (PEM). PEM has the advantage that any first principal knowledge on the system can be included during modeling[17]. But this is also the disadvantage of PEM algorithms—they are strongly dependent on the chosen parameterization. The main competitors of PEM are the subspace methods. One class of subspace methods is based on singular value decomposition (SVD). This means that these methods, as opposed to the PEM, are noniterative and require no other parameterization choice than the model order which can be estimated from the singular values of the input/output data. Two decisions should be made during subspace modeling: the size of the Hankel matrices and the model order n. A Hankel matrix is symmetric and has the same elements across the off-diagonals. Written out for the input series (u_0, u_1, u_2, …, u_{i+j-1}) and corresponding output series (y_0, y_1, y_2, …, y_{i+j-1}), the Hankel matrices would thus be[7]:

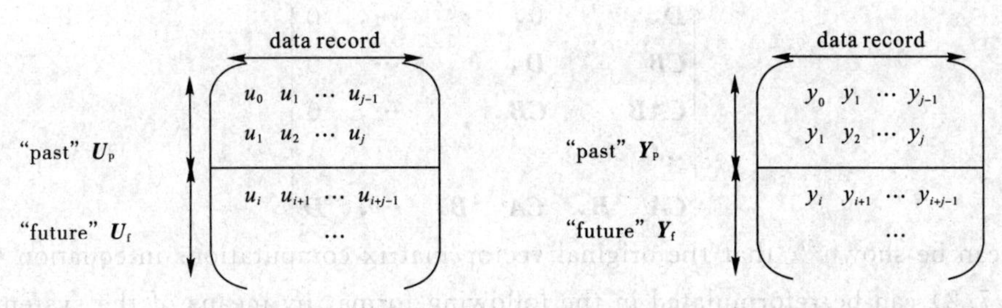

A similar Hankel matrix (effectively, a row and column time-shifted data representation) can be defined for the states series (x_0, x_1, x_2, ··· ,x_{i+j-1}), where each entry is a vector of length n (the rank of the system), instead of a scalar. The separation between "past" and "future" data reflects how future inputs, out-puts, and states can be regressed on past inputs, outputs, and states. The selection of the number of block rows (the "past" and "future" horizons) should be made so that i is larger than the expected system order n, whereas $i+j+1$ is determined by the length of the available training time series.

The input and output Hankel matrices can be combined in a block Hankel matrix $\boldsymbol{W}$. The "past" block Hankel matrix $\boldsymbol{W}_p$ would thereby, for example, be defined as[7]:

$$\boldsymbol{W}_p=\begin{pmatrix}\boldsymbol{U}_p\\ \boldsymbol{Y}_p\end{pmatrix}$$

The chemical/physical rank of $\boldsymbol{W}_p$ is an estimate of the true underlying number of dynamic components (which could be called eigenfrequencies) in the system. $\boldsymbol{W}_p$ can therefore be used to estimate the system order n. The block Hankel matrices for the observed data are closely related to the concepts of observablity and controlablity of the system states[7]. States can, in general terms, be said to be observable if they can be uniquely deter-mined from the output y_k of the system. A useful system-related matrix is the observability matrix $\boldsymbol{\Gamma}$, defined as:

$$\boldsymbol{\Gamma}=\begin{pmatrix}\boldsymbol{C}\\ \boldsymbol{CA}\\ \boldsymbol{CA}^2\\ \cdots\\ \boldsymbol{CA}^{j-1}\end{pmatrix}$$

If the rank of $\boldsymbol{\Gamma}$ is equal to n (number of elements in the state vector $\boldsymbol{x}_k$), then the system is observable. Another useful system-related matrix is the controllability matrix $\boldsymbol{\Delta}$. It is, as the name suggests, related to the controllability of the system. The system is controllable if it can be brought to any desired state by the in-put series $\boldsymbol{u}_k$. The controllability matrix is defined as:

$$\boldsymbol{\Delta}=(\boldsymbol{A}^{j-1}\boldsymbol{B}\quad \boldsymbol{A}^{j-2}\boldsymbol{B}\quad \cdots\quad \boldsymbol{AB}\quad \boldsymbol{B})$$

The last system-related matrix that needs to be defined is the lower block triangular toeplitz matrix $\boldsymbol{H}$:

$$\begin{bmatrix} \boldsymbol{D} & 0, & \cdots, & 0 \\ \boldsymbol{CB} & \boldsymbol{D}, & \cdots, & 0 \\ \boldsymbol{CAB} & \boldsymbol{CB}, & \cdots, & 0 \\ \cdots & & & \\ \boldsymbol{CA}^{j-2}\boldsymbol{B}, & \boldsymbol{CA}^{j-2}\boldsymbol{B}, & \cdots, & \boldsymbol{D} \end{bmatrix}$$

It can be shown[18] that the original vector/matrix computations in equation (3.7.1) and (3.7.2) can be reformulated in the following format by means of the system-related matrices as defined previously:

$$\boldsymbol{Y}_{\mathrm{p}} = \boldsymbol{\Gamma X}_{\mathrm{p}} + \boldsymbol{HU}_{\mathrm{p}}$$
$$\boldsymbol{Y}_{\mathrm{f}} = \boldsymbol{\Gamma X}_{\mathrm{f}} + \boldsymbol{HU}_{\mathrm{f}}$$
$$\boldsymbol{X}_{\mathrm{f}} = \boldsymbol{AX}_{\mathrm{p}} + \boldsymbol{\Delta U}_{\mathrm{p}}$$

The different subspace algorithms available essentially solve this set of equations from which the $\boldsymbol{A}$, $\boldsymbol{B}$, $\boldsymbol{C}$, and $\boldsymbol{D}$ matrices in equations (3.7.1) and (3.7.2) for a user-defined rank n of the system are estimated. The term "subspace" refers to the fact that the first step in the algorithms is an oblique (or nonorthogonal) projection O of the "future" outputs ($\boldsymbol{Y}_{\mathrm{f}}$) on the "past" block Hankel matrix $\boldsymbol{W}_{\mathrm{p}}$ along the future outputs $\boldsymbol{Y}_{\mathrm{f}}$. Singular SVD is then calculated on this weighted oblique projection: $\boldsymbol{G}_1\boldsymbol{O}\boldsymbol{G}_2 = \widetilde{\boldsymbol{U}}\boldsymbol{S}\boldsymbol{V}^{\mathrm{T}}$, where $\boldsymbol{G}_1$ and $\boldsymbol{G}_2$ are weights determined by the specific algorithm (where-in the case of a CVA solution-$\boldsymbol{G}_1$ contains the inverse square roots of the covariance estimate of the future outputs, and $\boldsymbol{G}_2$ is the identity[7]). $\widetilde{\boldsymbol{U}}$ and $\boldsymbol{S}$ are then used to determine the observability matrix ($\boldsymbol{\Gamma}$) by $\boldsymbol{\Gamma} = \boldsymbol{G}_1\widetilde{\boldsymbol{U}}\boldsymbol{S}^{1/2}$. Because the oblique projection is equal to the product of $\boldsymbol{\Gamma}$ and the states ($\boldsymbol{X}_k$), is it possible to determine the states by $\boldsymbol{X} = \boldsymbol{\Gamma}^{+}\boldsymbol{O}$, where the Moore-Penrose pseudo inverse of the observability matrix is used. The boundary be- tween "past" and "present" can then be shifted one step in order to determine the states at the next time step ($\boldsymbol{X}_{k+1}$), making the $\boldsymbol{A}$, $\boldsymbol{B}$, $\boldsymbol{C}$, and $\boldsymbol{D}$ matrices the only unknowns in the system of linear equations that can thus be solved by least squares.

For the batch situation without input signals discussed in this manuscript, CVA-based stochastic Algorithm 3 from the book "Subspace Identification for Linear Systems" by Peter van Overschee and Bart de Moor is used. The "past" block Hankel matrix is, in this case, equal to the "past" outputs ($\boldsymbol{Y}_{\mathrm{p}}$), and the algorithm then follows the same flow as in the deterministic case: determine $\boldsymbol{O}$ from "future" outputs and the block Hankel matrix (="future" outputs ($\boldsymbol{Y}_{\mathrm{f}}$)), determine the observability matrix ($\boldsymbol{\Gamma}$) from the weighted $\boldsymbol{O}$, determine the states by $\boldsymbol{X} = \boldsymbol{\Gamma}^{+}\boldsymbol{O}$, and solve the system of linear equations by least squares. The algorithm, furthermore, has the additional feature to produce positive real covariance sequences, making the solutions produced by the algorithm physically/chemically meaningful. The price to pay for this is a bias in the solution. The equivalent of controllability for the stochastic system, equation (3.7.3) and (3.7.4), is sometimes called reachability[12] (those latent states of the system that can be reached by the system dynamics and noise input), whereas the observability is some- times substituted by detectability (those latent states of the system that can be observed/detected or "are excited by" the system dynamics, plus noise input).

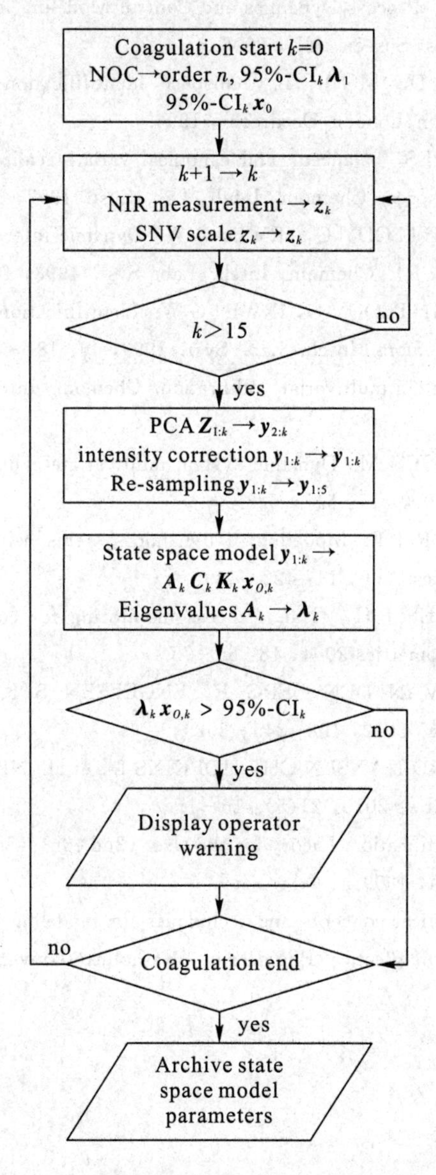

Fig. 3. 41 Dataflow for state space-based monitoring

References

[1] PERIS-VICENTE J, CARDA-BROCH S, ESTEVE-ROMERO J. Validation of Rapid Microbiological Methods[J]. Journal of Laboratory Automation, 20(3):259 - 264.

[2] LAURSEN K, RASMUSSEN MA, BRO R. Comprehensive control charting applied to chromatography. Chemom. Intell. Lab. Syst. 2011; 107: 215 - 225.

[3] QIN S J. Statistical process monitoring: basics and beyond. J. Chemometrics 2003; 17: 480 - 502.

[4] KOURTI T, MACGREGOR J F. Process analysis, monitoring and diagnosis, using multivariate projection methods. Chemom. Intell. Lab. Syst. 1995, 28: 3-21.

[5] SKAGERBERG B, MACGREGOR J F, KIPARISSIDES C. Multivariate data-analysis applied to low-density polyethylene reactors. Chemom. Intell. Lab. Syst. 1992, 14: 341-356.

[6] ROFFEL B, BETLEM B. Process Dynamics and Control Modeling for Control and Prediction. John Wiley & Sons, Ltd.: West Sussex, UK, 2006.

[7] VAN OVERSCHEE P, DE MOOR B. Subspace Identifification for Linear Systems, Kluwer Academic Publishers: oston/London/Dordrecht, 1996.

[8] NEGIZ A, CINAR A. PLS, balanced, and canonical variate realization techniques for identifying VARMA models in state space. Chemom. Intell. Lab. Syst. 1997, 38: 209-221.

[9] HARTNETT M K, LIGHTBODY G, IRWIN G W. Dynamic inferential estimation using principal components regression (PCR). Chemom. Intell. Lab. Syst. 1998, 40: 215-224.

[10] HARTNETT M K, LIGHTBODY G, IRWIN G W. Identifification of state models using principal components analysis. Chemom. Intell. Lab. Syst. 1999, 46: 181-196.

[11] ERGON R. Dynamic system multivariate calibration. Chemom. Intell. Lab. Syst. 1998, 44: 135-146.

[12] ERGON R, HALSTENSEN M. Dynamic system multivariate calibration with ow-sampling-rate y data. J. Chemometrics 2000, 14: 617-628.

[13] SHI R J, MACGREGOR J F. Modeling of dynamic systems using latent variable and subspace methods. J. Chemometrics 2000, 14: 423-439.

[14] PAN Y D, YOO C, LEE J H, et al. Process monitoring for continuous process with periodic characteristics. J. Chemometrics 2004, 18: 69-75.

[15] LYNDGAARD C H, VAN DEN BERG F, ENGELSEN S B. Real-time modeling of milk coagulation. J. Food Eng. 2012, 108: 345-352.

[16] DAHM D, LYNDGAARD HANSEN CH, HOPKINS D, et al. NIR discussion forum: analysis of coagulating milk. NIR News 2010, 21(5): 16-17.

[17] LJUNG L. System Identfification-Theory for the User (2nd edn). Prentice Hall PTR: Upper Saddle River, New Jersey, USA, 1999.

[18] DE MOOR B. Mathematical concepts and techniques for modeling of static and dynamic systems, PhD thesis, Department of Electical Engineering, Katholieke Universiteit Leuven, Belgium, 1988.

Chapter 4 Partial Least Squares Analysis

4.1 Basic Concept[1]

Partial Least Squares (PLS) is a wide class of methods for modeling relations between sets of observed variables by means of latent variables. It comprises of regression and classification tasks as well as dimension reduction techniques and modeling tools. The underlying assumption of all PLS methods is that the observed data is generated by a system or process which is driven by a small number of latent (not directly observed or measured) variables. Projections of the observed data to its latent structure by means of PLS was developed by Herman Wold and coworkers[26, 27].

PLS has received a great amount of attention in the field of chemometrics. The algorithm has become a standard tool for processing a wide spectrum of chemical data problems. The success of PLS in chemometrics resulted in a lot of applications in other scientific areas including bioinformatics, food research, medicine, pharmacology, social sciences, physiology—to name but a few[15, 16].

This chapter introduces the main concepts of PLS and provides an overview of its application to data analysis problems. Our aim is to present a concise introduction, that is, a valuable guide for anyone who is concerned with data analysis.

In its general form PLS creates orthogonal score vectors (also called latent vectors or components) by maximizing the covariance between different sets of variables. PLS dealing with two blocks of variables is considered in this chapter, although the PLS extensions to model relations among a higher number of sets exist [24, 25]. PLS is similar to Canonical Correlation Analysis (CCA) where latent vectors with maximal correlation are extracted [14]. There are different PLS techniques to extract latent vectors, and each of them gives rise to a variant of PLS.

PLS can be naturally extended to regression problems. The predictor and predicted (response) variables are each considered as a block of variables. PLS then extracts the score vectors which serve as a new predictor representation and regresses the response variables on these new predictors. The natural asymmetry between predictor and response variables is reflected in the way in which score vectors are computed. This variant is known under the names of PLS1 (one response variable) and PLS2 (at least two response variables). PLS regression used to be overlooked by statisticians and is still considered rather an algorithm than a rigorous statistical model [9]. Yet within the last years, interest in the statistical properties of PLS has risen. PLS has been related to other regression methods like Principal Component Regression (PCR) [16] and Ridge Regression (RR) [10] and all these methods can be cast under a unifying approach called continuum regression [22, 6]. The effectiveness of PLS has been studied theoretically in terms of its variance and its shrinkage properties [13, 12, 4]. The performance of PLS is investigated in several simulation studies [2].

PLS can also be applied to classification problems by encoding the class membership in an appropriate indicator matrix. There is a close connection of PLS for classification to Fisher Discriminant Analysis (FDA). PLS can be applied as a discrimination tool and dimension reduction method-similar to Principal Component Analysis (PCA). After relevant latent vectors are extracted, an appropriate classifier can be applied. The combination of PLS with Support Vector Machines (SVM) has been studied in [19].

Finally, the powerful machinery of kernel-based learning can be applied to PLS. Kernel methods are an elegant way of extending linear data analysis tools to nonlinear problems [21].

4.1.1 Partial Least Squares

Consider the general setting of a linear PLS algorithm to model the relation between two data sets (blocks of variables). Denote by $\boldsymbol{X}\subset\mathbb{R}^N$ an N-dimensional space of variables representing the first block and similarly by $\boldsymbol{y}\subset\mathbb{R}^N$ a space representing the second block of variables. PLS models the relations between these two blocks by means of score vectors. After observing n data samples from each block of variables, PLS decomposes the $(n\times N)$ matrix of zero-mean variables $\boldsymbol{X}$ and the $(n\times M)$ matrix of zero-mean variables $\boldsymbol{Y}$ into the form

$$\begin{aligned}\boldsymbol{X}&=\boldsymbol{T}\boldsymbol{P}^{\mathrm{T}}+\boldsymbol{E}\\ \boldsymbol{Y}&=\boldsymbol{U}\boldsymbol{Q}^{\mathrm{T}}+\boldsymbol{F}\end{aligned}\tag{4.1.1}$$

Graphically, it can be shown as Fig. 4.1, where the $\boldsymbol{T}$, $\boldsymbol{U}$ are $(n\times p)$ matrices of the p extracted score vectors (components, latent vectors), the $(N\times p)$ matrix $\boldsymbol{P}$ and the $(M\times p)$ matrix $\boldsymbol{Q}$ represent matrices of loadings and the $(n\times N)$ matrix $\boldsymbol{E}$ and the $(n\times M)$ matrix $\boldsymbol{F}$ are the matrices of residuals. The PLS method, which in its classical form is based on the nonlinear iterative partial least squares (NIPALS) algorithm [24], finds weight vectors $\boldsymbol{w}$, $\boldsymbol{c}$ such that

$$[\mathrm{cov}(\boldsymbol{t},\boldsymbol{u})]^2=[\mathrm{cov}(\boldsymbol{xW},\boldsymbol{Yc})]^2=\max_{|r|=|s|=1}[\mathrm{cov}(\boldsymbol{Xr},\boldsymbol{Ys})]^2\tag{4.1.2}$$

where $\mathrm{cov}(\boldsymbol{t},\boldsymbol{u})=\boldsymbol{t}^{\mathrm{T}}\boldsymbol{u}/n$ denotes the sample covariance between the score vectors $\boldsymbol{t}$ and $\boldsymbol{u}$. The NIPALS algorithm starts with random initialization of the $\boldsymbol{Y}$-space score vector $\boldsymbol{u}$ and repeats a sequence of the following steps until convergence.

(1) $\boldsymbol{w}=\dfrac{\boldsymbol{X}^{\mathrm{T}}\boldsymbol{u}}{\boldsymbol{u}^{\mathrm{T}}\boldsymbol{u}}$

(2) $\| w \| \rightarrow 1$

(3) $t = Xw$

(4) $c = \frac{Y^T t}{t^T t}$

(5) $\| c \| \rightarrow 1$

(6) $u = Yc$

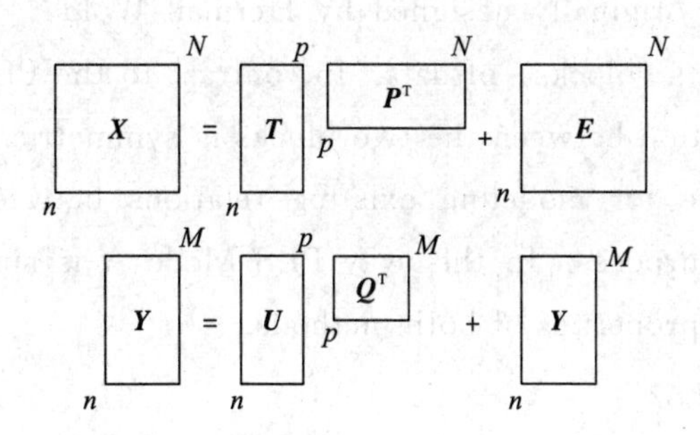

Fig. 4. 1 Where the T, U are $(n \times p)$ matrices of the p extracted score vectors (components, latent vectors), the $(N \times p)$ matrix P and the $(M \times p)$ matrix Q represent matrices of loadings and the $(n \times N)$ matrix E and the $(n \times M)$ matrix F are the matrices of residuals

Note that $u = y$ if $M = 1$, that is, Y is a one-dimensional vector that we denote by y. In this case the NIPALS procedure converges in a single iteration.

It can be shown that the weight vector w also corresponds to the first eigenvector of the following eigenvalue problem [10]

$$X^T Y Y^T X w = \lambda w \tag{4.1.3}$$

The X- and Y-space score vectors t and u are then given as

$$t = Xw \quad \text{and} \quad u = Yc \tag{4.1.4}$$

where the weight vector c is define in steps (4) and (5) of NIPALS. Similarly, eigenvalue problems for the extraction of t, u or c estimates can be derived [11]. The user then solves for one of these eigenvalue problems and the other score or weight vectors are readily computable using the relations defined in NIPALS.

4.1.2 Form of Partial Least Squares

PLS is an iterative process. After the extraction of the score vectors t, u the matrices X and Y are deflated by subtracting their rank-one approximations based on t and u. Different forms of deflation define several variants of PLS.

Using equation (4. 1. 1) the vectors of loadings p and q are computed as coefficients of regressing X on t and Y on u, respectively

$$p = \frac{X^T t}{t^T t} \quad \text{and} \quad q = \frac{Y^T u}{u^T u}$$

4.1.2.1 PLS Mode A

The PLS Mode A is based on rank-one deflation of individual block matrices using the corresponding score and loading vectors. In each iteration of PLS Mode A the $\boldsymbol{X}$ and $\boldsymbol{Y}$ matrices are deflated

$$\boldsymbol{X}=\boldsymbol{X}-\boldsymbol{t}\boldsymbol{p}^{\mathrm{T}} \text{ and } \boldsymbol{Y}=\boldsymbol{Y}-\boldsymbol{u}\boldsymbol{q}^{\mathrm{T}}$$

This approach was originally designed by Herman Wold [25] to model the relations between the different sets (blocks) of data. In contrast to the PLS regression approach, discussed next, the relation between the two blocks is symmetric. As such this approach seems to be appropriate for modeling existing relations between sets of variables in contrast to prediction purposes. In this way PLS Mode A is similar to CCA. Wegelin discusses and compares properties of both methods.

4.1.2.2 PLS1, PLS2

PLS1 (one of the block of data consists of a single variable) and PLS2 (both blocks are multidimensional) are used as PLS regression methods. These variants of PLS are the most frequently used PLS approaches. The relationship between $\boldsymbol{X}$ and $\boldsymbol{Y}$ is asymmetric. Two assumptions are made: i) the score vectors $\{\boldsymbol{t}_i\}_{i=1}^{p}$ are good predictors of $\boldsymbol{Y}$; p denotes the number of extracted score vectors—PLS iterations, ii) a linear inner relation between the scores vectors $\boldsymbol{t}$ and $\boldsymbol{u}$ exists; that is,

$$\boldsymbol{U}=\boldsymbol{T}\boldsymbol{D}+\boldsymbol{H} \tag{4.1.5}$$

where $\boldsymbol{D}$ is the $(p\times p)$ diagonal matrix and $\boldsymbol{H}$ denotes the matrix of residuals. The asymmetric assumption of the predictor-predicted variable(s) relation is transformed into a deflation scheme where the predictor space, say $\boldsymbol{X}$, score vectors $\{\boldsymbol{t}_i\}_{i=1}^{p}$ are good predictors of $\boldsymbol{Y}$. The score vectors are then used to deflate $\boldsymbol{Y}$, that is, a component of the regression of $\boldsymbol{Y}$ on $\boldsymbol{t}$ is removed from $\boldsymbol{Y}$ at each iteration of PLS

$$\boldsymbol{X}=\boldsymbol{X}-\boldsymbol{t}\boldsymbol{p}^{\mathrm{T}} \text{ and } \boldsymbol{Y}=\boldsymbol{Y}-\frac{\boldsymbol{t}\boldsymbol{t}^{\mathrm{T}}\boldsymbol{Y}}{\boldsymbol{t}^{\mathrm{T}}\boldsymbol{t}}=\boldsymbol{Y}-\boldsymbol{t}\boldsymbol{c}^{\mathrm{T}}$$

where we consider not scaled to unit norm weight vectors $\boldsymbol{c}$ defined in step 4 of NIPALS. This deflation scheme guarantees mutual orthogonality of the extracted score vectors $\{\boldsymbol{t}_i\}_{i=1}^{p}$ [10]. Note that in PLS1 the deflation of $\boldsymbol{y}$ is technically not needed during the iterations of PLS [10].

Singular values of the cross-product matrix $\boldsymbol{X}^{\mathrm{T}}\boldsymbol{Y}$ correspond to the sample covariance values [10]. Then the deflation scheme of extracting one component at a time has also the following interesting property. The first singular value of the deflated cross-product matrix $\boldsymbol{X}^{\mathrm{T}}\boldsymbol{Y}$ at iteration $i+1$ is greater or equal than the second singular value of $\boldsymbol{X}^{\mathrm{T}}\boldsymbol{Y}$ at iteration i[10]. This result can be also applied to the relation of eigenvalues of equation (4.1.3) due to the fact that (4.1.3) corresponds to the singular value decomposition of the transposed cross-product matrix $\boldsymbol{X}^{\mathrm{T}}\boldsymbol{Y}$. In particular, the PLS1 and PLS2 algorithms differ from the computation of all eigenvectors of equation (4.1.3) in one step.

4.1.2.3 PLS-SB

As outlined at the end of the previous paragraph the computation of all eigenvectors of equation (4.1.3) at once would define another form of PLS. This computation involves a sequence of implicit rank-one deflations of the overall cross-product matrix. This form of PLS was used in [19] and in accordance with it is denoted as PLS-SB. In contrast to PLS1 and PLS2, the extracted score vectors $\{\boldsymbol{t}_i\}_{i=1}^{p}$ are in general not mutually orthogonal.

SIMPLS: To avoid deflation steps at each iteration of PLS1 and PLS2, de Jong [5] has introduced another form of PLS denoted SIMPLS. The SIMPLS approach directly finds the weight vectors $\{\tilde{\boldsymbol{w}}\}_{i=1}^{p}$ which are applied to the original not deflated matrix $\boldsymbol{X}$. The criterion of the mutually orthogonal score vectors $\{\tilde{\boldsymbol{t}}\}_{i=1}^{p}$ is kept. It has been shown that SIMPLS is equal to PLS1 but differs from PLS2 when applied to the multidimensional matrix $\boldsymbol{Y}$ [5].

4.1.3 PLS Regression

As mentioned in the previous section, PLS1 and PLS2 can be used to solve linear regression problems. Combining assumption (4.1.5) of a linear relation between the scores vectors $\boldsymbol{t}$ and $\boldsymbol{u}$ with the decomposition of the $\boldsymbol{Y}$ matrix, equation (4.1.1) can be written as

$$\boldsymbol{Y}=\boldsymbol{T}\boldsymbol{D}\boldsymbol{Q}^{\mathrm{T}}+(\boldsymbol{H}\boldsymbol{Q}^{\mathrm{T}}+\boldsymbol{F})$$

This defines the equation

$$\boldsymbol{Y}=\boldsymbol{T}\boldsymbol{C}^{\mathrm{T}}+\boldsymbol{F}^{*} \tag{4.1.6}$$

where $\boldsymbol{C}^{\mathrm{T}}=\boldsymbol{D}\boldsymbol{Q}^{\mathrm{T}}$ now denotes the $(p\times M)$ matrix of regression coefficients and $\boldsymbol{F}^{*}=\boldsymbol{H}\boldsymbol{Q}^{\mathrm{T}}+\boldsymbol{F}$ is the residual matrix. Equation (4.1.6) is simply the decomposition of $\boldsymbol{Y}$ using ordinary least squares regression with orthogonal predictors $\boldsymbol{T}$.

We now consider orthonormalised score vectors $\boldsymbol{t}$, that is, $\boldsymbol{T}^{\mathrm{T}}\boldsymbol{T}=\boldsymbol{I}$, and the matrix $\boldsymbol{C}=\boldsymbol{Y}^{\mathrm{T}}\boldsymbol{T}$ of the not scaled to length one weight vectors $\boldsymbol{c}$. It is useful to redefine equation (4.1.6) in terms of the original predictors $\boldsymbol{X}$. To do this, we use the relationship [13]

$$\boldsymbol{T}=\boldsymbol{X}\boldsymbol{W}(\boldsymbol{P}^{\mathrm{T}}\boldsymbol{W})^{-1}$$

where $\boldsymbol{P}$ is the matrix of loading vectors defined in equation (4.1.1). Plugging this relation into equation (4.1.6), we yield

$$\boldsymbol{Y}=\boldsymbol{X}\boldsymbol{B}+\boldsymbol{F}^{*}$$

For a better understanding of these matrix equations, they are also given in graphical representation in Fig. 4.2.

where $\boldsymbol{B}$ represents the matrix of regression coefficients

$$\boldsymbol{B}=\boldsymbol{W}(\boldsymbol{P}^{\mathrm{T}}\boldsymbol{W})^{-1}\boldsymbol{C}^{\mathrm{T}}=\boldsymbol{X}^{\mathrm{T}}\boldsymbol{U}(\boldsymbol{T}^{\mathrm{T}}\boldsymbol{X}\boldsymbol{X}^{\mathrm{T}}\boldsymbol{U})^{-1}\boldsymbol{T}^{\mathrm{T}}\boldsymbol{Y}$$

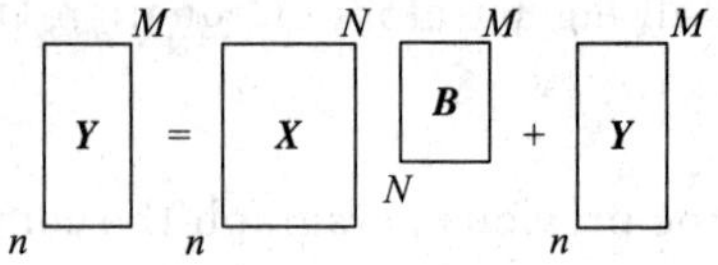

Fig. 4. 2 Where $\boldsymbol{B}$ represents the matrix of regression coefficients

For the last equality, the relations among $\boldsymbol{T}$, $\boldsymbol{U}$, $\boldsymbol{W}$ and $\boldsymbol{P}$ are used [12, 10, 17]. Note that different scalings of the individual score vectors $\boldsymbol{t}$ and $\boldsymbol{u}$ do not influence the $\boldsymbol{B}$ matrix. For training data the estimate of PLS regression is

$$\hat{\boldsymbol{Y}}=\boldsymbol{X}\boldsymbol{B}=\boldsymbol{T}\boldsymbol{T}^{\mathrm{T}}\boldsymbol{Y}=\boldsymbol{T}\boldsymbol{C}^{\mathrm{T}}$$

and for testing data we have

$$\hat{\boldsymbol{Y}}_{\mathrm{t}}=\boldsymbol{X}_{\mathrm{t}}\boldsymbol{B}=\boldsymbol{T}_{\mathrm{t}}\boldsymbol{T}^{\mathrm{T}}\boldsymbol{Y}=\boldsymbol{T}_{\mathrm{t}}\boldsymbol{C}^{\mathrm{T}}$$

where $\boldsymbol{X}_{\mathrm{t}}$ and $\boldsymbol{T}_{\mathrm{t}}=\boldsymbol{X}_{\mathrm{t}}\boldsymbol{X}^{\mathrm{T}}\boldsymbol{U}(\boldsymbol{T}^{\mathrm{T}}\boldsymbol{X}\boldsymbol{X}^{\mathrm{T}}\boldsymbol{U})^{-1}$ represent the matrices of testing data and score vectors, respectively.

4.1.4 Statistic

From the matrices of residuals $\boldsymbol{E}_{\mathrm{h}}$ and $\boldsymbol{F}_{\mathrm{h}}$ sums of squares can be calculated as follows: the total sum of squares over a matrix, the sums of squares over rows, and the sums of squares over columns. These sums of squares can be used to construct variance-like estimators. The statistical properties of these estimators have not undergone a rigorous mathematical treatment yet, but some properties can be understood intuitively.

The sum of squares of the $\boldsymbol{F}_{\mathrm{h}}$ is the indicator of how good the model. The sum of squares of $\boldsymbol{E}_{\mathrm{h}}$ is an indicator of how much of the X block is not used in the model. In some cases, a substantial part of the X block does not participate in the model, which means that the independent variables have unexpected properties or large errors.

Sums of squares over the columns indicate the importance of a variable for a certain component. Sums of squares over the rows indicate how well the objects fit the model. This can be used as an outlier detection criterion. Illustrations are given in Fig. 4. 3(a) for variable statistics and in Fig. 4. 3(b) for sample statistics.

An advantage of PLS is that these statistics can be calculated for every component. This is an ideal means of following the model-building process. The evolution of these statistics can be followed (as shown in Fig. 4. 3(a) and (b)) as more and more components are calculated so that an idea of how the different objects and variables fit can be obtained. In combination with a criterion for model dimensionality, the statistics can be used to estimate which objects and variables contribute mainly to the model and which contribute mainly to the residual.

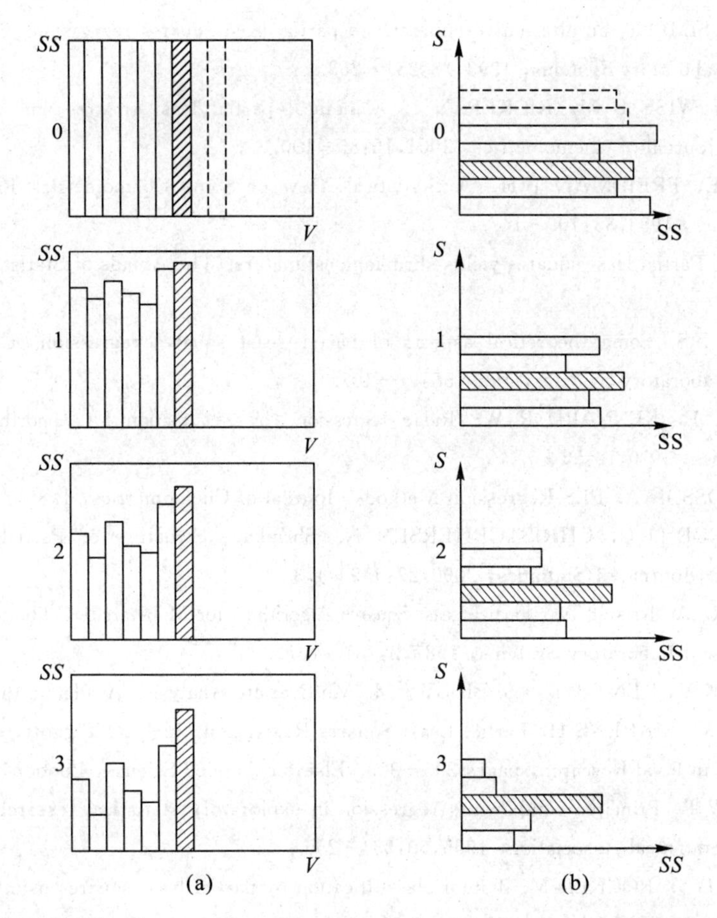

Fig. 4. 3(a) Statistics for the variables. The data are shown as bars representing the sum of squares per variable (for the model building both ***X*** and ***Y*** variables; for the prediction only ***X*** variables). After 0 PLS components, the data is in mean-centered and variance scaled form. As the number of PLS components increases, the information in each variable is exhausted. The hatched bar shows the behavior of a "special" variable, one that contributes little to the model

Fig. 4. 3(b) Statistics for the objects (samples). The data are shown as bars representing the sum of squares per object. As the number of PLS components increases, the sum of squares for each object decreases. The hatched bar shows the behavior of a "special" object, probably an outlier

Reference

[1] ROMAN R, KRÄMER N. Overview and Recent Advances in Partial Least Squares, 2005.

[2] ALMOY T. A simulation study on comparison of prediction models when only a few components are relevant. Computational Statistics and Data Analysis, 1996,21:87-107.

[3] BARKER M, RAYENS W S. Partial least squares for discrimination. Journal of Chemometrics, 2003,17:166-173.

[4] BUTLER N A, DENHAM M C. The peculiar shrinkage properties of partial least squares regression. Journal of the Royal Statistical Society B, 2000,62:585-593.

[5] JONG S D. SIMPLS: an alternative approach to partial least squares regression. Chemometrics and Intelligent Laboratory Systems, 1993,18:251 - 263.

[6] JONG S D, WISE B M, RICKER N L. Canonical partial least squares and continuum power regression. Journal of Chemometrics, 2001,15:85 - 100.

[7] FRANK I E, FREIDMAN J H. A Statistical View of Some Chemometrics Regression Tools. Technometrics, 1993,35:109 - 147.

[8] GOUTIS C. Partial least squares yields shrinkage estimators. The Annals of Statistics, 1996,24:816 - 824.

[9] HELLAND I S. Some theoretical aspects of partial least squares regression. Chemometrics and Intelligent Laboratory Systems, 1999,58:97 - 107.

[10] HORAL A E, KENNARD R W. Ridge regression: bias estimation for nonorthogonal problems. Technometrics, 1970,12:55 - 67.

[11] HOSKULDSSON A. PLS Regression Methods. Journal of Chemometrics, 1988,2:211 - 228.

[12] LINGJAERDE O C, CHRISTOPHERSEN N. Shrinkage Structure of Partial Least Squares. Scandinavian Journal of Statistics, 2000,27:459 - 473.

[13] MANNE R. Analysis of Two Partial-Least-Squares Algorithms for Multivariate Calibration. Chemometrics and Intelligent Laboratory Systems, 1987,2:187 - 197.

[14] MARDIA K V, KENT J T, and BIBBY J M. Multivariate Analysis. Academic Press, 1997.

[15] MATENS M, MARENS H. Partial Least Squares Regression. In J. R. Piggott, editor, Statistical Procedures in Food Research, pages 293 - 359. Elsevier Applied Science, London, 1986.

[16] MASSY W F. Principal components regression in exploratory statistical research. Journal of the American Statistical Association, 1965,60:234 - 256.

[17] NGUYEN D V, ROCKED M. Tumor classifification by partial least squares using microarray gene expression data. Bioinformatics, 2002,18:39 - 50.

[18] RANNAR S, LINDGREN F, GELADI P, et al. A PLS kernel algorithm for data sets with many variables and fewer objects. Part 1:Theory and algorithm. Chemometrics and Intelligent Laboratory Systems, 1994,8:111 - 125.

[19] ROSIPAL R, TREJO L J, MATTEWS B. Kernel PLS-SVC for Linear and Nonlinear Classifification. In Proceedings of the Twentieth International Conference on Machine Learning, 2008: 640 - 647.

[20] SAMPSON P D, STREISSGUTH, A P, BARR H M, et al. Neurobehavioral effffects of prenatal alcohol: Part II. Partial Least Squares analysis. Neurotoxicology and tetralogy, 1989,11:477 - 491.

[21] SCHOLKOPF B, SMOLA A J. Learning with Kernels-Support Vector Machines, Regularization, Optimization and Beyond. The MIT Press, 2002.

[22] STONE M, BROOKS R J. Continuum Regression: Cross-validated Sequentially Constructed Prediction Embracing Ordinary Least Squares, Partial Least Squares and Principal Components Regression. Journal of the Royal Statistical Society B, 1990,52:237 - 269.

[23] ANGEN L E, KOWALSKYB R. A multiblock partial least squares algorithm for investigating complex chemical systems. Journal of Chemometrics, 1989,3:3 - 20.

[24] WESTERHUIS J, KOUTI T, MACGREGOR J. Analysis of multiblock and hierarchical PCA and PLS models. Journal of Chemometrics, 1998,12:301 - 321.

[25] WOLD H. Path models with latent variables: The NIPALS approach. In H. M. Blalock et al., editor, Quantitative Sociology: International perspectives on mathematical and statistical model

building, pages 307 - 357. Academic Press, 1975.

[26] WOLD H. Soft modeling: the basic design and some extensions. In J.-K. Jöreskog and H. Wold, editor, Systems Under Indirect Observation, North Holland, Amsterdam, 1982,2:1 - 53.

[27] WOLD H. Partial least squares. In S. Kotz and N. L. Johnson, editors, "Encyclopedia of the Statistical Sciences". John Wiley & Sons, 1985,6:581 - 591.

4.2 NIPALS and SIMPLS Algorithm

This section continues to detail the two most popular methods of PLS (NIPALS and SIMPLS).

4.2.1 NIPALS

In this section, the transpose of the matrix is represented by the superscript.

4.2.1.1 Theory

The PLS algorithm as described in this section will be called the "standard" PLS algorithm. It has been presented in detail elsewhere [3 - 6]. For some alternative implementations of PLS see e.g. references [7 - 9]. The first step in standard PLS is to center the data matrices $\boldsymbol{X}$ and $\boldsymbol{Y}$, giving $\boldsymbol{X}_0$, and $\boldsymbol{Y}_0$, respectively. Then a set of $\boldsymbol{A}$ orthogonal $\boldsymbol{X}$ block factor scores $\boldsymbol{T}=[t_1, t_2, \cdots, t_A]$ and companion $\boldsymbol{Y}$ block factor scores $\boldsymbol{U}=[\boldsymbol{u}_1, \boldsymbol{u}_2, \cdots, \boldsymbol{u}_A]$ are calculated factor by factor. The first PLS factors $\boldsymbol{t}_1$, and $\boldsymbol{u}_1$, are weighted sums of the centered variables: $\boldsymbol{t}_1 = \boldsymbol{X}_0 \boldsymbol{w}_1$, and $\boldsymbol{u}_1 = \boldsymbol{Y}_0 \boldsymbol{q}_1$, respectively. Usually the weights are determined via the NIPALS algorithm. This is the iterative sequence:

$$\boldsymbol{w}_1 \propto \boldsymbol{X}_0^t \boldsymbol{u}_1 \tag{4.2.1}$$

$$\boldsymbol{t}_1 = \boldsymbol{X}_0 \boldsymbol{w}_1 \tag{4.2.2}$$

$$\boldsymbol{q}_1 \propto \boldsymbol{Y}_0^t \boldsymbol{t}_1 \tag{4.2.3}$$

$$\boldsymbol{u}_1 = \boldsymbol{Y}_0 \boldsymbol{q}_1 \tag{4.2.4}$$

Throughout the symbol α a not only indicates proportionality, but it also implies a subsequent normalization of the resultant vector. Thus the weight vectors $\boldsymbol{w}_1$ and $\boldsymbol{q}_1$ have length 1. (Different normalizations are possible, the specific choice being rather a matter of habit or of convenience.) The iteration sequence (4.2.1)-(4.2.4) starts by choosing for $\boldsymbol{u}_i$ some column of Y_0, e.g. the one having maximum variance, and it stops when $\boldsymbol{w}_1$ or $\boldsymbol{t}_1$ do not change given some prespecified tolerance. Manne[1] and Hoskuldsson [2] have shown that upon convergence the weight vectors $\boldsymbol{w}_1$ and $\boldsymbol{q}_1$ correspond to the first pair of left and right singular vectors obtained from a singular vector decomposition (SVD) [10] of the matrix of cross products $\boldsymbol{X}_0'\boldsymbol{Y}_0$,. Since the dominant singular value equals $\boldsymbol{w}_1 \boldsymbol{X}_0' \boldsymbol{Y}_0 \boldsymbol{q}_1 = \boldsymbol{t}_1' \boldsymbol{u}_1 = (n-1)\mathrm{cov}(\boldsymbol{t}_1, \boldsymbol{u}_1)$, the score vectors $\boldsymbol{t}_1$, and $\boldsymbol{u}_1$ have maximum covariance among all score vectors obtainable by applying normalized weights to $\boldsymbol{X}_0$ and $\boldsymbol{Y}_0$, respectively.

Once the first $\boldsymbol{X}$ block factor $\boldsymbol{t}_1$ is obtained one proceeds with deflating the data matrices. This yields new data sets $\boldsymbol{X}_1$ and $\boldsymbol{Y}_1$, which are the matrices of residuals obtained after regressing all variables on t_1

$$\boldsymbol{X}_1 = \boldsymbol{X}_0 - \frac{\boldsymbol{t}_1(\boldsymbol{t}'_1\boldsymbol{X}_0)}{\boldsymbol{t}'_1\boldsymbol{t}_1} \tag{4.2.5}$$

$$\boldsymbol{Y}_1 = \boldsymbol{Y}_0 - \frac{\boldsymbol{t}_1(\boldsymbol{t}'_1\boldsymbol{Y}_0)}{\boldsymbol{t}'_0\boldsymbol{t}_1} \tag{4.2.6}$$

Equation (4.2.5) may be rewritten as

$$\boldsymbol{X}_1 = \boldsymbol{X}_0 - \boldsymbol{t}_1\boldsymbol{p}'_0 \tag{4.2.7}$$

where $\boldsymbol{p}_1$ represents the vector of loadings of factor $\boldsymbol{t}_1$ on the $\boldsymbol{X}$ variables

$$\boldsymbol{p}_1 = \frac{\boldsymbol{X}\boldsymbol{t}_1}{\boldsymbol{t}'_1\boldsymbol{t}_1} \tag{4.2.8}$$

These loadings describe how strong the original $\boldsymbol{X}$ variables are related to this first PLS factor $\boldsymbol{t}_1$, similarly, equation (4.2.6) may be rewritten as

$$\boldsymbol{Y}_1 = \boldsymbol{Y}_b - b_1\boldsymbol{t}_1\boldsymbol{q}'_1 \tag{4.2.9}$$

The scalar b_1, is the estimated regression coefficient for the so-called inner relation between the two data sets relayed via their latent variables

$$\hat{\boldsymbol{u}} = b_1\boldsymbol{t}_1 \tag{4.2.10}$$

$$b_1 = \frac{\boldsymbol{u}'_1\boldsymbol{t}_1}{\boldsymbol{t}'_1\boldsymbol{t}_1} \tag{4.2.11}$$

The $\boldsymbol{Y}$ factor weights $\boldsymbol{q}_1$ and scores $\boldsymbol{u}_1$ and the inner-relation coefficient b_1 may aid interpretation of the latent structure underlying the data. The $\boldsymbol{u}$ scores, however, are not essential for establishing the multivariate linear regression model, $\boldsymbol{Y}=f(\boldsymbol{X})$.

The NIPALS algorithm now continues by repeating the steps described above, applying eqnuation (4.2.1)-(4.2.11) with all indices raised by 1. One starts with $\boldsymbol{X}_1$ and $\boldsymbol{Y}_1$, and identifies $\boldsymbol{w}_2$, $\boldsymbol{t}_2$, $\boldsymbol{q}_2$, $\boldsymbol{u}_2$, b_2, and $\boldsymbol{p}_2$ for the second dimension, $a=2$, analogous to the first dimension. Since $\boldsymbol{t}_1$ has been projected out of the system any new linear combination $\boldsymbol{t}_2=\boldsymbol{X}_1\boldsymbol{w}_2$ or $\boldsymbol{u}_2=\boldsymbol{Y}_1\boldsymbol{q}_2$ is orthogonal to $\boldsymbol{t}_1$ by construction. The normalized weights $\boldsymbol{w}_2$ and $\boldsymbol{q}_2$ when applied to the recall data sets, $\boldsymbol{X}_1$ and $\boldsymbol{Y}_1$, are chosen such as to maximize the covariance between the resulting score vectors $\boldsymbol{t}_2$ and $\boldsymbol{u}_2$.

Next, the data matrices are depleted further by projecting out $\boldsymbol{t}_2$, yielding $\boldsymbol{X}_2$ and $\boldsymbol{Y}_2$, and the third dimension is analyzed. The whole process proceeds until A factors have been determined. Choosing the right number A of factors is a crucial step in predictive modelling. However, in this paper the emphasis is on an alternative estimation of the PLS model given any dimensionality A.

4.2.1.2 NIPALS-PLS Factors in Terms of Original Variables

Each of the weight vectors $\boldsymbol{w}_a$, $a=2, 3, \cdots, A$, used for defining the associated factor scores, applies to a different matrix of residuals X_{a-1},

$$\boldsymbol{t}_a = \boldsymbol{X}_{a-1}\boldsymbol{w}_a \quad a=1, 2, \cdots, A \tag{4.2.12}$$

and not to the original centered data $\boldsymbol{X}_0$. This obscures the interpretation of the factors, mainly because one looses sight of what is in the depleted matrices $\boldsymbol{X}_a$, as one goes to higher dimensions, $a \geqslant 1$. Some $\boldsymbol{X}$ variables are used in the first factors, others only much later. The relation between factors and variables is better displayed by the loadings $\boldsymbol{p}_a (a = 1, 2, \cdots, A)$. Indeed, the weight vectors, collected in the $p \times A$ matrix $\boldsymbol{W}$, have found less use in interpreting PLS regression models than the loading vectors. It is therefore advantageous to re-express the NIPALS-PLS factors $\boldsymbol{t}_a$ in terms of the original centered data $\boldsymbol{X}_0$, say

$$\boldsymbol{t}_a = \boldsymbol{X}_0 \boldsymbol{T}_a \qquad a = 1, 2, \cdots, A \tag{4.2.13}$$

or, collecting the alternative weight vectors in a $p \times A$ matrix $\boldsymbol{R} = [r_1, r_2, \cdots, r_A]$,

$$\boldsymbol{T} = \boldsymbol{X}_0 \boldsymbol{R} \tag{4.2.14}$$

The factor scores $\boldsymbol{T}$ computed via NIPALS-PLS, i.e. via depleted $\boldsymbol{X}$ matrices, can be expressed exactly as linear combinations of the centered $\boldsymbol{X}$ variables, since all deflated matrices $\boldsymbol{X}_a$ and factor scores $\boldsymbol{t}_a$, $a = 1, 2, \cdots, A$, lie in the column space of $\boldsymbol{X}_0$. Thus, $\boldsymbol{R}$ can be computed from the regression of $\boldsymbol{T}$ on $\boldsymbol{X}_0$:

$$\boldsymbol{R} = \boldsymbol{X}_0^{+} \boldsymbol{T} = (\boldsymbol{X}'_0 \boldsymbol{X}_0)^{-} \boldsymbol{X}'_0 \boldsymbol{T} = (\boldsymbol{X}'_0 \boldsymbol{X}_0)^{-} \boldsymbol{P} (\boldsymbol{T}'\boldsymbol{T})^{-} \tag{4.2.15}$$

where $\boldsymbol{P} = [\boldsymbol{p}_1, \boldsymbol{p}_2, \cdots, \boldsymbol{p}_A]$ is the $(p \times A)$ matrix of factor loadings and the superscript $^{-}$ indicates any generalized inverse and $^{+}$ indicates the unique Moore-Penrose pseudo-inverse[16]. We also have the relation.

$$\boldsymbol{P}'\boldsymbol{R} = \boldsymbol{P}' (\boldsymbol{X}'_0 \boldsymbol{X}_0)^{-} \boldsymbol{P} (\boldsymbol{T}'\boldsymbol{T})^{-} = \boldsymbol{I}_A \tag{4.2.16}$$

Since $\boldsymbol{r}'_b \boldsymbol{p}_a = \boldsymbol{r}'_b \boldsymbol{X}'_0 \boldsymbol{t}_a / (\boldsymbol{t}'_a \boldsymbol{t}_a) = \boldsymbol{t}'_a \boldsymbol{t}_a / (\boldsymbol{t}'_a \boldsymbol{t}_a) = \boldsymbol{\delta}_{ab}$. Here $\boldsymbol{I}_A$ is the $(A \times A)$ identity matrix and $\boldsymbol{\delta}_{ab}$ is Kronecker's delta. Thus $\boldsymbol{R}$ is a generalized inverse of $\boldsymbol{P}'$. Another expression for $\boldsymbol{R}$ is [17, 18].

$$\boldsymbol{R} = \boldsymbol{W} (\boldsymbol{P}'\boldsymbol{W})^{-1} \tag{4.2.17}$$

which follows from the observation that $\boldsymbol{R}$ and $\boldsymbol{W}$ share the same column space and that $\boldsymbol{P}'\boldsymbol{R}$ should be equal to the identity matrix.

The explicit computation of the (pseudo-)inverse matrices in eqnuation (4.2.15) and (4.2.17) detracts somewhat from the PLS-NIPALS algorithm, that is otherwise very straightforward. Hiiskuldsson[2] gives the following recurrent relation.

$$\boldsymbol{r}_a = \boldsymbol{w}_a - (\boldsymbol{p}'_{a-1} \boldsymbol{w}_a) \boldsymbol{r}_{a-1} \quad \text{for } a > 1 \tag{4.2.18}$$

starting with $\boldsymbol{r}_1 = \boldsymbol{w}_1$, However, this relation depends on the tridiagonal structure of $\boldsymbol{P}'\boldsymbol{P}$ and is only correct for univariate $\boldsymbol{Y} = \boldsymbol{y}$ ($m = 1$, PLS1). Equation (4.2.19) and (4.2.20) form a set of updating formulas that is generally applicable:

$$\boldsymbol{r}_a = \boldsymbol{G}_a \boldsymbol{w}'_a \quad a = 1, 2, \cdots, A \tag{4.2.19}$$

$$\boldsymbol{G}_{a+1} = \boldsymbol{G}_a - \boldsymbol{r}_a \boldsymbol{p}'_a \quad a = 1, 2, \cdots, A-1 \tag{4.2.20}$$

starting with $\boldsymbol{G}_1 = \boldsymbol{I}_p$[19]. Note that the vectors $\boldsymbol{r}_a$ are not normalized, in contrast to the weight vectors $\boldsymbol{w}_a$, Thus in equation (4.2.13), neither $\boldsymbol{t}_a$ nor $\boldsymbol{r}_a$ are normalized.

4.2.1.3 Prediction

The main application of PLS modelling in multivariate calibration is to use the final regression model for predictive purposes. For prediction from a newly measured row vector $\boldsymbol{x}^*$, one extracts factor scores $\boldsymbol{t}_a$ using calibration weights $\boldsymbol{W}$ and loadings $\boldsymbol{P}$ while building up $\hat{\boldsymbol{y}}^*$ starting from $\overline{\boldsymbol{y}}$ and using $\boldsymbol{b}$ and $\boldsymbol{Q}=[\boldsymbol{q}_1, \boldsymbol{q}_2, \cdots, \boldsymbol{q}_A]$:

$$\boldsymbol{x}_0^* = \boldsymbol{x}^* - \overline{\boldsymbol{x}} \tag{4.2.21}$$

$$\boldsymbol{t}_a^* = \boldsymbol{x}_{a-1}^* \boldsymbol{w}_a \quad a=1, \cdots, A \tag{4.2.22}$$

$$\boldsymbol{x}_a^* = \boldsymbol{x}_{a-1}^* - \boldsymbol{t}_a^* \boldsymbol{p}'_a \quad a=1, \cdots, A \tag{4.2.23}$$

$$\hat{\boldsymbol{y}}^* = \overline{\boldsymbol{y}} + b_1 \boldsymbol{t}_1^* \boldsymbol{q}'_1 + b_2 \boldsymbol{t}_2^* \boldsymbol{q}'_2 + \cdots + b_A \boldsymbol{t}_A^* \boldsymbol{q}'_A \tag{4.2.24}$$

When the $\boldsymbol{R}$ weights are available, a closed form multiple regression-type prediction model can be obtained more readily:

$$\hat{\boldsymbol{Y}}_0 = \boldsymbol{T}\mathrm{diag}(\boldsymbol{b})\boldsymbol{Q}' = \boldsymbol{X}\boldsymbol{R}\mathrm{diag}(\boldsymbol{b})\boldsymbol{Q}' = \boldsymbol{X}_0 \boldsymbol{B}_{\mathrm{PLS}} \tag{4.2.25}$$

Here, $\boldsymbol{B}_{\mathrm{PLS}} = \boldsymbol{R}\mathrm{diag}(\boldsymbol{b})\boldsymbol{Q}' = \boldsymbol{W}(\boldsymbol{P}'\boldsymbol{W})^{-1}\mathrm{diag}(\boldsymbol{b})\boldsymbol{Q}'$ is the $p \times m$ set of biased multivariate regression coefficients obtained via PLS regression.

4.2.2 SIMPLS

4.2.2.1 Theory

Our alternative approach to the PLS method is, in some sense, opposite to standard PLS: first we specify our objective, then derive an optimizing criterion, next we try and optimize the criterion and, finally, we build an algorithm. This is the classical approach in multivariate statistics, e.g. in principal component analysis (PCA), discriminant analysis and canonical correlation analysis.

Our objective is to find a predictive linear model $\hat{\boldsymbol{Y}} = \boldsymbol{X}\boldsymbol{B}$. We opt for a biased regression method in order to stabilize the parameter estimate, which hopefully leads to more reliable predictions. In order to stay in line with the standard PLS method we choose for extracting successive orthogonal factors of $\boldsymbol{X}$, $\boldsymbol{t}_a = \boldsymbol{X}_0 \boldsymbol{r}_a$, that are determined by maximizing their covariance with corresponding factors of $\boldsymbol{Y}$, $\boldsymbol{u}_a = \boldsymbol{Y}_0 \boldsymbol{q}_a$ ($a=1, 2, \cdots, A$). Hoskuldsson [14] has advanced many plausible reasons why this covariance measure is such a sensible criterion.

The modification we propose leads to the direct computation of the weights $\boldsymbol{R}$. In this way we avoid the construction of deflated data matrices $\boldsymbol{X}_1, \boldsymbol{X}_2, \cdots, \boldsymbol{X}_A$ and $\boldsymbol{Y}_1, \boldsymbol{Y}_2, \cdots, \boldsymbol{Y}_A$ and by-pass the calculation of weights $\boldsymbol{W}$. The explicit computation of matrix inverses as in equation (4.2.15) or (4.2.17) is also circumvented. The newly defined $\boldsymbol{R}$ is similar, but not identical, to the "standard" $\boldsymbol{R}$ introduced in equation (4.2.14). In fact, our new $\boldsymbol{R}$ contains normalized weight vectors just as $\boldsymbol{W}$ in standard PLS.

Thus, the task we face is to compute weight vectors $\boldsymbol{r}_a$ and $\boldsymbol{q}_a$ ($a = 1, 2, \cdots, A$), which can be applied directly to the centered data:

$$\boldsymbol{t}_a = \boldsymbol{X}_0 \boldsymbol{r}_a \quad a=1, 2, \cdots, A \tag{4.2.26}$$

$$\boldsymbol{u}_a = \boldsymbol{Y}_0 \boldsymbol{q}_a \quad a=1, 2, \cdots, A \tag{4.2.27}$$

The weights should be determined such as to maximize the covariance of score vectors $\boldsymbol{t}_a$ and $\boldsymbol{u}_a$ under some constraints. (The term covariance will be used somewhat loosely and interchangeably with the terms cross-product or inner product; they merely differ by a scalar factor $n-1$). Specifically, four conditions control the solution:

(1) maximization of covariance:

$$\boldsymbol{u}'_a\boldsymbol{t}_a=\boldsymbol{q}'_a(\boldsymbol{Y}_0\boldsymbol{X}_0)\boldsymbol{r}_a=\max!$$

(2) normalization of weights $\boldsymbol{r}_a$:

$$\boldsymbol{r}'_a\boldsymbol{r}_a=1$$

(3) normalization of weights $\boldsymbol{q}_a$:

$$\boldsymbol{q}'_a\boldsymbol{q}_a=1$$

(4) orthogonality of $\boldsymbol{t}$ scores:

$$\boldsymbol{t}'_b\boldsymbol{t}_a=0 \quad \text{for} \quad a>b$$

Without the last constraint there is only one, straightforward solution: $\boldsymbol{r}_1$ and $\boldsymbol{q}_1$ are the first left and right singular vectors of the $p\times m$ cross-product matrix $\boldsymbol{S}_0=\boldsymbol{X}'_0\boldsymbol{Y}_0$. However, in order to get more than one solution and to generate a set of orthogonal factors of $\boldsymbol{X}$, the orthogonality restriction condition (4) has to be added. Thus we require

$$\boldsymbol{t}'_b\boldsymbol{t}_a=\boldsymbol{t}'_b\boldsymbol{X}_0\boldsymbol{r}_a=\boldsymbol{t}'_b\boldsymbol{t}_b\boldsymbol{p}'_b\boldsymbol{r}_a=0 \tag{4.2.28}$$

for $a>b$. Here, $\boldsymbol{p}_b$ is a loading vector expressing the relation between the original $\boldsymbol{X}$ variables and the b-th PLS factor (cf. equation (4.2.8)). Equation (4.2.28) stipulates that any new weight vector $\boldsymbol{r}_a(a>1)$ should be orthogonal to all preceding loading vectors, i.e., to the columns of $\boldsymbol{P}_{a-1}=[\boldsymbol{p}_1, \boldsymbol{p}_2, \cdots, \boldsymbol{p}_{a-1}]$. Letting $\boldsymbol{P}^{\perp}_{a-1}$ be the required orthogonal projector

$$\boldsymbol{P}^{\perp}_{a-1}=\boldsymbol{I}_p-\boldsymbol{P}_{a-1}(\boldsymbol{P}'_{a-1}\boldsymbol{P}_{a-1})^{-1}\boldsymbol{P}'_{a-1} \tag{4.2.29}$$

we therefore demand

$$\boldsymbol{r}_a=\boldsymbol{P}^{\perp}_{a-1}\boldsymbol{r}_a \quad \text{for } a>1 \tag{4.2.30}$$

Equation (4.2.29) and (4.2.30) concisely account for the fourth orthogonality restriction. The solution for $\boldsymbol{q}_a$ and $\boldsymbol{r}_a$ is now given by the first pair of singular vectors from the singular value decomposition (SVD) of $\boldsymbol{S}_0$ projected on a subspace orthogonal to $\boldsymbol{P}_{a-1}$, i.e. the SVD of $\boldsymbol{P}^{\perp}_{a-1}\boldsymbol{S}_0$. In general, we will indicate the cross product after a loading vectors have been projected out as $\boldsymbol{S}_a$, where

$$\boldsymbol{S}_a=\boldsymbol{P}^{\perp}_a(\boldsymbol{X}'_0\boldsymbol{Y}_0)=\boldsymbol{P}^{\perp}_a\boldsymbol{S}_0 \quad \text{for } a\geqslant 1 \tag{4.2.31}$$

4.2.2.2 SIMPLS Algorithm

It is expedient to compute $\boldsymbol{S}_{a+1}$ from its predecessor $\boldsymbol{S}_a$. To achieve the projection onto the column space of $\boldsymbol{P}_a$ will be carried out as a sequence of orthogonal projections. For this we need an orthonormal basis of $\boldsymbol{P}_a$, say $\boldsymbol{V}_a=[\boldsymbol{v}_1, \boldsymbol{v}_2, \cdots, \boldsymbol{v}_a]$. $\boldsymbol{V}_a$ may be obtained from a GramSchmidt orthonormalization of $\boldsymbol{P}_a$, i.e.,

$$\boldsymbol{v}_a\propto\boldsymbol{p}_a-\boldsymbol{V}_{a-1}(\boldsymbol{V}'_{a-1}\boldsymbol{p}_a) \quad a=2,3,\cdots,A \tag{4.2.32}$$

starting with $\boldsymbol{V}_1=\boldsymbol{v}_1\propto\boldsymbol{p}_1$. An additional simplification is possible when the response is univariate ($m=1$, PLS1). In this case, one may employ the orthogonality properties of $\boldsymbol{P}$, viz., $\boldsymbol{p}'_b\boldsymbol{p}_a=0$, for $b\leqslant a-2$. These properties carry over to the orthonormalized loadings

$\boldsymbol{V}$, i. e., $\boldsymbol{p}'_b\boldsymbol{V}_a = 0$, for $b \leqslant a-2$. Thus, orthogonality of $\boldsymbol{p}_a$ with respect to $\boldsymbol{V}_{a-2}$ is automatically taken care of and equation (4.2.32) simplifies to

$$\boldsymbol{V}_a \propto \boldsymbol{p}_a - \boldsymbol{V}_{a-1}(\boldsymbol{V}'_{a-1}\boldsymbol{p}_a) \quad a>1, m=1 \tag{4.2.33}$$

The projection onto the subspace spanned by the first a loading vectors, $\boldsymbol{P}_a(\boldsymbol{P}'_a\boldsymbol{P}_a)^{-1}\boldsymbol{P}'_a$, can now be replaced by $\boldsymbol{V}_a\boldsymbol{V}'_a$ and the projection on the orthogonal implement $\boldsymbol{P}_a^{\perp}$ by $\boldsymbol{I}_p - \boldsymbol{V}_a\boldsymbol{V}'_a = \prod_1^a(\boldsymbol{I}_p\boldsymbol{V}_b\boldsymbol{V}'_b)$. Thus, utilizing the orthonormality of $\boldsymbol{V}$, the product matrices $\boldsymbol{S}_a(a=1, 2, \cdots)$, are steadily depleted by projecting out the perpendicular directions $\boldsymbol{v}_a$:

$$\boldsymbol{S}_a = \boldsymbol{S}_{a-1} - \boldsymbol{v}_a(\boldsymbol{V}'_a\boldsymbol{S}_{a-1}) \quad a>1 \tag{4.2.34}$$

The main difference with the standard PLS algorithm is that the deflation process applies to the cross-product $\boldsymbol{S}_0$ and not to the larger data matrices $\boldsymbol{X}_0$ and $\boldsymbol{Y}_0$. The first pair of singular vectors of each $\boldsymbol{S}_a$ may be calculated using the iterative power method. This has the advantage of extracting only the pair of singular vectors of interest, i. e., the ones corresponding to the dominant singular value, which equals the maximum attainable covariance. Usually, the number of $\boldsymbol{Y}$ variables is smaller than the number of $\boldsymbol{X}$ variables, $m < p$. For example, in multivariate calibration, $\boldsymbol{p}$ may be a few hundreds and m may be quite low, say < 5. Then it will be efficient to calculate $\boldsymbol{q}_a$ as the dominant eigenvector of the small $m \times m$ symmetric matrix $\boldsymbol{S}'_{a-1}\boldsymbol{S}_{a-1}$ and finding $\boldsymbol{r}_a$ as

$$\boldsymbol{r}_a \propto \boldsymbol{S}_{a-1}\boldsymbol{q}_a \tag{4.2.35}$$

For univariate $\boldsymbol{Y}(=\boldsymbol{y})$, $\boldsymbol{S}_{a-1}(=\boldsymbol{s}_{a-1})$ is a vector of covariances of $\boldsymbol{y}$ with the $\boldsymbol{X}$ variables, hence $\boldsymbol{S}'_{a-1}\boldsymbol{S}_{a-1} = \boldsymbol{s}'_{a-1}\boldsymbol{s}_{a-1}$ is scalar, $q_a = 1$, and $\boldsymbol{r}_a \propto \boldsymbol{s}_{a-1}$. Thus, in this case the solution is non-iterative, as for standard PLS1 ($m=1$). It may be shown that the algorithm for univariate $\boldsymbol{y}$ is closely related to the conjugated gradient algorithms Bidiag2 and LSQR [1, 15].

Another simplification concerns the centering of $\boldsymbol{X}$ and $\boldsymbol{Y}$. It is not necessary to carry out this very first step of the standard deflation process: $\boldsymbol{S}_0$ may be computed as $\boldsymbol{S}_0 = \boldsymbol{X}'\boldsymbol{Y} - n\overline{\boldsymbol{x}}'\overline{\boldsymbol{y}}$. Alternatively, one might center just $\boldsymbol{Y}$, since $\boldsymbol{S}_0 = \boldsymbol{X}'_0\boldsymbol{Y}_0 = \boldsymbol{X}'\boldsymbol{Y}_0$. Usually, $\boldsymbol{Y}$ is the smaller of the two data sets, often a single column, $\boldsymbol{y}$. When $\boldsymbol{X}$ is centered, giving $\boldsymbol{X}_0$, the $\boldsymbol{t}_a = \boldsymbol{X}_0\boldsymbol{r}_a$ scores are automatically centered. When $\boldsymbol{X}$ is not centered, the scores, computed initially as $\boldsymbol{X}\boldsymbol{r}_a$, will have to be centered explicitly. The $\boldsymbol{u}_a = \boldsymbol{Y}\boldsymbol{q}_a$ scores may be centered as well, although this is not mandatory.

It may also be useful to orthogonalize the $\boldsymbol{u}_a$ scores to the preceding $\boldsymbol{t}$ scores, $\boldsymbol{t}_1, \boldsymbol{t}_2, \cdots, \boldsymbol{t}_{a-1}$. Again this is not necessary, but it removes some collinearity among the $\boldsymbol{u}$ scores, and it provides for equivalence with the standard PLS results as well as for a better interpretation. The Appendix gives a more elaborate pseudo-code incorporating all of these improvements. We have coined the name SIMPLS for our PLS algorithm, since it is a straightforward implementation of a statistically inspired modification of the PLS method according to the simple concept. Detailed codes of the SIMPLS algorithm for use with MATLAB [16] or SAS/IML [17] are available on request. The algorithm can be extended to deal with missing values in the same approximate way as done in the standard NIPALS procedure, i. e., accumulating sums of products over available data and correcting such sums for the number of entries.

4.2.2.3 Fitting, Prediction and Residual Analysis

For the development of the theory and algorithm of SIMPLS it was convenient to choose normalized weight vectors $\boldsymbol{r}_a$. This choice, however, is in no way essential. We will now switch to a normalization of the scores $\boldsymbol{t}_a$ instead, since this considerably simplifies some of the ensuing formulas. The code given in the Appendix already uses the latter normalization scheme. Thus we redefine $\boldsymbol{r}_a = \boldsymbol{r}_a / |\boldsymbol{X}_0 \boldsymbol{r}_a|$ and $\boldsymbol{t}_a = \boldsymbol{t}_a / |\boldsymbol{t}_a|$, giving unit-length score vectors $\boldsymbol{t}_a$ and orthonormal $\boldsymbol{T}: \boldsymbol{T}'\boldsymbol{T} = \boldsymbol{I}A$.

Predicted values of the calibration samples are now obtained as

$$\hat{\boldsymbol{Y}}_0 = \boldsymbol{T}\boldsymbol{T}\boldsymbol{Y}_0 = \boldsymbol{X}_0\boldsymbol{R}\boldsymbol{R}'\boldsymbol{X}_0'\boldsymbol{Y}_0, = \boldsymbol{X}_0\boldsymbol{R}\boldsymbol{R}'\boldsymbol{S}_0 \tag{4.2.36}$$

giving

$$\boldsymbol{B}_{\text{PLS}} = \boldsymbol{R}(\boldsymbol{R}'\boldsymbol{S}_0) = \boldsymbol{R}(\boldsymbol{T}'\boldsymbol{Y}_0) = \boldsymbol{R}(\boldsymbol{T}'\boldsymbol{Y}) = \boldsymbol{R}\boldsymbol{Q}' \tag{4.2.37}$$

with non-normalized $\boldsymbol{Y}$ loadings $\boldsymbol{Q} \equiv \boldsymbol{Y}_0'\boldsymbol{T}$. Thus, the regression coefficients are obtained directly without the explicit computation of matrix inverses (equation (4.2.15) and (4.2.17)) or singular value decomposition [9]. This is one of the advantages of the SIMPLS algorithm. Comparing $\boldsymbol{B}_{\text{PLS}}$ with the least-squares estimator $(\boldsymbol{X}_0'\boldsymbol{X}_0)^- \boldsymbol{S}_0$ we find that $\boldsymbol{R}\boldsymbol{R}'$ plays the role of a generalized inverse of the cross product $\boldsymbol{X}_0'\boldsymbol{X}_0$. Thus, $\boldsymbol{R}\boldsymbol{R}'$ may be interpreted as being proportional to the variance-covariance matrix of the parameter estimates $\boldsymbol{B}_{\text{PLS}}$. Likewise, $\boldsymbol{R}\boldsymbol{T}'$ can be seen as a generalized inverse of $\boldsymbol{X}_0$ in the PLS subspace. Recently, a better approximation to the variance-covariance matrix of $\boldsymbol{B}_{\text{PLS}}$ has been proposed by Phatak, et al. [18].

The variation of $\boldsymbol{X}$ around the mean $\bar{\boldsymbol{x}}$ can be analysed in terms of the PLS factors. using $\boldsymbol{X}_0 = \boldsymbol{T}\boldsymbol{P}' + \boldsymbol{X}_A$, the cross-product $\boldsymbol{X}_0'\boldsymbol{X}_0$, can be decomposed as

$$\boldsymbol{X}_0'\boldsymbol{X}_0 = \boldsymbol{P}\boldsymbol{T}'\boldsymbol{T}\boldsymbol{P}' + \boldsymbol{X}_A'\boldsymbol{X}_A = \boldsymbol{P}\boldsymbol{P}' + \boldsymbol{X}_A'\boldsymbol{X}_A \tag{4.2.38}$$

The total sum of squares of $\boldsymbol{X}$, $\text{tr}(\boldsymbol{X}_0'\boldsymbol{X}_0)$, is then accounted for by the A PLS factors to the extent $\text{tr}(\boldsymbol{P}\boldsymbol{P}') = \text{tr}(\boldsymbol{P}'\boldsymbol{P}) = \sum \boldsymbol{p}_a'\boldsymbol{p}_a$. Each term in the summation gives the contribution of a particular PLS dimension a. Similar expressions are obtained for the variance of $\boldsymbol{Y}$. Using $\boldsymbol{Y}_0 = \boldsymbol{T}\boldsymbol{Q} + \boldsymbol{Y}A$, the total sum of squares of the $\boldsymbol{Y}$ variables explained by the PLS factors equals $\text{tr}(\boldsymbol{Q}\boldsymbol{Q}') = \sum \boldsymbol{q}_a'\boldsymbol{q}_a$.

The $n \times n$ hat matrix [19] $\boldsymbol{H}$ is given by $\boldsymbol{T}\boldsymbol{T}' + \boldsymbol{I}_n/n$ (the constant $1/n$ on the diagonal accounts for the intercept term). The leverages for training objects are on the diagonal of H, $h_1 \equiv \frac{1}{n} + \sum t_{ia}^2$. These leverages are of interest in their own right and may also be used to transform the vector of residuals for the i-th multivariate observation, $\boldsymbol{e} \equiv \boldsymbol{y}_i - \hat{\boldsymbol{y}}_i$, into a vector of prediction errors $\boldsymbol{e}_{(i)}$,

$$e_{(i)} \equiv \frac{e_i}{1-h_i} \tag{4.2.39}$$

i. e., as if the observation had been left out during the model building. Equation (4. 2. 39) has been borrowed from ordinary least-squares regression theory [18], where it is exact. In the context of PLS regression [25], however, it is only a quick-and-dirty approximation, since $\boldsymbol{T}$ is not a fixed design matrix.

For new objects we employ the straightforward prediction formula

$$\hat{y}^* = \bar{y} + (\boldsymbol{x}^* - \bar{\boldsymbol{x}})\boldsymbol{B}_{\mathrm{PLS}} \tag{4.2.40}$$

The factor scores $\boldsymbol{t}_a^* = \boldsymbol{x}_0^* \boldsymbol{r}_a$ and leverage $h^* = \sum \boldsymbol{t}_a^{*2}$ may be computed for diagnostic purposes, e. g., to assess whether or not the new object lies within the region covered by the training objects.

4. 2. 2. 4 Detailed SIMPLS Algorithm

INPUT: $n \times p$ matrix $\boldsymbol{X}$,

$n \times m$ matrix $\boldsymbol{Y}$,

number of factors A.

OUTPUT: $\boldsymbol{R}$, $\boldsymbol{T}$, $\boldsymbol{P}$, $\boldsymbol{Q}$, $\boldsymbol{U}$ and $\boldsymbol{V}$

1) $\boldsymbol{Y}_0 = \boldsymbol{Y} - \mathrm{MEAN}(\boldsymbol{Y})$	center $\boldsymbol{Y}$
2) $\boldsymbol{S} = \boldsymbol{X}' * \boldsymbol{Y}_0$	cross-product
3) For $a = 1, \cdots, A$	per dimension
4) $\boldsymbol{q}$ = dominant eigenvectorof $\boldsymbol{S}' * \boldsymbol{S}$	$\boldsymbol{Y}$ block factor weight
5) $\boldsymbol{r} = \boldsymbol{S} * \boldsymbol{q}$	$\boldsymbol{X}$ block factor weights
6) $\boldsymbol{t} = \boldsymbol{X} * \boldsymbol{r}$	$\boldsymbol{X}$ block factor scores
7) $\boldsymbol{t} = \boldsymbol{t} - \mathrm{MEAN}(\boldsymbol{t})$	center scores
8) normt = SQRT(t′ * t)	compute norm
9) $\boldsymbol{t} = \boldsymbol{t}/\mathrm{normt}$	normalize scores
10) $\boldsymbol{r} = \boldsymbol{r}/\mathrm{normt}$	adapt weights accordingIy
11) $\boldsymbol{p} = \boldsymbol{X}' * \boldsymbol{t}$	$\boldsymbol{X}$ block factor loadings
12) $\boldsymbol{q} = \boldsymbol{Y}_0' * \boldsymbol{t}$	$\boldsymbol{Y}$ block factor loadings
13) $\boldsymbol{u} = \boldsymbol{Y}_0 * \boldsymbol{q}$	$\boldsymbol{Y}$ block factor scores
14) $\boldsymbol{v} = \boldsymbol{p}$	initialize orthogonal loadings
15) if $a > 1$ then	
16) $\boldsymbol{v} = \boldsymbol{v} - \boldsymbol{V} * (\boldsymbol{V}' * \boldsymbol{p})$	make $\boldsymbol{v} \perp$ perevious loadings
17) $\boldsymbol{u} = \boldsymbol{u} - \boldsymbol{T} * (\boldsymbol{T}' * \boldsymbol{p})$	make $\boldsymbol{u} \perp$ perveious $\boldsymbol{t}'$ values
18) end	
19) $\boldsymbol{v} = \boldsymbol{v}/\mathrm{SQRT}(\boldsymbol{v}' * \boldsymbol{v})$	normalize orthogonal loadings
20) $\boldsymbol{S} = \boldsymbol{S} - \boldsymbol{v}' * (\boldsymbol{v}' * \boldsymbol{S})$	deflate $\boldsymbol{S}$ with respect to current loadings
21) Store $\boldsymbol{r}$, $\boldsymbol{t}$, $\boldsymbol{p}$, $\boldsymbol{q}$, $\boldsymbol{u}$, and $\boldsymbol{v}$ into o	
22) into $\boldsymbol{R}$, $\boldsymbol{T}$, $\boldsymbol{P}$, $\boldsymbol{Q}$, $\boldsymbol{U}$, and $\boldsymbol{V}$, respectively	

End

23) $\boldsymbol{B}=\boldsymbol{R}^{*}\boldsymbol{Q}'$ regression coefficients

24) $\boldsymbol{h}=\mathrm{DIAG}(T^{*}T')+1/n$ leverages of objects

25) $\mathrm{van}\boldsymbol{X}=\mathrm{DIAG}(P'*P)/(n-1)$ variance explained for $\boldsymbol{X}$ variables

26) $\mathrm{van}\boldsymbol{Y}=\mathrm{DIAG}(\boldsymbol{Q}'*\boldsymbol{Q})/(n-1)$ variance explained for $\boldsymbol{Y}$ variables

References

[1] MANNE R. Analysis of two partial-least-squares algorithms for multivariate calibration, Chemometrics and Intelligent Laboratory Systems, 1987,2:283 - 290.

[2] HIISKULASSON A. PLS regression methods, Journal of Chemometrics, 2 (1988) 211 - 228.

[3] WOLD S, ALBANO C, DUNNn W J III, et al. Pattern recognition: Finding and using regularities in multi-variate data, in H. Martens and H. Russwurm, Jr. (Editors), Food Research and Data Analysis, Applied Science Publishers, London, 1983: 147 - 188.

[4] GELADI P, KOWALSKI B R. Partial least-squares regression: A tutorial, Analytica Chimica Acta, 185 (1986) 1 - 17.

[5] NAES T, IRGENS C, MARTENS H. Comparison of linear statistical methods for calibration of NIR instruments, Applied Statistics, 1986,35: 195 - 206.

[6] HAALAND D M, THOMAS E V. Partial least-squares methods for spectral analyses. 1. Relation to other quantitative calibration methods and the extraction of qualitative information, Analytical Chemistry, 1998,60: 1193 - 1202.

[7] WOLD S, RUHE A, WOLD H, et al. The collinearity problem in linear regression. The partial least squares (PLS) approach to generalized inverses, SIAM Journal of Scientific and Statistical Compuling, 1984,5: 735 - 743.

[8] NAES T, MARTENS H. Comparison of prediction methods for multicollinear data, Communications in Statistics-Simularions and Computations, 1985,14: 545 - 576.

[9] LORBER A, KOWALSKI B R. A note on the use of the partial least-squares method for multivariate calibration, Applied Spectroscopy, 1998,42: 1572 - 1574.

[10] LAWSONC L, HANSON R J. Solcing Least Squares Problems, Prentice-Hall, Englewood Cliffs, NJ, 1974.

[11] SEARLE S R. Linear Models, Wiley, New York, 1971.

[12] MARTENS H, NAZS T. Multivariate calibration by data compression, in P. C. Williams and K. Norris (Editors), Near-Infrared Technology in the Agricultural and Food Industries, American Association of Cereal Chemists, St. Paul, MN, 1987: 57 - 87.

[13] HELLAND I S. On the structure of partial least squares regression, Communications in Statistics-Simulations and Computations, 1998,17: 581 - 607.

[14] HSSKULDSSON A. The H-principle in modelling with applications to chemometrics, Chemometrics and Intelligent Laboratory Systems, 1992,14: 139 - 153.

[15] PAIGE C, SAUNDERS M A. LSQR: an algorithm for sparse linear equations and sparse least squares, ACM Transactions on Mathematical Software, 1982,8: 43 - 47, 195 - 209.

[16] MATLABTM User's Guide, The Math Works Inc. , South Natick, MA, 1989.

[17] SAS/IML'. u User's Guide for Personal Computers, Version 6, SAS Institute Inc. , Cary, NC, 1985.

[18] PHATAK A, REILLY P M, PENLIDIS A. An approach to interval estimation in partial least

squares regression, Analyrica Chimica Acta, in press.

[19] WEISBERG S. Applied Regression Analysis, Wiley, New York, 2nd ed. , 1985.

[20] NAES T, MARTENS H. Principal component regression in NIR analysis: viewpoints, background details and selection of components, Journal of Chemometrics, 1988,2: 155 - 167.

4.3 Programming Method of Standard Partial Least Squares

4.3.1 Cross-validation

Learning the parameters of a prediction function and testing it on the same data is a methodological mistake: a model that would just repeat the labels of the samples that it has just seen would have a perfect score but would fail to predict anything useful on yet-unseen data. This situation is called overfitting. To avoid it, it is common practice when performing a (supervised) machine learning experiment to hold out part of the available data as a test set X_test, Y_test. Note that the word "experiment" is not intended to denote academic use only, because even in commercial settings machine learning usually starts out experimentally.

When evaluating different settings ("hyperparameters") for estimators, there is still a risk of overfitting on the test set because the parameters can be tweaked until the estimator performs optimally. This way, knowledge about the test set can "leak" into the model and evaluation metrics no longer report on generalization performance. To solve this problem, yet another part of the dataset can be held out as a so-called "validation set": training proceeds on the training set, after which evaluation is done on the validation set, and when the experiment seems to be successful, final evaluation can be done on the test set.

However, by partitioning the available data into three sets, we drastically reduce the number of samples which can be used for learning the model, and the results can depend on a particular random choice for the pair of (train, validation) sets.

A solution to this problem is a procedure called cross-validation (CV for short). A test set should still be held out for final evaluation, but the validation set is no longer needed when doing CV.

4.3.1.1 Cross-validation Iterators For i. i. d. Data

Assuming that some data is Independent and Identically Distributed (i. i. d.) is making the assumption that all samples stem from the same generative process and that the generative process is assumed to have no memory of past generated samples.

The following cross-validators can be used in such cases.

Note

While i. i. d. data is a common assumption in machine learning theory, it rarely holds in practice. If one knows that the samples have been generated using a time-dependent process, it's safer to use a time-series aware cross-validation scheme. Similarly if we know that the generative process has a group structure (samples from collected from different subjects, experiments, measurement devices) it safer to use group-wise cross-validation.

1. K-Fold

In the basic approach, called K-Fold CV, the training set is split into k smaller sets (other approaches are described below, but generally follow the same principles). The following procedure is followed for each of the K "folds":

(1) A model is trained using $k-1$ of the folds as training data;

(2) the resulting model is validated on the remaining part of the data.

The performance measure reported by K-Fold cross-validation is then the average of the values computed in the loop. This approach can be computationally expensive, but does not waste too much data (as is the case when fixing an arbitrary validation set), which is a major advantage in problems such as inverse inference where the number of samples is very small.

K-Fold divides all the samples in k groups of samples, called folds (if $k=n$, this is equivalent to the Leave One Out strategy), of equal sizes (if possible). The prediction function is learned using $k-1$ folds, and the fold left out is used for test.

Example of 2-fold cross-validation on a dataset with 4 samples:

```
>>>import numpy as np
>>>from sklearn.model_selection import KFold

>>>X = ["a", "b", "c", "d"]
>>>kf = KFold(n_splits=2)
>>>for train, test in kf.split(X):
...     print("%s %s" % (train, test))
[2 3] [0 1]
[0 1] [2 3]
```

Here is a visualization of the cross-validation behavior in Fig. 4.4. Note that K-Fold is not affected by classes or groups.

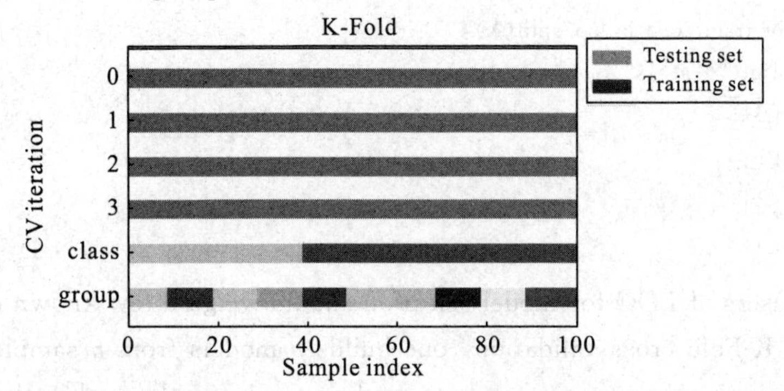

Fig. 4.4 Each fold is constituted by two arrays: the first one is related to the training set, and the second one to the test set. Thus, one can create the training/test sets using numpy indexing

```
>>>X = np.array([[0., 0.], [1., 1.], [-1., -1.], [2., 2.]])
>>>y = np.array([0, 1, 0, 1])
>>>X_train, X_test, y_train, y_test = X[train], X[test], y[train], y[test]
```

2. Repeated K-Fold

Repeated K-Fold repeats K-Fold n times. It can be used when one requires to run K-Fold n times, producing different splits in each repetition

Example of 2-fold K-Fold repeated 2 times:

```
>>>import numpy as np
>>>from sklearn.model_selection import RepeatedKFold

>>>X = np.array([[1, 2], [3, 4], [1, 2], [3, 4]])
>>>random_state = 12883823
>>>rkf = RepeatedKFold(n_splits=2, n_repeats=2, random_state=random_state)
>>>for train, test in rkf.split(X):
...     print("%s %s" % (train, test))
[2 3] [0 1]
[0 1] [2 3]
[0 2] [1 3]
[1 3] [0 2]
```

3. Leave One Out (LOO)

Leave One Out (or LOO) is a simple cross-validation. Each learning set is created by taking all the samples except one, the test set being the sample left out. Thus, for n samples, we have n different training sets and n different tests set. This cross-validation procedure does not waste much data as only one sample is removed from the training set:

```
>>>from sklearn.model_selection import LeaveOneOut

>>>X = [1, 2, 3, 4]
>>>loo = LeaveOneOut()
>>>for train, test in loo.split(X):
...     print("%s %s" % (train, test))
[1 2 3] [0]
[0 2 3] [1]
[0 1 3] [2]
[0 1 2] [3]
```

Potential users of LOO for model selection should weigh a few known caveats. When compared with K-Fold cross validation, one builds n models from n samples instead of k models, where $n>k$. Moreover, each is trained on $n-1$ samples rather than $(k-1)n/k$. In both ways, assuming k is not too large and $k<n$, LOO is more computationally expensive than K-Fold cross validation.

In terms of accuracy, LOO often results in high variance as an estimator for the test error. Intuitively, since $n-1$ of the n samples are used to build each model, models constructed from folds are virtually identical to each other and to the model built from the entire training set.

However, if the learning curve is steep for the training size in question, then 5- or 10-fold cross validation can overestimate the generalization error.

As a general rule, most authors, and empirical evidence, suggest that 5-fold or 10-fold cross validation should be preferred to LOO.

4. Leave P Out (LPO)

Leave P Out is very similar to LeaveOneOut as it creates all the possible training/test sets by removing p samples from the complete set. For n samples, this produces $\binom{n}{p}$ train-test pairs. Unlike Leave One Out and K-Fold, the test sets will overlap for $p>1$.

Example of Leave-2-Out on a dataset with 4 samples:

```
>>>from sklearn.model_selection import LeavePOut

>>>X = np.ones(4)
>>>lpo = LeavePOut(p=2)
>>>for train, test in lpo.split(X):
...     print("%s %s" % (train, test))
[2 3] [0 1]
[1 3] [0 2]
[1 2] [0 3]
[0 3] [1 2]
[0 2] [1 3]
[0 1] [2 3]
```

5. Leave P Out (LPO)Random Permutations Cross-validation a.k.a. Shuffle & Split

The Shuffle Split iterator will generate a user defined number of independent train / test dataset splits. Samples are first shuffled and then split into a pair of train and test sets.

It is possible to control the randomness for reproducibility of the results by explicitly seeding the random state pseudo random number generator.

Here is a usage example:

```
>>>from sklearn.model_selection import ShuffleSplit

>>>X = np.arange(10)
>>>ss = ShuffleSplit(n_splits=5, test_size=0.25,
...     random_state=0)
>>>for train_index, test_index in ss.split(X):
...     print("%s %s" % (train_index, test_index))
```

```
[9 1 6 7 3 0 5] [2 8 4]
[2 9 8 0 6 7 4] [3 5 1]
[4 5 1 0 6 9 7] [2 3 8]
[2 7 5 8 0 3 4] [6 1 9]
[4 1 0 6 8 9 3] [5 2 7]
```

Here is a visualization of the cross-validation behavior in Fig. 4. 5. Note that Shuffle Split is not affected by classes or groups.

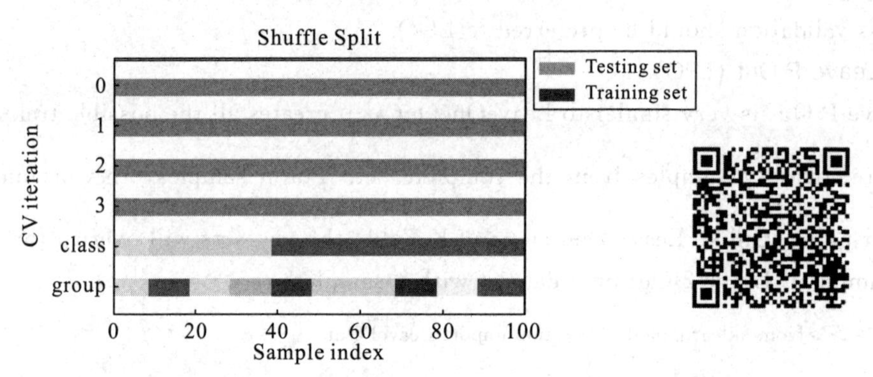

Fig. 4. 5　Shuffle Split is thus a good alternative to K-Fold cross validation that allows a finer control on the number of iterations and the proportion of samples on each side of the train/test split

4. 3. 1. 2　Cross-validation Iterators with Stratification Based on Class Labels

Some classification problems can exhibit a large imbalance in the distribution of the target classes: for instance there could be several times more negative samples than positive samples. In such cases it is recommended to use stratified sampling as implemented in Stratified K-Fold and Stratified Shuffle Split to ensure that relative class frequencies is approximately preserved in each train and validation fold.

1. Stratified K-Fold

Stratified K-Fold is a variation of K-Fold which returns stratified folds: each set contains approximately the same percentage of samples of each target class as the complete set.

Example of stratified 3-fold cross-validation on a dataset with 10 samples from two slightly unbalanced classes:

```
>>>from sklearn.model_selection import StratifiedKFold

>>>X = np.ones(10)
>>>y = [0, 0, 0, 0, 1, 1, 1, 1, 1, 1]
>>>skf = StratifiedKFold(n_splits=3)
>>>for train, test in skf.split(X, y):
...     print("%s %s" % (train, test))
[2 3 6 7 8 9] [0 1 4 5]
```

[0 1 3 4 5 8 9] [2 6 7]
[0 1 2 4 5 6 7] [3 8 9]

Here is a visualization of the cross-validation behavior in Fig. 4. 6.

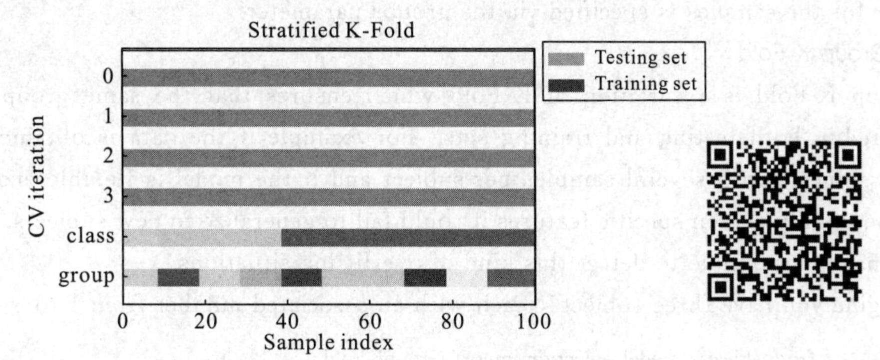

Fig. 4. 6 Repeated Stratified K-Fold can be used to repeat Stratified K-Fold n times with different randomization in each repetition

2. Stratified Shuffle Split

Stratified Shuffle Split is a variation of Shuffle Split, which returns stratified splits, i. e, which creates splits by preserving the same percentage for each target class as in the complete set.

Here is a visualization of the cross-validation behavior in Fig. 4. 7.

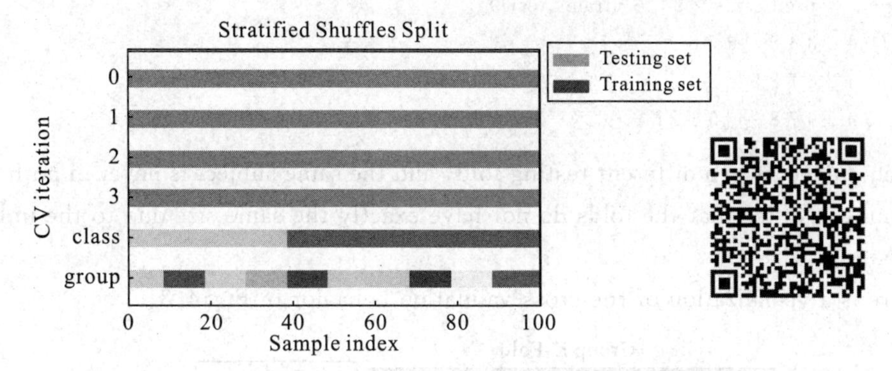

Fig. 4. 7 Stratified Shuffle Split

4. 3. 1. 3 Cross-validation Iterators for Grouped Data

The i. i. d. assumption is broken if the underlying generative process yield groups of dependent samples.

Such a grouping of data is domain specific. An example would be when there is medical data collected from multiple patients, with multiple samples taken from each patient. And such data is likely to be dependent on the individual group. In our example, the patient id for each sample will be its group identifier.

In this case we would like to know if a model trained on a particular set of groups generalizes well to the unseen groups. To measure this, we need to ensure that all the samples in the validation fold come from groups that are not represented at all in the paired training fold.

The following cross-validation splitters can be used to do that. The grouping identifier for the samples is specified via the groups parameter.

1. Group K-Fold

Group K-Fold is a variation of K-Fold which ensures that the same group is not represented in both testing and training sets. For example if the data is obtained from different subjects with several samples per-subject and if the model is flexible enough to learn from highly person specific features it could fail to generalize to new subjects. Group K-Fold makes it possible to detect this kind of overfitting situations.

Imagine you have three subjects, each with an associated number from 1 to 3:

```
>>>from sklearn.model_selection import GroupKFold

>>>X = [0.1, 0.2, 2.2, 2.4, 2.3, 4.55, 5.8, 8.8, 9, 10]
>>>y = ["a", "b", "b", "b", "c", "c", "c", "d", "d", "d"]
>>>groups = [1, 1, 1, 2, 2, 2, 3, 3, 3, 3]

>>>gkf = GroupKFold(n_splits=3)
>>>for train, test in gkf.split(X, y, groups=groups):
...     print("%s %s" % (train, test))
[0 1 2 3 4 5] [6 7 8 9]
[0 1 2 6 7 8 9] [3 4 5]
[3 4 5 6 7 8 9] [0 1 2]
```

Each subject is in a different testing fold, and the same subject is never in both testing and training. Notice that the folds do not have exactly the same size due to the imbalance in the data.

Here is a visualization of the cross-validation behavior in Fig. 4.8.

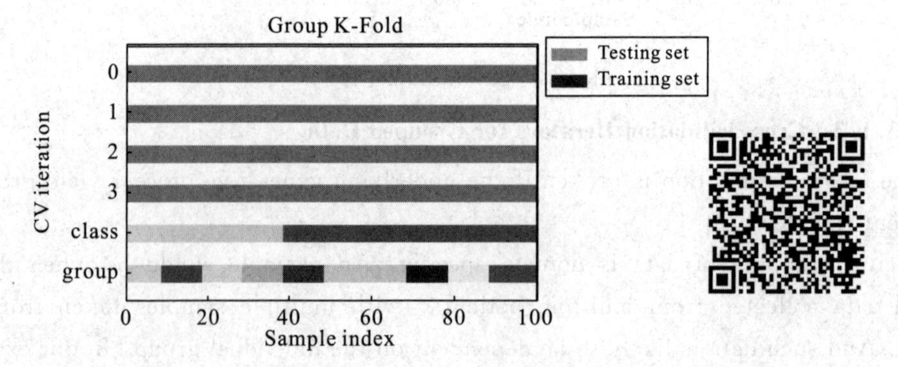

Fig. 4.8　Group K-Fold

2. Leave One Group Out

Leave One Group Out is a cross-validation scheme which holds out the samples according to a third-party provided array of integer groups. This group information can be used to encode arbitrary domain specific pre-defined cross-validation folds.

Each training set is thus constituted by all the samples except the ones related to a specific group.

For example, in the cases of multiple experiments, Leave One Group Out can be used to create a cross-validation based on the different experiments: we create a training set using the samples of all the experiments except one:

```
>>>from sklearn. model_selection import Leave One Group Out

>>>X = [1, 5, 10, 50, 60, 70, 80]
>>>y = [0, 1, 1, 2, 2, 2, 2]
>>>groups = [1, 1, 2, 2, 3, 3, 3]
>>>logo = LeaveOneGroupOut()
>>>for train, test in logo. split(X, y, groups=groups):
...     print("%s %s" % (train, test))
[2 3 4 5 6] [0 1]
[0 1 4 5 6] [2 3]
[0 1 2 3] [4 5 6]
```

Another common application is to use time information: for instance the groups could be the year of collection of the samples and thus allow for cross-validation against time-based splits.

3. Leave P Group Out

Leave P Groups Out is similar as Leave One Group Out, but removes samples related to P groups for each training/test set.

```
Example of Leave-2-Group Out:
>>>from sklearn. model_selection import LeavePGroupsOut

>>>X = np. arange(6)
>>>y = [1, 1, 1, 2, 2, 2]
>>>groups = [1, 1, 2, 2, 3, 3]
>>>lpgo = LeavePGroupsOut(n_groups=2)
>>>for train, test in lpgo. split(X, y, groups=groups):
...     print("%s %s" % (train, test))
[4 5] [0 1 2 3]
[2 3] [0 1 4 5]
[0 1] [2 3 4 5]
```

4. Group Shuffle Split

The Group Shuffle Split iterator behaves as a combination of Shuffle Split and Leave P Groups Out, and generates a sequence of randomized partitions in which a subset of groups are held out for each split.

Here is a usage example:

```
>>>from sklearn.model_selection import GroupShuffleSplit

>>>X = [0.1, 0.2, 2.2, 2.4, 2.3, 4.55, 5.8, 0.001]
>>>y = ["a", "b", "b", "b", "c", "c", "c", "a"]
>>>groups = [1, 1, 2, 2, 3, 3, 4, 4]
>>>gss = GroupShuffleSplit(n_splits=4, test_size=0.5, random_state=0)
>>>for train, test in gss.split(X, y, groups=groups):
...     print("%s %s" % (train, test))
...
[0 1 2 3] [4 5 6 7]
[2 3 6 7] [0 1 4 5]
[2 3 4 5] [0 1 6 7]
[4 5 6 7] [0 1 2 3]
```

Here is a visualization of the cross-validation behavior in Fig. 4.9.

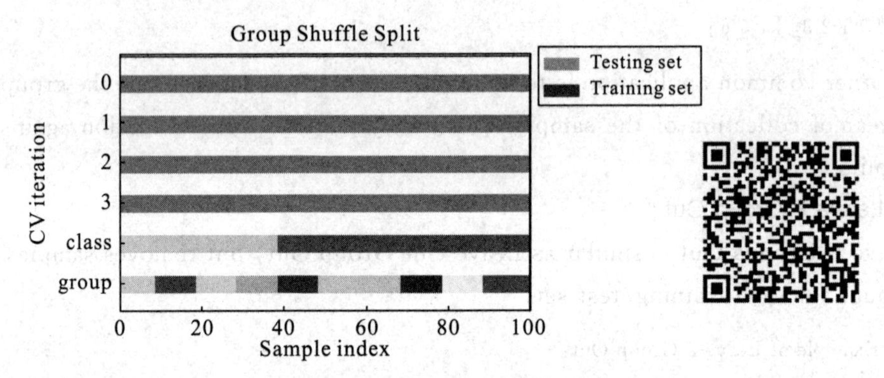

Fig. 4.9 Group Shuffle Split

This class is useful when the behavior of Leave P Groups Out is desired, but the number of groups is large enough that generating all possible partitions with P groups withheld would be prohibitively expensive. In such a scenario, Group Shuffle Split provides a random sample (with replacement) of the train / test splits generated by Leave P Groups Out.

4.3.2 Procedure of NIPALS

4.3.2.1 Inner Loop of The Iterative NIPALS Algorithm

Provides an alternative to the svd($\boldsymbol{X}'\boldsymbol{Y}$); returns the first left and right singular vectors of $\boldsymbol{X}'\boldsymbol{Y}$. See PLS for the meaning of the parameters. It is similar to the Power method for determining the eigenvectors and eigenvalues of a $X'Y$.

```
def _nipals_twoblocks_inner_loop(X, Y, max_iter=500, tol=1e-06, ):
    y_score = Y[:, [0]]
    x_weights_old =0
    ite =1

    while True:
        # 1.1 Update u: the X weights
        # regress each X column on y_score
        # w=X.T*Y[:, 0]/||Y[:, 0]||
        x_weights = np.dot(X.T, y_score) / np.dot(y_score.T, y_score)
        # 1.2 Normalize u
        # w=w/||w||
        x_weights /= np.sqrt(np.dot(x_weights.T, x_weights))
        # 1.3 Update x_score: the X latent scores
        # t=X*w
        x_score = np.dot(X, x_weights)
        # 2.1  regress each Y column on x_score
        # q=Y*t/(t.T*t)
        y_weights = np.dot(Y.T, x_score) / np.dot(x_score.T, x_score)

        # 2.2 Update y_score: the Y latent scores
        # u=Y*q/(q.T, q)
        y_score = np.dot(Y, y_weights) / np.dot(y_weights.T, y_weights)
        x_weights_diff = x_weights - x_weights_old
        if np.dot(x_weights_diff.T, x_weights_diff) < tol :
            break

        if ite == max_iter:
            warnings.warn('Maximum number of iterations reached')
            break
        x_weights_old = x_weights
        ite +=1

    return x_weights, y_weights
```

4.3.2.2 Center X and Y

```
def _center_xy(X, Y):

    # center
```

```
x_mean = X.mean(axis=0)
X_center = np.subtract(X, x_mean)
y_mean = Y.mean(axis=0)
    Y_center = np.subtract(Y, y_mean)

    return X_center, Y_center, x_mean, y_mean
```

4.3.2.3 NIPALS

This class implements the generic PLS algorithm, constructors'parameters allow to obtain a specific implementation such as:

This implementation uses the PLS Wold 2 blocks algorithm based on two nested loops:

(i) The outer loop iterate over components.

(ii) The inner loop estimates the weights vectors. This can be done with two algo. (a) the inner loop of the original NIPALS algo, or (b) a SVD on residuals cross-covariance matrices.

```
class _NIPALS():
    '''
    Parameters
    ----------
    X: array-like of predictors, shape = [n_samples, p], Training vectors, where n_samples in
 the number of samples and p is the number of predictors.
    Y: array-like of response, shape = [n_samples, q], Training vectors, where n_samples in
 the number of samples and q is the number of response variables.
    n_components: int, number of components to keep. (default 2).
    max_iter: an integer, the maximum number of iterations (default 500) of the NIPALS inner
 loop (used only if algorithm="nipals")
    tol: non-negative real, default 1e-06, The tolerance used in the iterative algorithm.

    Attributes
    ----------
    'x_weights_' : array, [p, n_components]
    X block weights vectors.
    'y_weights_' : array, [q, n_components]
    Y block weights vectors.
    'x_loadings_' : array, [p, n_components]
    X block loadings vectors.
    'y_loadings_' : array, [q, n_components]
    Y block loadings vectors.
    'x_scores_' : array, [n_samples, n_components]
    'y_scores_' : array, [n_samples, n_components]
    'x_rotations_' : array, [p, n_components]
```

```
X block to latents rotations.
'y_rotations_' : array, [q, n_components]
Y block to latents rotations.
coefs: array, [p, q]
The coefficients of the linear model: Y = X coefs + Err
'''
def __init__(self, n_components, max_iter=500, tol=1e-06, copy=True):
    self.n_components = n_components
    self.max_iter = max_iter
    self.tol = tol
    self.copy = copy

def fit(self, X, Y, n_components):
    n = X.shape[0]
    p = X.shape[1]
    q = Y.shape[1]

    if n != Y.shape[0]:
        'Incompatible shapes: X has %s samples, while Y '
        'has %s' % (X.shape[0], Y.shape[0])
    if self.n_components < 1 or self.n_components > p:
        raise ValueError('invalid number of components')

    Xcenter, Ycenter, self.x_mean_, self.y_mean_ = _center_xy(X, Y)
    # Residuals (deflated) matrices
    Xk = Xcenter
    Yk = Ycenter
    # Results matrices
    self.x_scores_ = np.zeros((n, self.n_components))
    self.y_scores_ = np.zeros((n, self.n_components))
    self.x_weights_ = np.zeros((p, self.n_components))
    self.y_weights_ = np.zeros((q, self.n_components))
    self.x_loadings_ = np.zeros((p, self.n_components))
    self.y_loadings_ = np.zeros((q, self.n_components))

    # NIPALS algo: outer loop, over components
    for k in range(self.n_components):
        x_weights, y_weights = _nipals_twoblocks_inner_loop(
                X=Xk, Y=Yk, max_iter=self.max_iter, tol=self.tol, )
        # compute scores
        x_scores = np.dot(Xk, x_weights)
        y_ss = np.dot(y_weights.T, y_weights)
        y_scores = np.dot(Yk, y_weights) / y_ss
```

```
            x_loadings = np.dot(Xk.T, x_scores) / np.dot(x_scores.T, x_scores)
            # - substract rank-one approximations to obtain remainder matrix
            Xk -= np.dot(x_scores, x_loadings.T)

            y_loadings = (np.dot(Yk.T, x_scores) / np.dot(x_scores.T, x_scores))
            Yk -= np.dot(x_scores, y_loadings.T)
            self.x_scores_[:, k] = x_scores.ravel()  # T
            self.y_scores_[:, k] = y_scores.ravel()  # U
            self.x_weights_[:, k] = x_weights.ravel()  # W
            self.y_weights_[:, k] = y_weights.ravel()  # C
            self.x_loadings_[:, k] = x_loadings.ravel()  # P
            self.y_loadings_[:, k] = y_loadings.ravel()  # Q

        lists_coefs = []
        for i in range(n_components):
            self.x_rotations_ = np.dot(self.x_weights_[:, :i + 1], linalg.inv(np.dot(self.
x_loadings_[:, :i + 1].T, self.x_weights_[:, :i + 1])))
            self.coefs = np.dot(self.x_rotations_, self.y_loadings_[:, :i + 1].T)

            lists_coefs.append(self.coefs)

        return lists_coefs

    def predict(self, x_test, coefs_B, xtr_mean, ytr_mean):

        xte_center = np.subtract(x_test, xtr_mean)
        y_pre = np.dot(xte_center, coefs_B)
        y_predict = np.add(y_pre, ytr_mean)

        return y_predict
```

4.4 Example Application

4.4.1 Demo of PLS

Software version python 2.7, and a Microsoft Windows 7 operating system. Cross-validation and train test split are performed using the sklearn package, respectively. Dataset loading is done using the scipy package, and other programs can be implemented by individuals.

```
# -*- coding: utf-8 -*-
```

```
from sklearn.cross_validation import train_test_split
from scipy.io.matlab.mio import loadmat
fromPLS.PLS import PLS

if __name__ == '__main__':
    fname = loadmat('NIRcorn.mat')
    x = fname['cornspect']
    y = fname['cornprop'][:, 0:1]
    print x.shape, y.shape

    x_train, x_test, y_train, y_test = train_test_split(x, y, test_size=0.2, random_state=0)

    demo = PLS(x_train, y_train, x_test, y_test, n_fold=10, max_components=9)
    RMSECV, min_RMSECV, comp_best, RMSEC, RMSEP = demo.pls()


    print 'RMSECV', RMSECV
    print 'min_RMSECV', min_RMSECV
    print  'comp_best', comp_best
    print 'RMSEP:', RMSEP

# -*- coding: utf-8 -*-

from cross_validation import Cross_Validation
from NIPALS import _NIPALS
import numpy as np

class PLS():
    def __init__(self, x_train, y_train, x_test, y_test, n_fold=10, max_components=10):
        self.x_train = x_train
        self.x_test = x_test
        self.y_train = y_train
        self.y_test = y_test
        self.n_fold = n_fold
        self.max_components = max_components

    def pls(self):
        # Select the optimal principal component number
        pls_cv = Cross_Validation(self.x_train, self.y_train,
                                  self.n_fold, self.max_components)
```

```
        y_allPredict, y_measure = pls_cv.predict_cv()
        RMSECV, min_RMSECV, comp_best = pls_cv.mse_cv(y_allPredict, y_measure)
        #  Modeling by optimal principal component number
        pls = _NIPALS(comp_best)
        List_coef_B = pls.fit(self.x_train, self.y_train, comp_best)
        coef_B = List_coef_B[comp_best -1]

        x_trainMean = np.mean(self.x_train, axis=0)
        y_trainMean = np.mean(self.y_train, axis=0)
        y_trainPredict = pls.predict(self.x_train, coef_B, x_trainMean, y_trainMean)
        # compute RMSEC
        press = np.square(np.subtract(self.y_train, y_trainPredict))
        all_press = np.sum(press, axis=0)
        RMSEC = np.sqrt(all_press /self.x_train.shape[0])
        # compute RMSEP
        y_predict = pls.predict(self.x_test, coef_B, x_trainMean, y_trainMean)
        press = np.square(np.subtract(self.y_test, y_predict))
        all_press = np.sum(press, axis=0)
        RMSEP = np.sqrt(all_press /self.x_test.shape[0])

        return RMSECV, min_RMSECV, comp_best, RMSEC, RMSEP

# -*- coding: utf-8 -*-

import numpy as np
from sklearn import cross_validation
from NIPALS import _NIPALS

class Cross_Validation():    # Variable initialization

    def __init__(self, x, y, n_fold, max_components):
        self.x = x
        self.y = y
        self.n = x.shape[0]
        self.n_fold = n_fold
        self.max_components = max_components

    def cv(self):    # Divide training sets and test sets
        kf = cross_validation.KFold(self.n, self.n_fold)
        x_train = []
```

```
        y_train = []
        x_test = []
        y_test = []

        for train_index, test_index in kf:
            xtr, ytr = self.x[train_index], self.y[train_index]
            xte, yte = self.x[test_index], self.y[test_index]
            x_train.append(xtr)
            y_train.append(ytr)
            x_test.append(xte)
            y_test.append(yte)

        return x_train, x_test, y_train, y_test

    def predict_cv(self):
        x_train, x_test, y_train, y_test = self.cv()
        y_allPredict = np.ones((1, self.max_components))
        pls = _NIPALS(self.max_components)

        for i in range(self.n_fold):
            y_predict = np.zeros((y_test[i].shape[0], self.max_components))
            x_trainMean = np.mean(x_train[i], axis=0)
            y_trainMean = np.mean(y_train[i], axis=0)
            x_testCenter = np.subtract(x_test[i], x_trainMean)
            list_coef_B = pls.fit(x_train[i], y_train[i], self.max_components)
            for j in range(self.max_components):
                y_pre = np.dot(x_testCenter, list_coef_B[j])
                y_pre = y_pre + y_trainMean
                y_predict[:, j] = y_pre.ravel()
            y_allPredict = np.vstack((y_allPredict, y_predict))
        y_allPredict = y_allPredict[1:]

        return y_allPredict, self.y

    def mse_cv(self, y_allPredict, y_measure):

        PRESS = np.square(np.subtract(y_allPredict, y_measure))
        all_PRESS = np.sum(PRESS, axis=0)

        RMSECV = np.sqrt(all_PRESS / self.n)
```

```
        min_RMSECV = min(RMSECV)
        comp_array = RMSECV.argsort()
        comp_best = comp_array[0] + 1

return RMSECV, min_RMSECV, comp_best

# -*-coding: utf-8-*-

import matplotlib.pyplot as plt

def draws_pre_pharm(Y_test, Y_predict, y_trainPredict, y_train):

    plt.figure(figsize=(8, 8), facecolor='white')
    plt.subplot(321)
    plt.title('spectrometer1: weight')
    plt.plot([min(y_train[0]), max(y_train[0])], [min(y_train[0]), max(y_train[0])],
            'black', label='y=x')
    plt.scatter(y_train[0], y_trainPredict[0], s=20, c='r', marker='o',
              label='calibration set')
    plt.scatter(Y_test[0], Y_predict[0], s=30, c='b', marker='o', label='test set')
    plt.xlabel('Measured value')
    plt.ylabel(' Predicted value')

    plt.subplot(322)
    plt.title("spectrometer1: hardness")
    plt.plot([min(y_train[1]), max(y_train[1])], [min(y_train[1]), max(y_train[1])],
            'black', label='y=x')
    plt.scatter(y_train[1], y_trainPredict[1], s=20, c='r', marker='o',
              label='calibration set')
    plt.scatter(Y_test[1], Y_predict[1], s=30, c='b', marker='o', label='test set')
    plt.xlabel('Measured value')
    plt.ylabel(' Predicted value')

    plt.subplot(323)
    plt.title("spectrometer1: assay")
    plt.plot([min(y_train[2]), max(y_train[2])], [min(y_train[2]), max(y_train[2])],
            'black', label='y=x')
    plt.scatter(y_train[2], y_trainPredict[2], s=20, c='r', marker='o',
              label='calibration set')
    plt.scatter(Y_test[2], Y_predict[2], s=30, c='b', marker='o', label='test set')
```

```
    plt.xlabel('Measured value')
    plt.ylabel(' Predicted value')

    plt.subplot(324)
    plt.title("spectrometer2: weight")
    plt.plot([min(y_train[3]), max(y_train[3])], [min(y_train[3]), max(y_train[3])],
             'black', label='y=x')
    plt.scatter(y_train[3], y_trainPredict[3], s=20, c='r', marker='o',
                label='calibration set')
    plt.scatter(Y_test[3], Y_predict[3], s=30, c='b', marker='o', label='test set')
    plt.xlabel('Measured value')
    plt.ylabel(' Predicted value')

    plt.subplot(325)
    plt.title("spectrometer2: hardness")
    plt.plot([min(y_train[4]), max(y_train[4])], [min(y_train[4]), max(y_train[4])],
             'black', label='y=x')
    plt.scatter(y_train[4], y_trainPredict[4], s=20, c='r', marker='o',
                label='calibration set')
    plt.scatter(Y_test[4], Y_predict[4], s=30, c='b', marker='o', label='test set')
    plt.xlabel('Measured value')
    plt.ylabel(' Predicted value')

    plt.subplot(326)
    plt.title("spectrometer2: assay")
    plt.plot([min(y_train[5]), max(y_train[5])], [min(y_train[5]), max(y_train[5])],
             'black', label='y=x')
    plt.scatter(y_train[5], y_trainPredict[5], s=20, c='r', marker='o',
                label='calibration set')
    plt.scatter(Y_test[5], Y_predict[5], s=30, c='b', marker='o', label='test set')
    plt.xlabel('Measured value')
    plt.ylabel(' Predicted value')

    plt.tight_layout()
    plt.show()

def rmsecv_comp_line_pharm(max_components, rmsecv_list):

    plt.figure(figsize=(8, 8), facecolor='white')
    plt.subplot(321)
```

```
plt.title('spectrometer1: weight')
plt.plot(range(1, max_components + 1), rmsecv_list[0], '-o')
plt.xlabel('num_components')
plt.ylabel('RMSECV')

plt.subplot(322)
plt.title("spectrometer1: hardness")
plt.plot(range(1, max_components + 1), rmsecv_list[1], '-o')
plt.xlabel('num_components')
plt.ylabel('RMSECV')

plt.subplot(323)
plt.title("spectrometer1: assay")
plt.plot(range(1, max_components + 1), rmsecv_list[2], '-o')
plt.xlabel('num_components')
plt.ylabel('RMSECV')

plt.subplot(324)
plt.title("spectrometer2: weight")
plt.plot(range(1, max_components + 1), rmsecv_list[3], '-o')
plt.xlabel('num_components')
plt.ylabel('RMSECV')

plt.subplot(325)
plt.title("spectrometer2: hardness")
plt.plot(range(1, max_components + 1), rmsecv_list[4], '-o')
plt.xlabel('num_components')
plt.ylabel('RMSECV')

plt.subplot(326)
plt.title("spectrometer2: assay")
plt.plot(range(1, max_components + 1), rmsecv_list[5], '-o')
plt.xlabel('num_components')
plt.ylabel('RMSECV')

plt.tight_layout()
plt.show()
```

4.4.2 Corn Dataset

In this section the corn dataset was used for experiments. Latent variables of PLS are allowed to take values in the set [1, 15], and it is determined by the 10-fold cross-validation. No pre-processing methods were used other than mean-centering. Table 4.1 shows the training error, cross-validation error, prediction error, and principal component number of the PLS model for moisture, oil, protein, and starch content directly using the corn data set.

Table 4.1 Summary of the PLS models and properties

Instrument	Reference values	RMSEC	RMSEP	$RMSECV_{min}$	Latent Variable (LV)
m5spec	moisture	0.00505	0.00799	0.01017	15
m5spec	oil	0.02680	0.03934	0.06355	15
m5spec	protein	0.05979	0.09843	0.09710	14
m5spec	starch	0.08879	0.14584	0.19731	15
mp5spec	moisture	0.09826	0.11225	0.14085	11
mp5spec	oil	0.08043	0.07055	0.10078	7
mp5spec	protein	0.09945	0.15854	0.144019	10
mp5spec	starch	0.24319	0.35500	0.36011	11
mp6spec	moisture	0.08464	0.11239	0.16117	15
mp6spec	oil	0.06239	0.07214	0.10260	12
mp6spec	protein	0.10276	0.14496	0.14700	10
mp6spec	starch	0.21746	0.42008	0.34952	13

RMSEC: Root Mean Square Error of calibration set

RMSEP: Root Mean Square Error of test set

$RMSECV_{min}$: Minimum Root Mean Square Error of Cross-Validation

LV: The optimal number of latent variables is selected only when the lowest RMSECV

It can be seen from Table 4.1 that there is no significant difference in RMSEC, RMSECV, and RMSEP of each component in corn, indicating that there is no over-fitting phenomenon, and the RMSEP is small, indicating that there is no under-fitting phenomenon. The selection of the number of principal components is reasonable.

In this paper, the principal component of the PLS algorithm is selected by the 10-fold cross-validation method. The RMSECV of the PLS model is given in Fig. 4.10 – Fig. 4.12, respectively.

In order to more intuitively compare the fitting effect of the PLS algorithm, Fig. 4.13– Fig. 4.15 show the plots of measured vs predicted values for the calibration set and the test set of various components in corn. If the model predicts better, the corresponding point is closer to the line $y=x$. Therefore, the prediction performance of the PLS algorithm can be judged based on the degree of concentration of the data point near the line $y=x$. Further, it is possible to observe their fitting effects more intuitively.

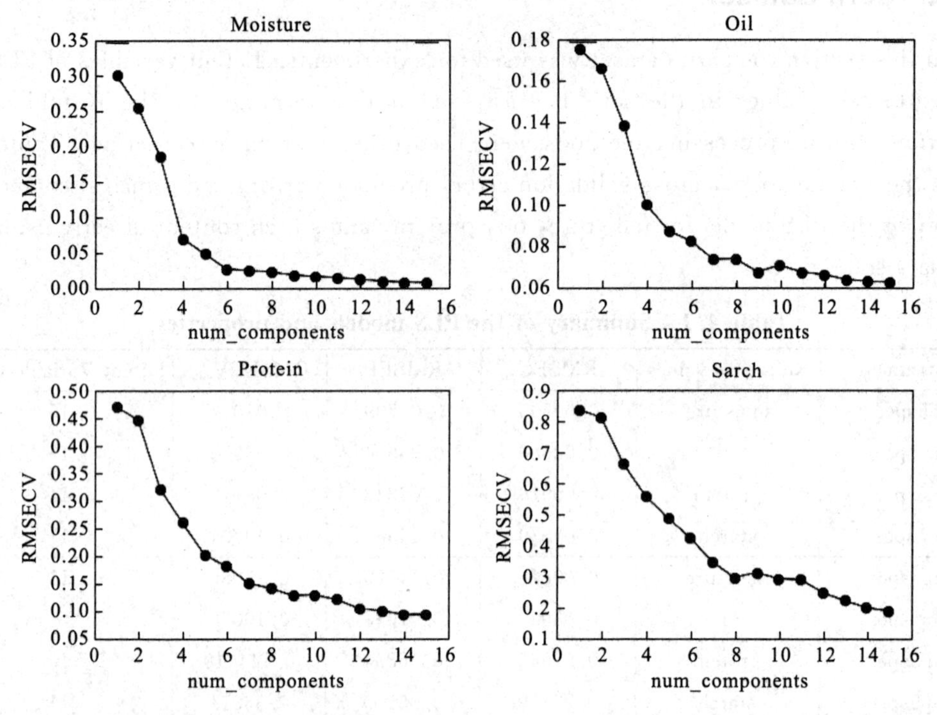

Fig. 4. 10　The selection process of the optimal latent variables number from PLS model about the m5 spec instrument

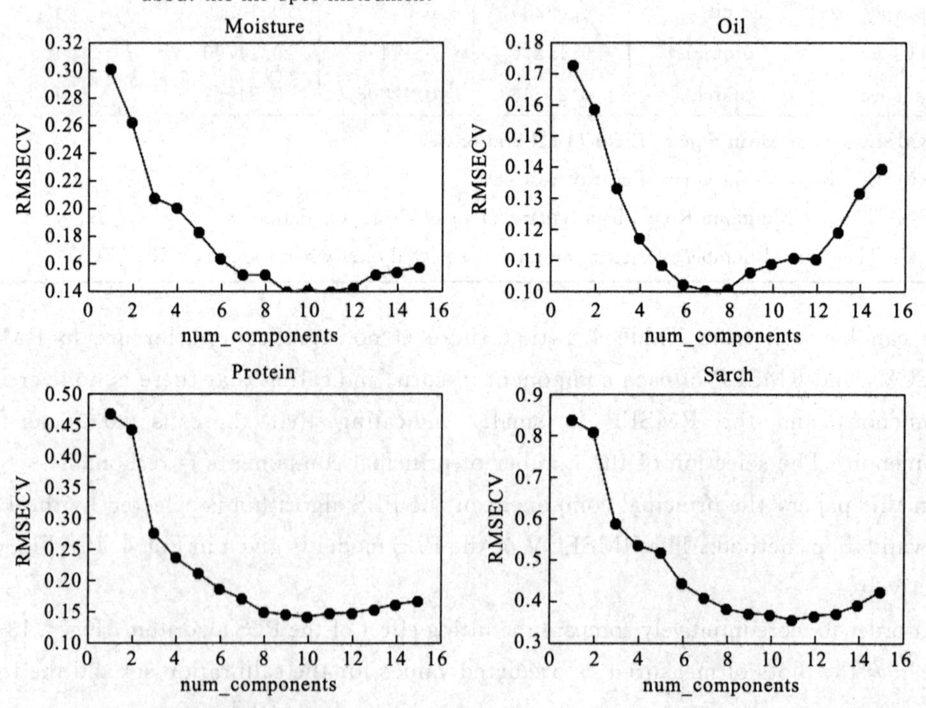

Fig. 4. 11　The selection process of the optimal latent variables number from PLS model about the mp5 spec instrument

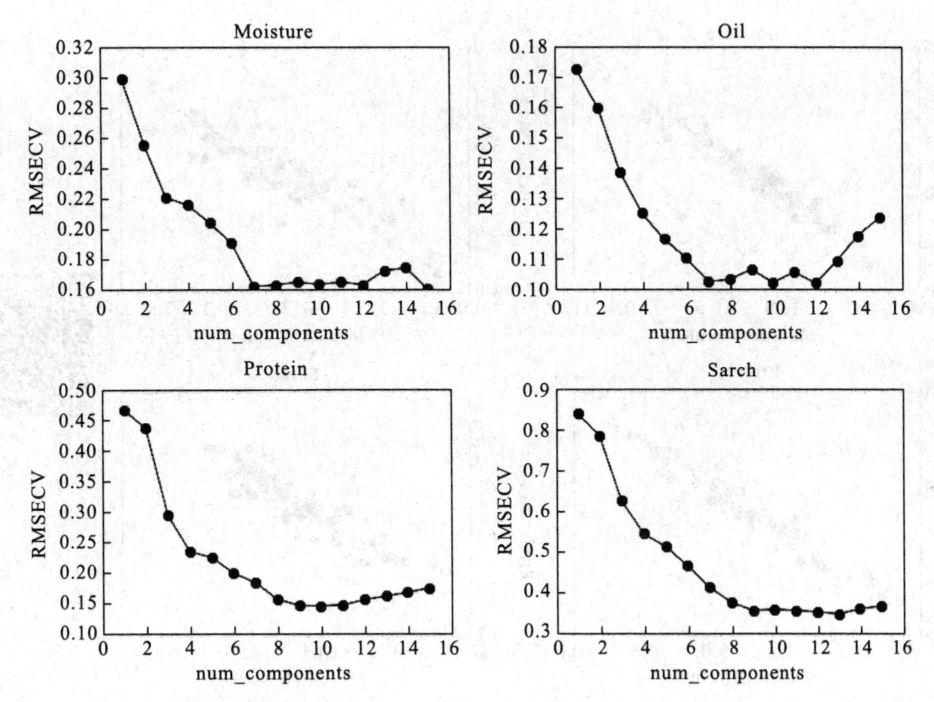

Fig. 4. 12 The selection process of the optimal latent variables number from PLS model about the mp6 spec instrument

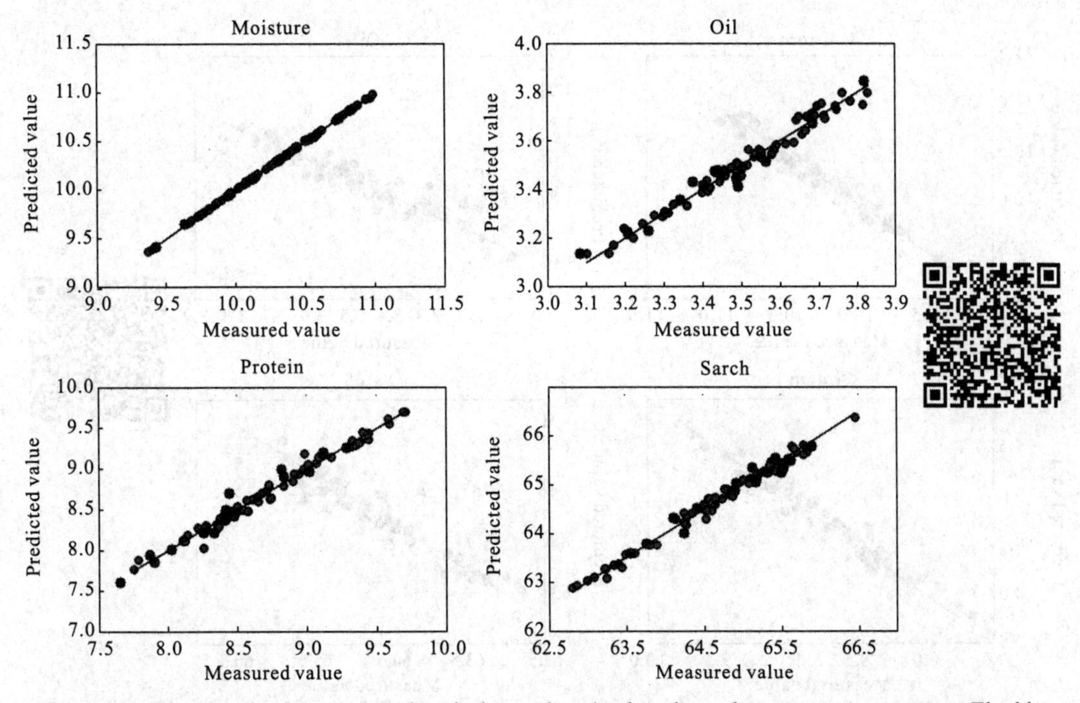

Fig. 4. 13 The actual value compared with the predicted value about the m5 spec instrument. The blue and red dots represent the results for each sample in the train set and test set, respectively

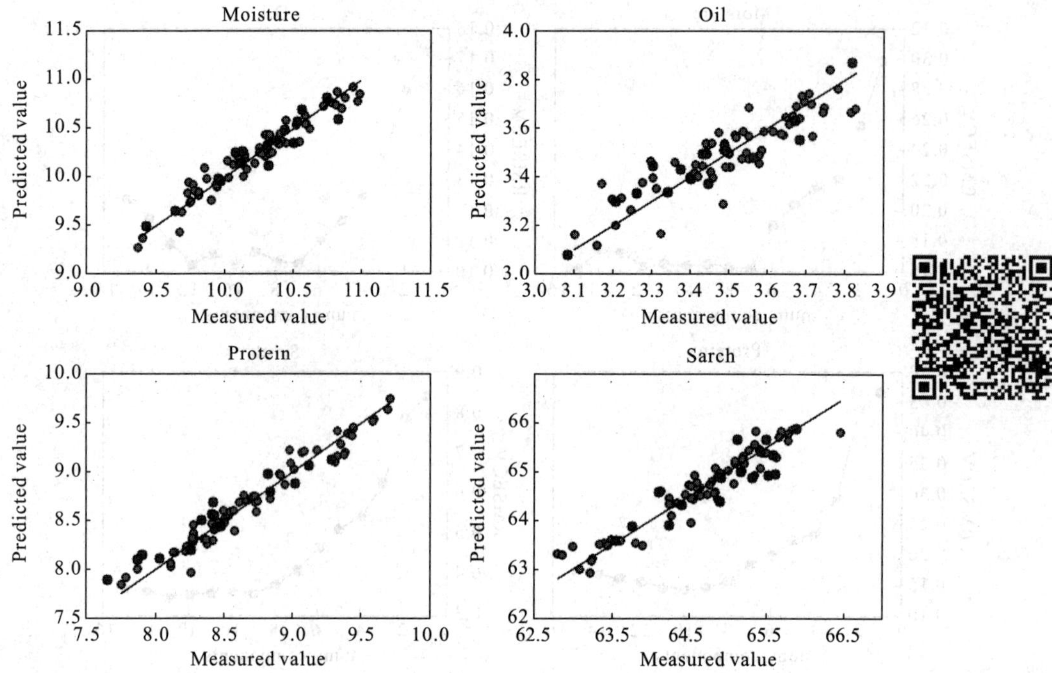

Fig. 4. 14　The actual value compared with the predicted value about the mp5 spec instrument. The blue and red dots represent the results for each sample in the train set and test set, respectively

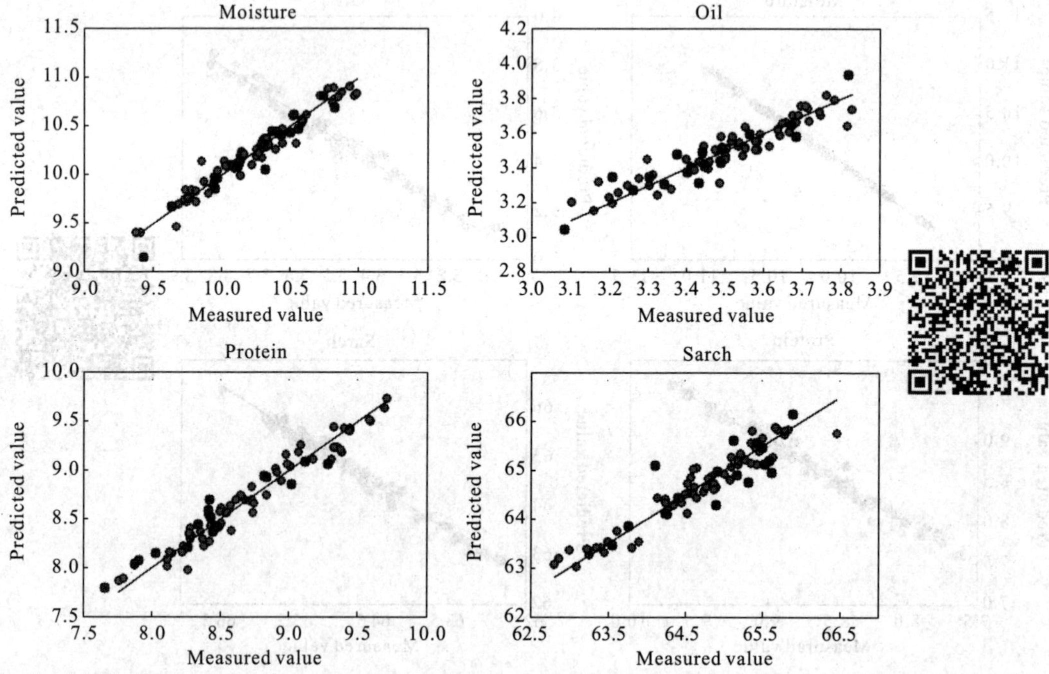

Fig. 4. 15　The actual value compared with the predicted value about the mp6 spec instrument. The blue and red dots represent the results for each sample in the train set and test set, respectively

4.4.3 Wheat Dataset

Table 4.2 also lists the RMSE of different PLS models for cross-validation set, calibration set, and test set. According to minimum RMSECV criterion, the optimal number of latent variable is determined and listed in Table 4.2. For PLS model, the optimal number of latent variables are 15, 15, 15, 15, 8 and 15, respectively.

Table 4.2 Summary of the PLS Models and Properties

Instrument	Reference Values	RMSEC	RMSEP	$RMSECV_{min}$	Latent Variable (LV)
B1	protein	0.31905	0.37823	0.52004	15
B2	protein	0.37818	0.39704	0.55749	15
B3	protein	0.34824	0.41466	0.48623	15
A1	protein	0.17773	0.29443	0.23551	15
A2	protein	0.53254	0.64587	0.93716	8
A3	protein	0.17221	0.28497	0.23113	15

RMSEC: Root Mean Square Error of calibration set

RMSEP: Root Mean Square Error of test set

$RMSECV_{min}$: Minimum Root Mean Square Error of Cross-Validation

LV: The optimal number of latent variables is selected only when the lowest RMSECV

In order to achieve the best performance for each model, the optimal number of latent variable was pre-optimized by performing 10-fold cross-validation. The RMSECV of the PLS model is given in Fig. 4.16, and the minimum cross-validation error at the lowest point of each figure is the number of optimal latent variables.

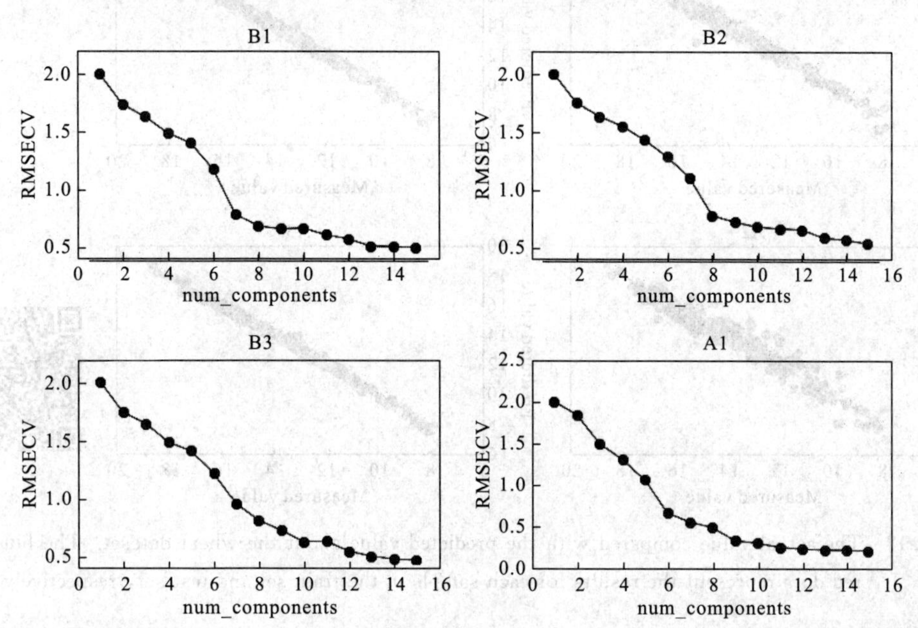

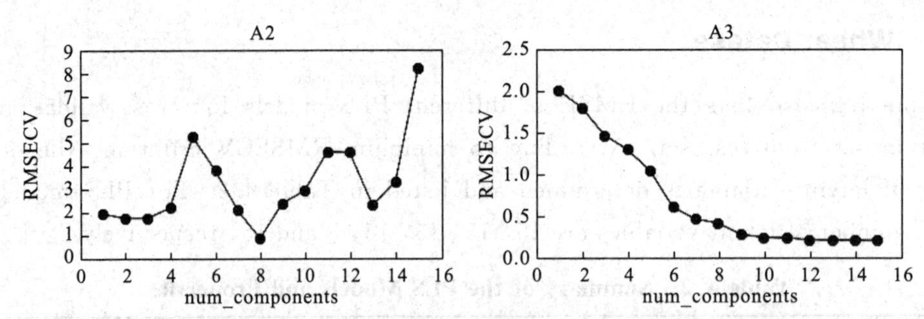

Fig. 4. 16 The selection process of the optimal latent variables number from PLS model about the wheat dataset

Moreover, Fig. 4. 17 exhibits the relationship between the measured values and the prediction values of the training and test sets by different PLS models. As can be seen, good correlations are found between expected and predicted concentrations which confirm the good performance of the PLS model.

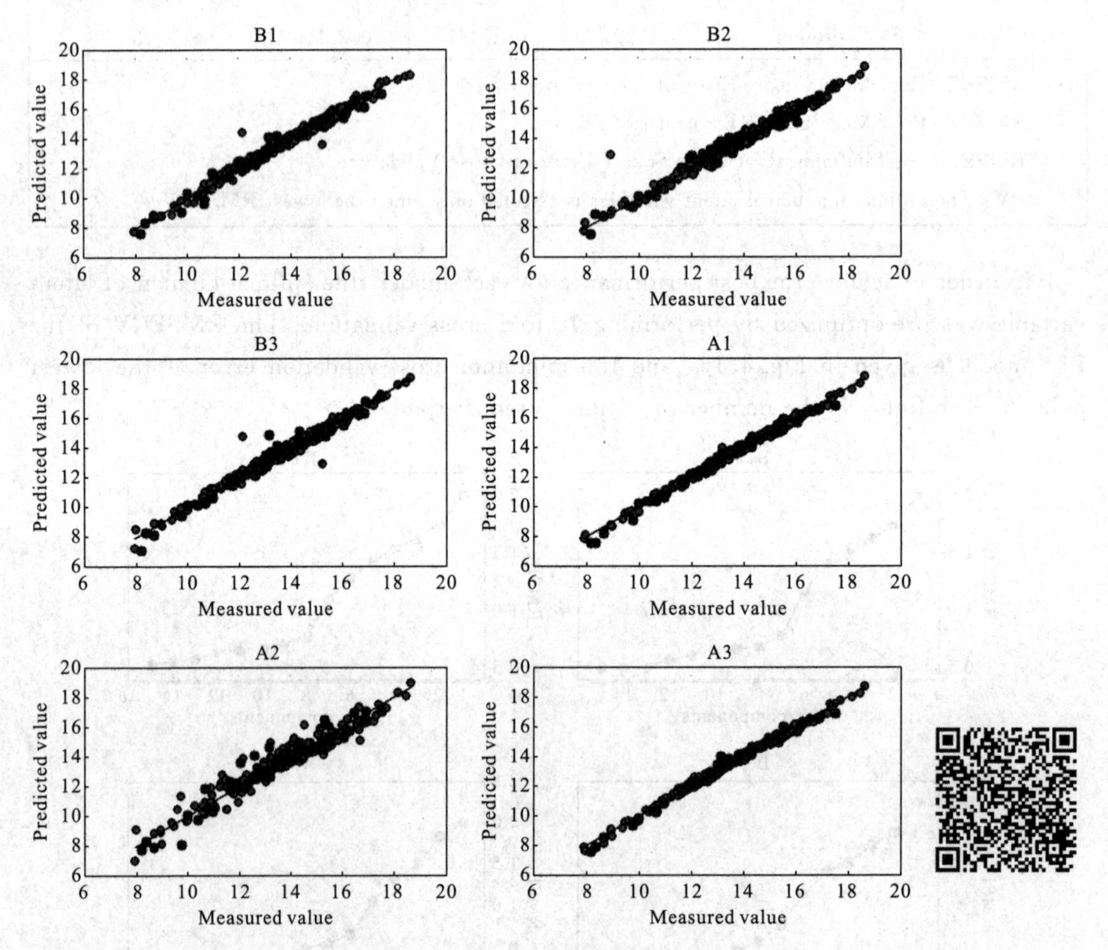

Fig. 4. 17 The actual value compared with the predicted value about the wheat dataset. The blue and red dots represent the results for each sample in the train set and test set, respectively

4.4.4 Pharmaceutical Tablet Dataset

This section uses the pharmaceutical tablet data set for experiments. Table 4.3 shows the training error, cross-validation error, prediction error, and principal component number of the PLS model by the pharmaceutical tablet training set. As can be seen from Table 4.3, the RMSEC, RMSECV, and RMSEP of each component in the pharmaceutical tablet data set are all on the same order of magnitude, indicating that there is no over-fitting phenomenon, and the RMSEP is small, indicating that there is no under-fitting phenomenon. Explain that the selection of the number of principal components is reasonable.

Table 4.3 Summary of the PLS Models and Properties

Instrument	Reference values	RMSEC	RMSEP	$RMSECV_{min}$	Latent Variable (LV)
spectrometer1	weight	3.47831	4.12929	3.81790	3
spectrometer1	hardness	1.28772	0.72098	1.39317	3
spectrometer1	assay	4.19221	5.54743	4.97014	5
spectrometer2	weight	2.73142	4.40738	3.44580	6
spectrometer2	hardness	1.35162	0.61732	1.40090	2
spectrometer2	assay	4.28022	5.84714	4.92193	5

RMSEC: Root Mean Square Error of calibration set

RMSEP: Root Mean Square Error of test set

$RMSECV_{min}$: Minimum Root Mean Square Error of Cross-Validation

LV: The optimal number of latent variables is selected only when the lowest RMSECV

In this section, the principal component of the PLS algorithm is selected by a 10-fold cross-validation method, and the maximum principal component is set to 15. The Fig. 4.18 shows the variation of the RMSECV of the PLS model with the content of the three active ingredients in the pharmaceutical tablet with the number of principal components. When the number of principal components is 3, 3, 5, 6, 2 and 5, the minimum value of RMSECV is obtained.

In order to more intuitively compare the prediction stability of various methods, Fig. 4.19 shows the comparative graphs between the expected and predicted concentrations with different PLS models for pharmaceutical tablet dataset, respectively.

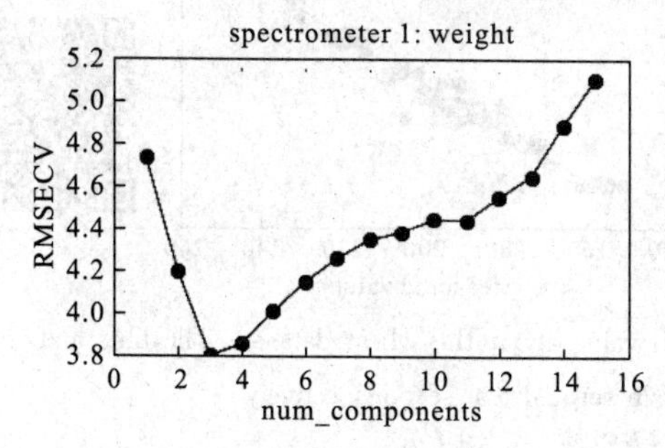

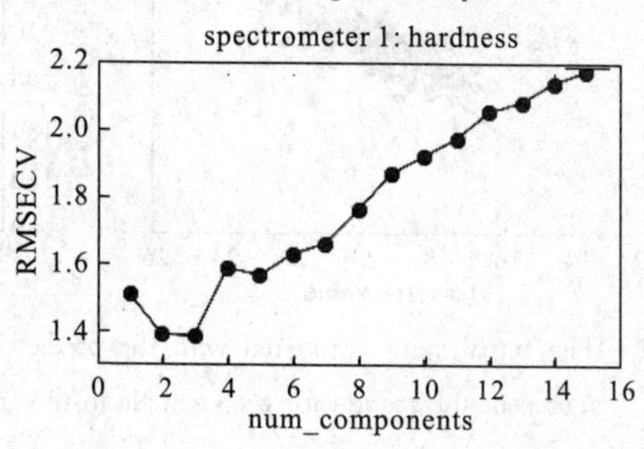

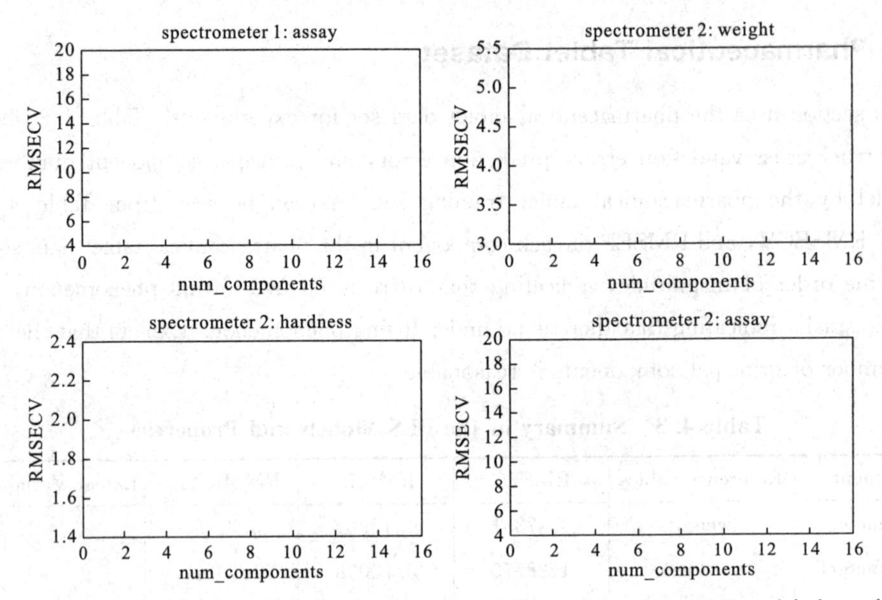

Fig. 4. 18 The selection process of the optimal latent variables number from PLS model about the about the wheat dataset

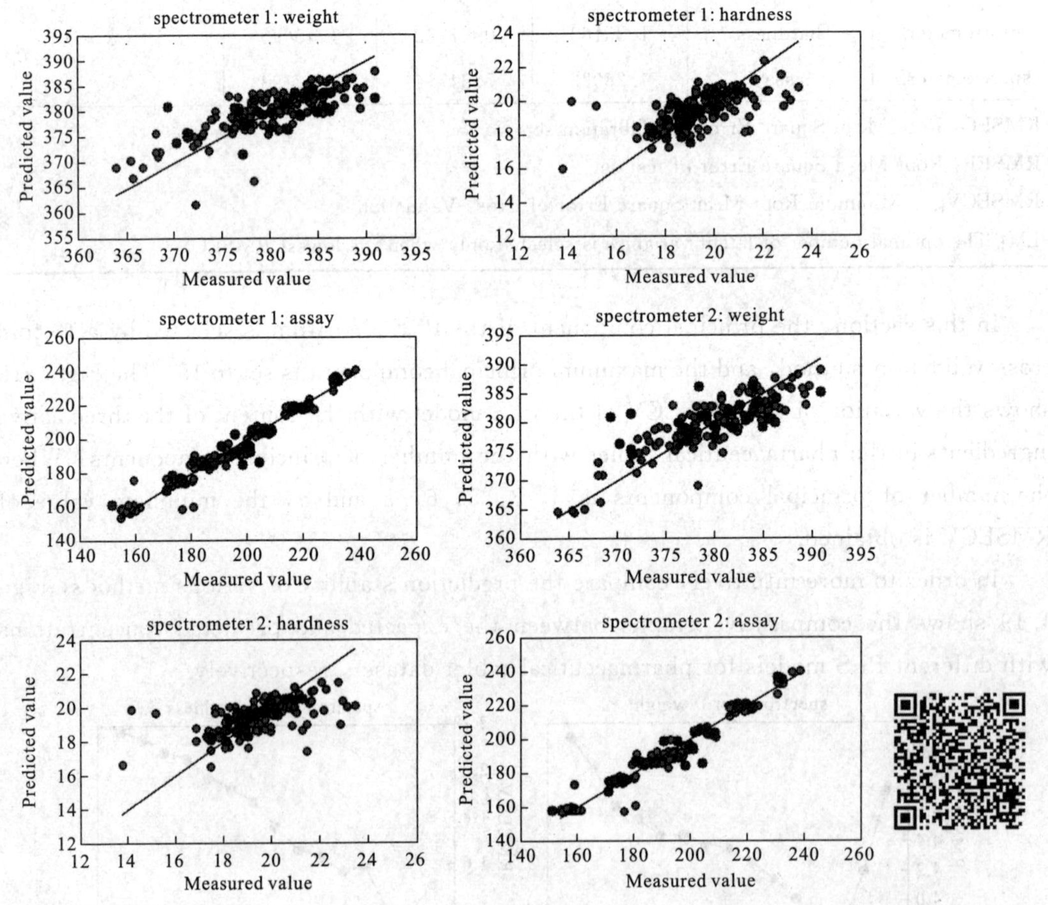

Fig. 4. 19 The actual value compared with the predicted value about the wheat dataset. The blue and red dots represent the results for each sample in the train set and test set, respectively

4.5 Stack Partial Least Squares

4.5.1 Introduction

A spectrum, especially a near-infrared (NIR) spectrum, usually consists of hundreds or even thousands of measurements or channels. The spectrum collected from instruments contains beneficial information for a calibration but often contains redundant and irrelevant information. Theoretical and experimental evidence indicates that some form of wavelength selection can improve the predictive performance from a PLS regression model by avoiding the irrelevant information embedded in the response data [1, 2]. Wavelength selection methods have been shown to improve the stability of the predictive model and increase the interpretability of the relationship between the response and property [3, 4]. In some methods, the goal is to extract the spectral channels strongly associated with the property by stepwise elimination of wavelengths with high uncertainty [5, 6]. Other methods rely on genetic algorithms [7, 8] to optimize and select subsets or intervals of the spectrum.

An interval PLS (iPLS) algorithm has been proposed by Nørgaard [3] to select the best single interval of the spectrum, based on cross-validation performance of PLS models built on a series of user-selected intervals. This is an attractive and simple approach to wavelength selection, but improper selection of the size of the interval used in iPLS can corrupt the predictive performance from iPLS regression model, so Xu and Schechter [1] proposed a modified wavelength selection algorithm optimizing both interval position and size. Despite the improvement seen in models using the best interval or even several of the better intervals, there is still the risk that eliminating all other intervals, as is done in iPLS, might result in loss of useful information [9].

Model fusion offers an alternative to variable or interval selection. Fusion methods rely on the idea of combining predictors instead of selecting only the single best predictor or best set of predictors to improve model robustness [9-11]. In a stacked regression, a series of separate regression models built on subsets of the full data set are combined by determining a set of weights for each subset or sub-model, and these weights are used to produce a fused prediction from the set of predictions generated from the set of sub-models. These weights can be obtained from cross-validation [9, 10] of each model or subset, or from use of a Bayesian criterion [12]. Stacking does not involve a conventional variable selection, and so a stacked model reduces the information loss associated with wavelength selection while gaining many of the benefits demonstrated from focusing on relevant portions of the data [9]. While a stacked model may not always predict better than one using a single best variable or subset, a stacked model may offer other advantages, including improved precision and better resistance to the effects of outliers [11, 12]. These possible benefits can offset the additional time and work needed to generate the sub-models that must be combined to generate the stacked model. Stacked regressions are increasingly used in the area of sensor fusion and design [13] and in forecasting [14]. In Monte Carlo cross-validation stacked regression (MCCVSR) modeling reported by Xu [15], fixed window subsets of the response spectrum are stacked to try to overcome the disadvantages of wavelength selection inherent in iPLS modeling.

In this section, we report a novel stacking algorithms developed to take full advantage of the information in an entire spectral response while emphasizing intervals that are highly correlated to the target property. The first, the stacked PLS (SPLS) algorithm, splits a spectrum into a set of equally sized, disjoint sub-intervals similar to that done in iPLS regression and then performs stacking, using errors from cross-validation to weight the PLS regression models obtained on the sub-intervals. The two-dimensional cross-validation used in SPLS regression determines the number of intervals to be used in the stacked regression model as well as the number of latent variables (LVs) needed. We show that predictions from the stacked PLS regression are more robust than those obtained from conventional PLS modeling regression performed on the whole spectrum, and are never worse than those from a iPLS regression performed on the single best sub-interval of the spectrum. The performance of the SPLS algorithms is evaluated using a simulation and NIR spectra datasets. The results obtained from these data sets show that the proposed methods not only yield superior performance compared to that of regular PLS and iPLS algorithms, but also have a potential use in predicting spectral data including an outlier sample.

4.5.2 Theory of Stack Partial Least Squares

4.5.2.1 Interval Partial Least Squares Algorithm (iPLS)

Nørgaard et al. [3] developed a wavelength selection process driven by a local regression method called interval partial least squares (iPLS) regression to improve the predictive power and to enhance the interpretation of PLS models. The approach used in the interval PLS algorithm splits the response matrix ($\boldsymbol{X}$ for m samples measured at p spectral wavelengths) into n disjoint intervals ($\boldsymbol{X}_1$, $\boldsymbol{X}_2$, $\cdots$, $\boldsymbol{X}_n$) of equal width (of p/n channels), where each sub-interval may contain different information and variance that is related to the property of interest. RMSECV values and number of LVs obtained by PLS regression on each sub-interval are employed to focus on the important spectral regions and to eliminate the other regions. The best regression model based on sub-intervals should require the smallest number of PLS components and produce the lowest RMSECV values. Jiang [4] showed the inherent relationship between conventional PLS regression and iPLS regression theoretically and experimentally, and found that adding spectral channels to the calibration model led to an increase in root mean square error of prediction (RMSEP). Selection of suitable regions or channels in the spectrum could enhance the predictive performance of the resulting calibration model, suggesting that

$$\text{RMSEP (iPLS)} < \text{RMSEP (PLS)}$$

4.5.2.2 Stack Partial Least Squares Algorithm (SPLS)

Although the iPLS algorithm focuses the modeling on the most important interval in the spectra and removes other intervals, the intervals eliminated by the iPLS algorithm might contain information beneficial to the calibration. In order to avoid information leakage, a new algorithm named stacked partial least squares (SPLS), combining predictors from all interval models by application of conventional PLS regression, is developed here. Regression analysis used in SPLS consists of two steps, calibration and prediction, similar to that in regular PLS and iPLS; however, several steps are involved to find the stacked regression vector associated with the calibration model. First, sets of sub-interval spectra are created. As in iPLS regression, the set of calibration spectra $\boldsymbol{X}$ is split into n disjoint intervals $\boldsymbol{X}_k$ of equal width ($k=1, 2, \cdots, n$). For the calibration step, n PLS models are developed between the $m\times 1$ target property vector $\boldsymbol{y}$ and each of the n intervals $\boldsymbol{X}_k$, and a set of n PLS regression vectors, $\hat{\boldsymbol{\beta}}_{\text{PLS},k}$, are obtained. Unlike the approach taken in iPLS, all n of the interval PLS model regression vectors $\hat{\boldsymbol{\beta}}_{\text{PLS},k}$ are retained and are used to generate a prediction from new spectral data with the regression vector from each of the n PLS interval models weighted according to the model's predictive performance in cross validation[9]. The aim is to have a set of weights that combine predicted values from all intervals by application of regular PLS and that minimize the cross-validated error in the stacked regression model, as indicated in the following equation: [10, 11]

$$\hat{w}=\text{ARG}\min_{w\in R}(\boldsymbol{y}-\sum_{k=1}^{n}w_k\hat{y}_k)^2$$

where $\hat{y}_k$ is the prediction of the property from the PLS model developed on the k-th interval and w_k is the stacking weight of the PLS model developed for the k-th interval. The relationship between the regression vector of the PLS model in the k-th interval and the regression vector for the SPLS model can be expressed by

$$\hat{\boldsymbol{\beta}}_{k,\text{SPLS}}=w_k\hat{\boldsymbol{\beta}}_{\text{PLS},k}$$

where $\hat{\boldsymbol{\beta}}_{\text{PLS},k}$ is the regression vector for the k-th interval as estimated by regular PLS and $\hat{\boldsymbol{\beta}}_{k,\text{SPLS}}$ is the regression vector for the k-th interval for the SPLS regression model.

The components w_k of the weight vector $\boldsymbol{w}$ used to set stacking weights for combining predicted values of the target property from all n of the PLS interval models can be obtained by cross-validation performed on individual interval models using the calibration set, according to the following equation:

$$w_k=\frac{s_k^2}{\sum_{k=1}^{n}s_k^2}$$

Here s_k is the reciprocal of the cross-validation error of the PLS model developed on the k-th interval, determined for the calibration set. The stacking weights w_k are constrained between 0 and 1, and their sum is normalized to 1. The stacking weights are one measure of how well each of the subsets of the spectrum correlate with the target property.

The external prediction set $\boldsymbol{X}_u(\boldsymbol{X}_{1,u}\cdots\boldsymbol{X}_{n,u})$ can then be used to predict the values for the dependent property $\boldsymbol{y}_u$ by direct application of the SPLS regression vector

$$\hat{\boldsymbol{y}}_u=\sum_{k=1}^{n}\boldsymbol{X}_{k,u}\hat{\boldsymbol{\beta}}_{k,\mathrm{SPLS}}$$

As above equation indicates, the weighted predictions from each interval PLS model obtained in the calibration step are summed to produce a fused prediction based on the predictive results of all interval models. In this way, all n interval PLS models contribute to a prediction, though their contributions need not be equal or even similar.

The stacking—combining all subsets of the spectrum with a set of non-negative, constrained weights—makes the predictive performance from the SPLS algorithm at least equal to that from any single interval selected, for example, by the iPLS algorithm. Given that $w_k(k=1, 2, \cdots, n)$ is the k-th weight used to combine $\boldsymbol{y}_{\mathrm{pred},k}$ obtained from the k-th PLS interval model, and that $\sum w_k=1$, the theoretical relationship between the stacking PLS and iPLS algorithms can be demonstrated by a comparison of the norms of their prediction residuals. The mean-squared error of prediction from the stacked PLS model is

$$\mathrm{RMSEP(SPLS)}=E(\boldsymbol{y}-\sum w_k\hat{\boldsymbol{y}}_k)^2$$

where $\boldsymbol{y}$ is the reference value of the property and $\hat{\boldsymbol{y}}_k$ is predicted P value of the dependent property from the k-th interval. For $\sum w_k=1$, $\sum w_k\hat{\boldsymbol{y}}_k$ is the vector of expected values from predicted values on each individual interval. Therefore,

$$\sum w_k\hat{\boldsymbol{y}}_k=E(\hat{\boldsymbol{y}}_k)\quad k=1, 2, \cdots, n$$

which indicates that the predicted values from the stacked PLS model are closer to the reference values than those from any of the individual models. Then,

$$\|\boldsymbol{y}-\sum w_k\hat{\boldsymbol{y}}_k\|\leqslant\|\boldsymbol{y}-\hat{\boldsymbol{y}}_k\|$$

where $\| \quad \|$ is the Frobenius norm. How large the inequality in above equation depends on how much difference there is among the predicted values from the individual PLS models; no significant difference between the performance individual models will make the two sides of the equation nearly equal. However, because of the weight vector, the mean-squared error of prediction from the stacked PLS model will be less than a PLS prediction made from a model based on any single interval (region) of the spectrum, so that

$$E(\boldsymbol{y}-\sum w_k\hat{\boldsymbol{y}}_k)^2\leqslant E(\boldsymbol{y}-\hat{\boldsymbol{y}}_k)^2$$

This result suggests that RMSEP (SPLS)$\leqslant$ RMSEP (iPLS), and from above

$$\mathrm{RMSEP(SPLS)}\leqslant\mathrm{RMSEP\ (iPLS)}\leqslant\mathrm{RMSEP\ (PLS)}$$

These results imply that predictions from a stacked PLS model created from a data set are at least as precise as those from the iPLS model based on the best single subset of that data set, and that the stacked model will always predict at least as well as the conventional PLS model built on the same data set. There are two cases where the predictive performance from a model developed from the SPLS algorithm will be identical to that from a model developed using the iPLS algorithm. The first case is one where all interval models have the same predictive performance, so that all intervals are best subsets and all have equal stacking weights ($w_1 = w_2 = \cdots = w_n$, where $\sum w_k = 1$). The other case is when only one interval of the spectrum is correlated to the target property and the others are entirely irrelevant to the target property. In this case, the most important subset has a stacking weight near 1 and the others have weights near 0, and SPLS regression reduces to iPLS regression.

4.5.2.3 Choosing the Intervals for Stacking

The SPLS algorithm includes a cross-validation procedure to determine how many intervals the stacked model should optimally include. We generate a cross-validation surface to optimize the predictive ability of the stacked regression methods by performing the cross-validation as a function of the number of intervals and the number of LVs. The number of intervals and appropriate number of LVs within the range of the number of intervals and the number of LVs considered is selected to minimize the error of cross-validation. In general, the larger the number of intervals tested in the surface, the better the results, but in fact, 25 or 30 disjoint intervals were sufficient to discriminate the narrowest target property in the data examined here.

The cross-validation method (Fig. 4.20) reported for iPLS chooses one optimal interval from a pre-determined number of intervals. In this work, however, we use the same cross-validation surface used to optimize SPLS models to determine the number of intervals into which the spectra are partitioned for iPLS regression to allow a direct comparison of the methodology.

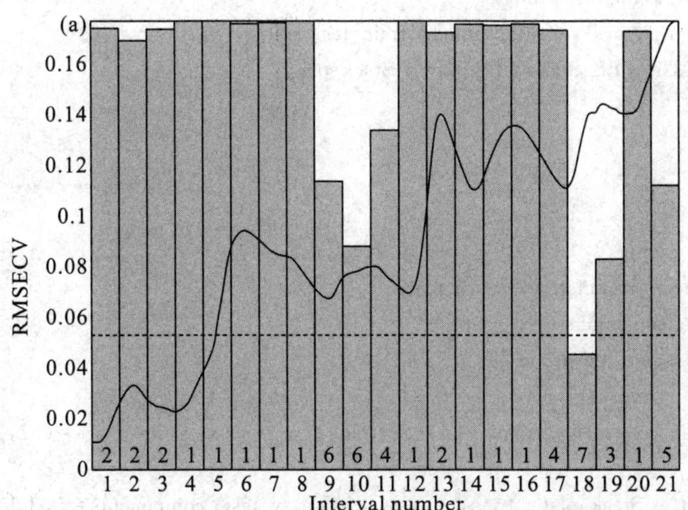

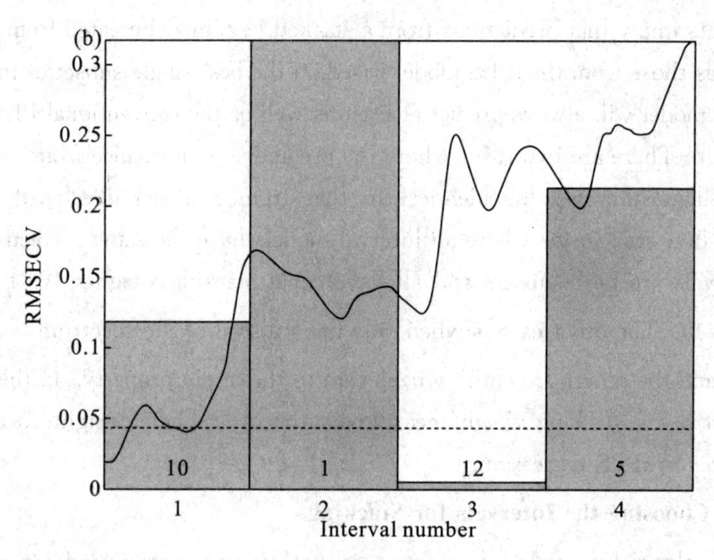

Fig. 4.20　Cross-validation (RMSECV) of iPLS model for corn data (m5spec): (a) for oil and (b) for moisture. Numbers shown in each bar represent the latent variables in each interval, the line is the mean spectrum of the calibration data and the broken line is the RMSECV value obtained from the PLS regression model generated from the full spectrum

4.5.3 Demo of SPLS

Software version python 2.7 and a Microsoft Windows 7 operating system. Cross-validation, MinMaxScaler and train test split are performed using the sklearn package, respectively. Dataset loading is done using the scipy package, and other programs can be implemented by individuals.

```
# -*- coding: utf-8 -*-

from scipy.io import loadmat
from sklearn.cross_validation import train_test_split
from STACK_PLS.Stack_PLS import Stack_pls


if __name__ == '__main__':


    fname = loadmat('NIRcorn.mat')
    x = fname['m5spec']['data'][0][0]
    y = fname['cornprop'][:, 3:4]

    x_train, x_test, y_train, y_test = train_test_split(x, y, test_size=0.2, random_state=0)

    demo = Stack_pls(x_train, x_test, y_train, y_test, components=5, folds=5)
```

```
    RMSEP = demo.stackPls(start=3, end=12, intervals=2)
    print RMSEP


# -*- coding: utf-8 -*-
'''
Created on 2019.3.10
@author: zzh
'''
import numpy as np
from sklearn.preprocessing import MinMaxScaler
from sklearn import cross_validation

def Weight(RMSECV, INTERVAL):
    s = []
    sum_s =0
    for i in range(INTERVAL):
        s.append(1.0 / RMSECV[i])
        sum_s = sum_s + s[i] * s[i]

    Weight_mat = np.zeros((INTERVAL, 1))
    for i in range(INTERVAL):
        w = s[i] * s[i] / sum_s
        Weight_mat[i] = w.ravel()
    Weight_minmax = MinMaxScaler((0, 1)).fit_transform(Weight_mat)  #
    Weight_minmax_sum = np.sum(Weight_minmax)
    W_mat = Weight_minmax / Weight_minmax_sum
    return W_mat


def cv(n_fold, x, y):    #
        kf = cross_validation.KFold(x.shape[0], n_fold)

        x_train = []
        y_train = []
        x_test = []
        y_test = []
        for train_index, test_index in kf:
            xtr, ytr = x[train_index], y[train_index]
            xte, yte = x[test_index], y[test_index]
            x_train.append(xtr)
            y_train.append(ytr)
            x_test.append(xte)
            y_test.append(yte)
```

```
        return x_train, x_test, y_train, y_test


# -*- coding: utf-8 -*-
'''
Created on 2019.3.10
@author: zzh
'''
import numpy as np
from select_interAndLv import select_interAndLv
from pls_split import pls_split
from PLS.NIPALS import _NIPALS
from function_cvAndWeight import cv, Weight


class Stack_pls():
    def __init__(self, x_train, x_test, y_train, y_test, components, folds):
        self.x_train = x_train
        self.y_train = y_train
        self.x_test = x_test
        self.y_test = y_test
        self.components = components
        self.folds = folds

    def stackPls(self, start, end, intervals):

        '''Choosing the optimal parameters, the optimal interval and the optimal latent variable '''
        demo = select_interAndLv(self.x_train, self.y_train, start, end, intervals)
        better_components, better_intervals = demo.select_interAndLv(self.components,
self.folds)
        print 'better: ', better_components, better_intervals

        ''' Calculate cross validation error based on optimal parameters'''

        RMSECV_list =self.get_rmsecv(better_components, better_intervals)

        ''' Find the weight based on the cross validation error'''
        W_mat = Weight(RMSECV_list, better_intervals)

        '''The final RMSEP is calculated'''
        better_intervals_demo = pls_split(self.x_train, self.y_train)
        better_split_list, better_intervals_list =
    better_intervals_demo.split(better_intervals)
        y_predict =0
```

```
        for j in range(better_intervals):

            xTrain = better_split_list[j]
            Intervals = better_intervals_list[j]
            xTest =self.x_test[:, Intervals[0]: Intervals[1]]

            xtrMean = np.mean(xTrain, axis=0)
            ytrMean = np.mean(self.y_train, axis=0)

            better_demo = _NIPALS(better_components)
            coef_list = better_demo.fit(xTrain, self.y_train, better_components)
            coef_B = coef_list[better_components -1]
            yte_predict = better_demo.predict(xTest, coef_B, xtrMean, ytrMean)
            y_predict = np.add(y_predict, W_mat[j] * yte_predict)

        press = np.square(np.subtract(self.y_test, y_predict))
        all_press = np.sum(press, axis=0)
        RMSEP = np.sqrt(all_press /self.x_test.shape[0])

        return RMSEP


    def get_rmsecv(self, better_components, better_intervals):

        rmsecv_list = []

        x_train, x_test, y_train, y_test = cv(self.folds, self.x_train, self.y_train)

        for j in range(better_intervals):

            RMSECV =0
            for i in range(self.folds):

                better_intervals_demo = pls_split(x_train[i], y_train[i])
                better_split_list, better_intervals_list =
better_intervals_demo.split(better_intervals)

                xTrain = better_split_list[j]
                Intervals = better_intervals_list[j]
                xTest = x_test[i][:, Intervals[0]: Intervals[1]]

                xtrMean = np.mean(xTrain, axis=0)
                ytrMean = np.mean(y_train[i], axis=0)
```

```
                better_demo = _NIPALS(better_components)
                coef_list = better_demo.fit(xTrain, y_train[i], better_components)
                coef_B = coef_list[better_components -1]
                yte_predict = better_demo.predict(xTest, coef_B, xtrMean, ytrMean)

                PRESS = np.square(np.subtract(yte_predict, y_test[i]))
                all_PRESS = np.sum(PRESS, axis=0)
                rmsecv = np.sqrt(all_PRESS / y_test[i].shape[0])

                RMSECV = RMSECV + rmsecv
            rmsecv_list.append(RMSECV /10)

        return rmsecv_list


# -*- coding: utf-8 -*-
'''
Created on 2019.3.10

@author: ZZH
'''

from pls_split import pls_split
import numpy as np
from PLS.NIPALS import _NIPALS
from function_cvAndWeight import cv, Weight


class select_interAndLv():
    def __init__(self, x_cal, y_cal, start, end, intervals):
        self.x_cal = x_cal
        self.y_cal = y_cal
        self.start = start
        self.end = end
        self.intervals = intervals


    def select_interAndLv(self, components, folds):

        x_train, x_test, y_train, y_test = cv(folds, self.x_cal, self.y_cal)

        length = (self.end - self.start) / self.intervals  #
        if (self.end - self.start) % self.intervals != 0 :
```

```
    length = length +1

error = np.zeros((components, length))

for k in range(folds):
  i2 =0
  for i in range(self.start, self.end, self.intervals):

      demo = pls_split(x_train[k], y_train[k])
      split_list, intervals = demo.split(i)
      for j in range(0, components):

          rmsecv_list = []
          Y_predict = np.zeros((y_test[k].shape[0], i))

          for i1 in range(i):
              better_components = j +1
              xTrain = split_list[i1]
              Intervals = intervals[i1]
              xTest = x_test[k][:, Intervals[0]:Intervals[1]]

              xtrMean = np.mean(xTrain, axis=0)
              ytrMean = np.mean(y_train[k], axis=0)

              better_demo = _NIPALS(better_components)
              coef_list = better_demo.fit(xTrain, y_train[k], better_components)
              coef_B = coef_list[better_components -1]
              yte_predict = better_demo.predict(xTest, coef_B, xtrMean, ytrMean)

                Y_predict[:, i1] = yte_predict.ravel()
                rmsecv =self.rmsecv(yte_predict, y_test[k])
                rmsecv_list.append(rmsecv)

            W_mat = Weight(rmsecv_list, i)

            y_predict =0
            for i1 in range(i):
                y_predict = np.add(y_predict, W_mat[i1] * Y_predict[:, i1])

            y_predict = y_predict.reshape(-1, 1)
            RMSECV =self.rmsecv(y_predict, y_test[k])

            error[j, i2] = error[j, i2] + RMSECV
```

```
                i2 = i2 +1

            component_op, interval_op =self.get_cv_parameter(error)
            component_op = component_op +1
            interval_op =self.start + interval_op * self.intervals

            return component_op, interval_op

        def rmsecv(self, y_predict, y_measure):
            '''  calculate   RMSECV'''
            PRESS = np.square(np.subtract(y_predict, y_measure))
            all_PRESS = np.sum(PRESS, axis=0)
            RMSECV = np.sqrt(all_PRESS / y_measure.shape[0])
            return   RMSECV


        def get_cv_parameter(self, error):
            '''Find the subscript of the minimum cross validation error in the two-dimensional error list'''

            min_value = np.amin(error)
            component, interval = error.shape

            for i in range(component):
                for j in range(interval):
                    if min_value == error[i, j]:
                        component_op, interval_op = i, j

            return component_op, interval_op


# -*- coding: utf-8 -*-
'''
Created on 2019.3.10
@author: Administrator
'''

import numpy as np
import scipy.io as sio
from sklearn.cross_validation import train_test_split


class pls_split():
    def __init__(self, x_cal, y_cal):
        self.x_cal = x_cal
```

```
        self.y_cal = y_cal
    def split(self, intervals):   #
        n, m = np.shape(self.x_cal)
#
        num = m / intervals#
        mod = m % intervals#
        before = num +1
        split_list = []
        intervals_list = []
        for i in range(mod):
            self.intervals = self.x_cal[:, i * before: (i + 1) * before]
            split_list.append(self.intervals)
            before_intervals = (i * before, (i +1) * before)
            intervals_list.append(before_intervals)
        before_num = mod * before

        behind = intervals - mod

        for i in range(behind):
        self.intervals = self.x_cal[:, before_num + i * num: before_num + (i + 1) * num]
            split_list.append(self.intervals)
            behind_intervals = (before_num + i * num, before_num + (i +1) * num)
            intervals_list.append(behind_intervals)

        return split_list, intervals_list
```

4.5.4 Experiments

4.5.4.1 Corn Dataset

The corn data set was chosen to show the effects of the SPLS algorithms in this study, and to demonstrate how SPLS takes full advantage of the information contained in the spectrum. No pre-processing methods were used other than mean-centering. The result shown in Table 4.4 indicates that RMSEP of SPLS model.

The maximum number of latent variable of PLS is set to 15, the number of interval is set in the range [2, 12], and the interval is 2. Two-way cross-validation was applied to the training set, to optimize both the number of LVs and the number of disjoint intervals. Fig. 4.21 – Fig. 4.23 is the error surface of the parameter selection in the three instruments of corn data set under different components, respectively.

To more intuitively compare the fitting effects of the SPLS algorithm, Fig. 4.24 – Fig. 4.26 shows the relationship between the measured values and the predicted values of the calibration set and test set. If the model predicts better, the corresponding point is closer to the line $y=x$. The prediction performance of the SPLS algorithm can be judged based on the degree of concentration of data points near the line $y=x$.

Table 4.4 Summary of the SPLS models and properties

Instrument	Reference Values	$RMSECV_{min}$	RMSEP	Latent Variable (LV)	Interval
m5spec	moisture	0.00374	0.00221	14	4
m5spec	oil	0.04423	0.04460	11	7
m5spec	protein	0.04912	0.04042	12	5
m5spec	starch	0.07798	0.09491	15	3
mp5spec	moisture	0.11079	0.13279	9	3
mp5spec	oil	0.07205	0.07178	9	11
mp5spec	protein	0.10096	0.16395	12	9
mp5spec	starch	0.24694	0.34657	14	9
mp6spec	moisture	0.10513	0.13043	14	9
mp6spec	oil	0.06874	0.09032	9	8
mp6spec	protein	0.10562	0.15136	8	5
mp6spec	starch	0.22725	0.32890	14	9

RMSEP: Root Mean Square Error of test set

$RMSECV_{min}$: Minimum Root Mean Square Error of Cross-Validation

LV: The optimal number of latent variables is selected only when the lowest RMSECV

Interval: The optimal number of interval is selected only when the lowest RMSECV

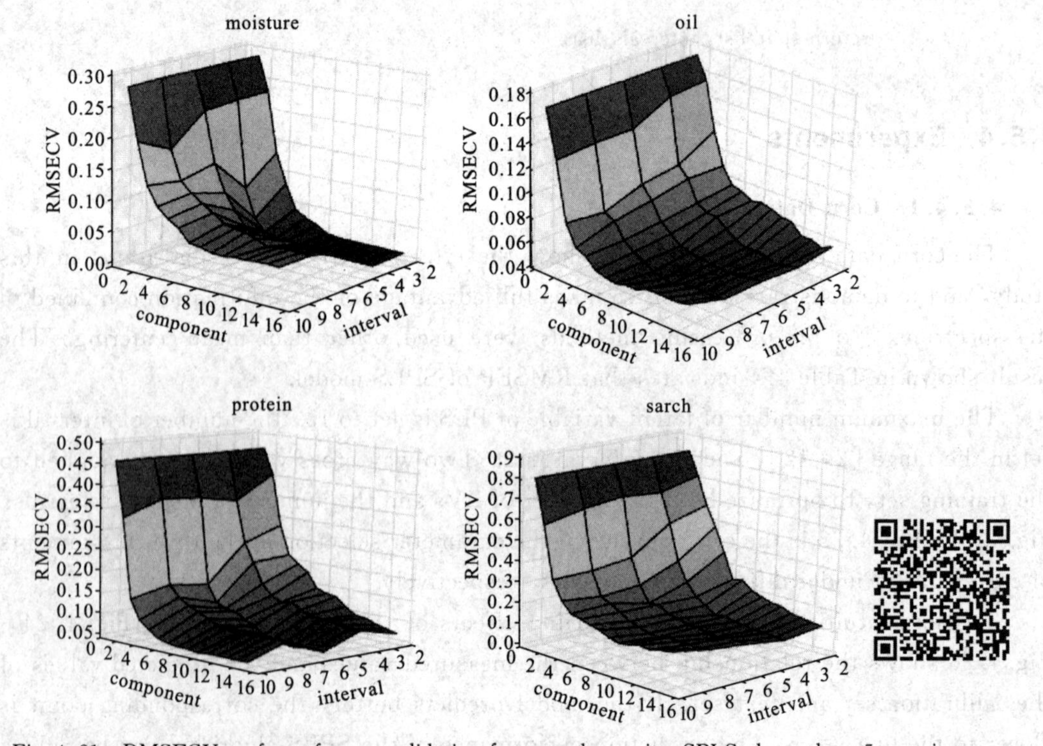

Fig. 4.21 RMSECV surface of cross-validation for corn data using SPLS about the m5 spec instrument

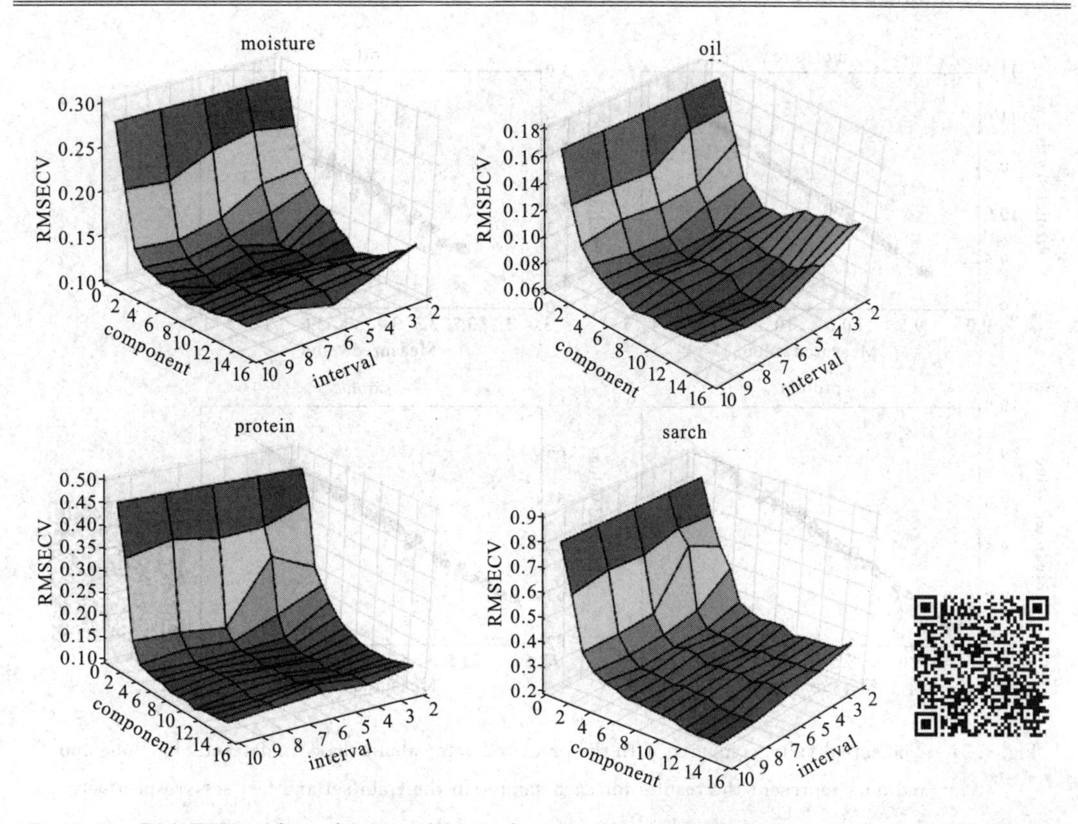

Fig. 4. 22 RMSECV surface of cross-validation for corn data using SPLS about the mp5 spec instrument

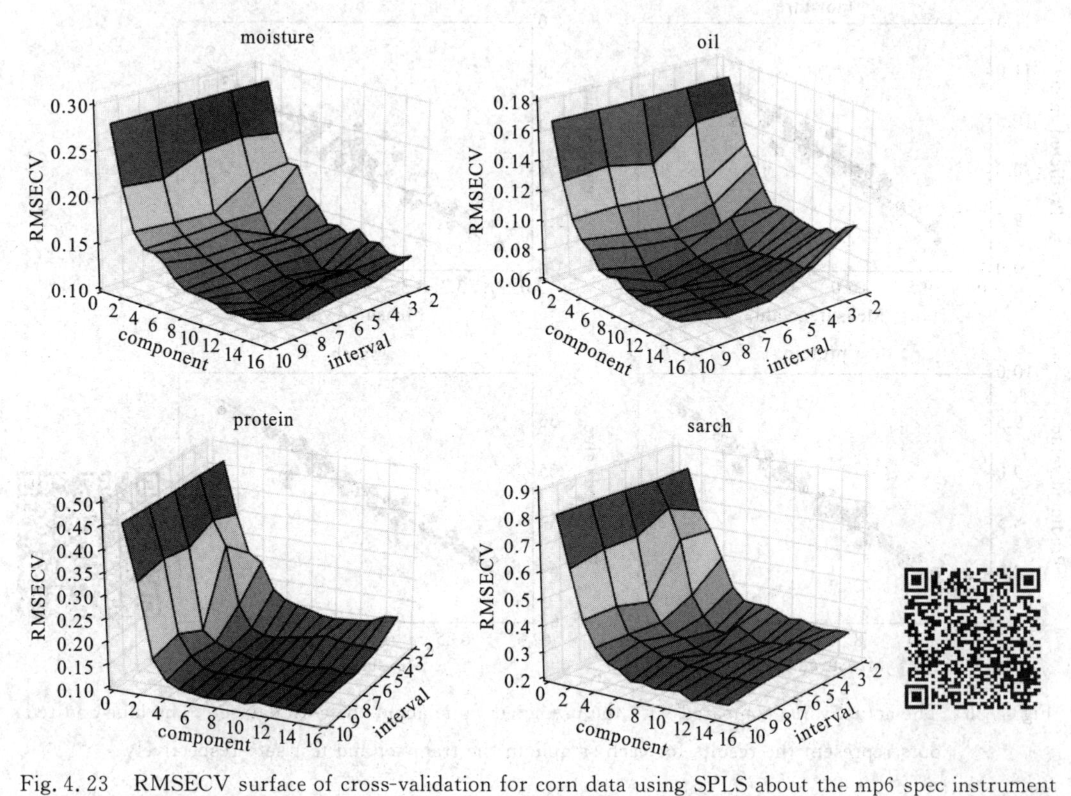

Fig. 4. 23 RMSECV surface of cross-validation for corn data using SPLS about the mp6 spec instrument

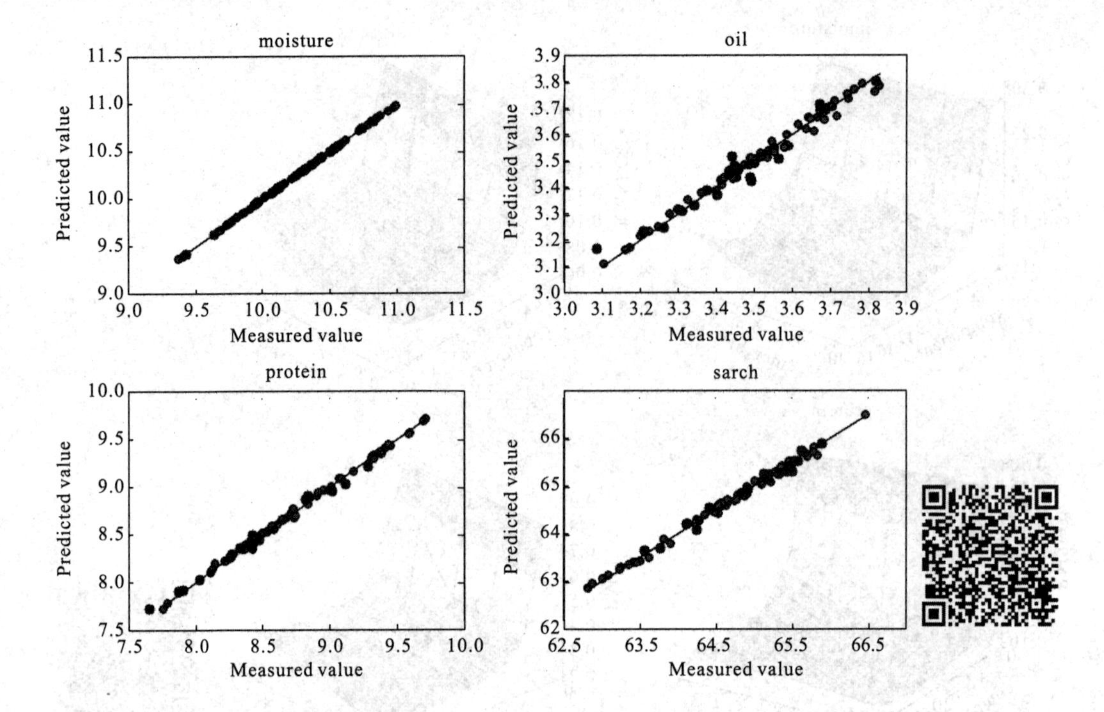

Fig. 4. 24 The actual value compared with the predicted value about the corn dataset. The blue and red dots represent the results for each sample in the train set and test set, respectively

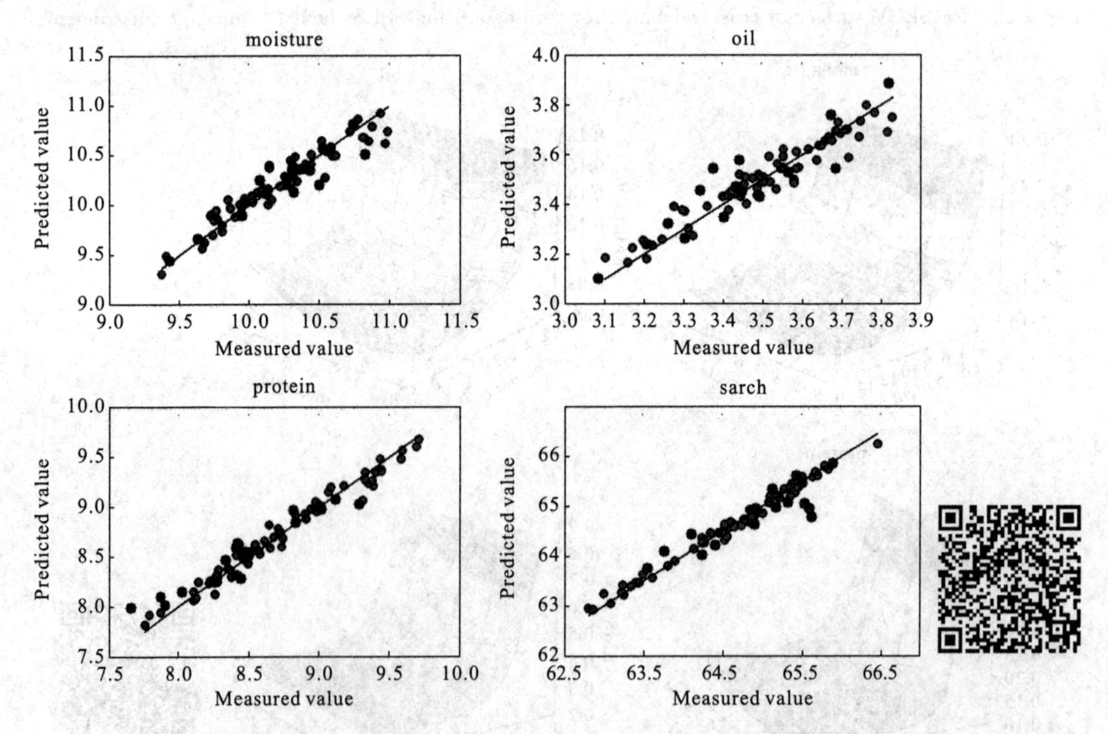

Fig. 4. 25 The actual value compared with the predicted value about the corn dataset. The blue and red dots represent the results for each sample in the train set and test set, respectively

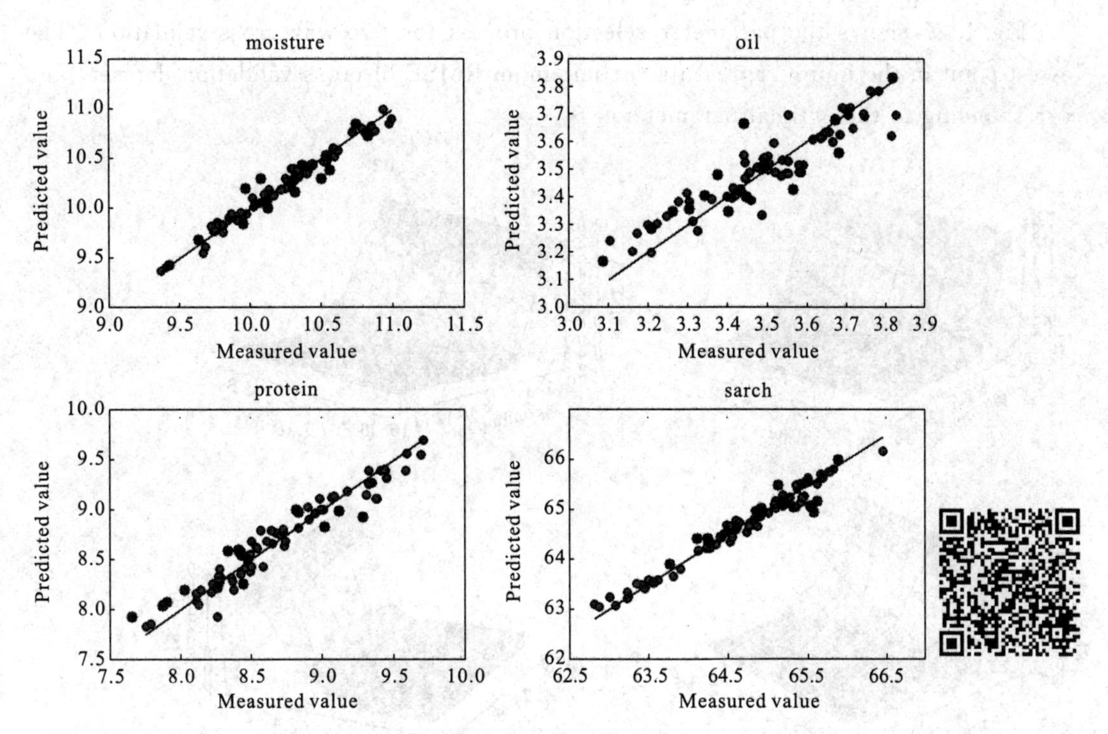

Fig. 4. 26 The actual value compared with the predicted value about the corn dataset. The blue and red dots represent the results for each sample in the train set and test set, respectively

4. 5. 4. 2 Wheat Dataset

Table 4. 5 also lists the RMSE of the different SPLS models for the cross-validation set and calibration set. The number of optimal latent variables and optimal intervals is determined according to the minimum RMSECV standard and is listed in Table 4. 5. It can be seen from that the RMSEPs obtained from SPLS regressions of the different instrument are significantly better than those produced from the regular PLS algorithm.

Table 4. 5 Summary of the PLS Models and Properties

Instrument	Reference Values	$RMSECV_{min}$	RMSEP	Latent Variable (LV)	Interval
B1	protein	0. 34014	0. 25129	9	4
B2	protein	0. 34483	0. 28004	9	4
B3	protein	0. 34914	0. 25440	8	4
A1	protein	0. 23726	0. 28232	7	6
A2	protein	0. 24144	0. 23129	7	6
A3	protein	0. 22690	0. 24835	7	6

RMSEP: Root Mean Square Error of test set

$RMSECV_{min}$: Minimum Root Mean Square Error of Cross-Validation

LV: The optimal number of latent variables is selected only when the lowest RMSECV

Interval: The optimal number of interval is selected only when the lowest RMSECV

Fig. 4. 27 shows the parameter selection process for two-way cross-validation. The lowest point in the figure represents the minimum RMSE of cross-validation dataset, and corresponding to the optimal parameters.

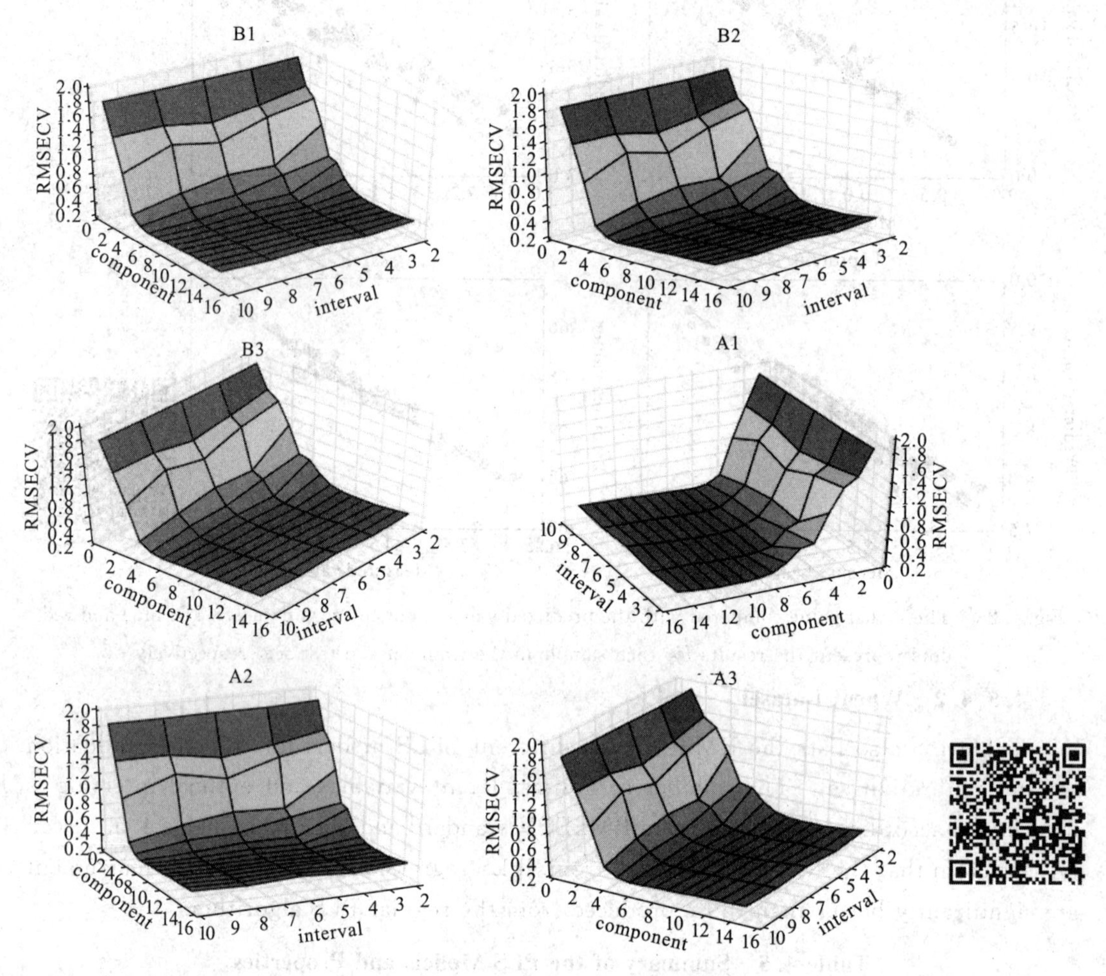

Fig. 4. 27　RMSECV surface of cross-validation using SPLS about the wheat dataset

In addition, Fig. 4. 28 shows the relationship between the measured values and predicted values of the training set and test set for different SPLS models. We can see that each instrument has a good fitting effect in wheat dataset.

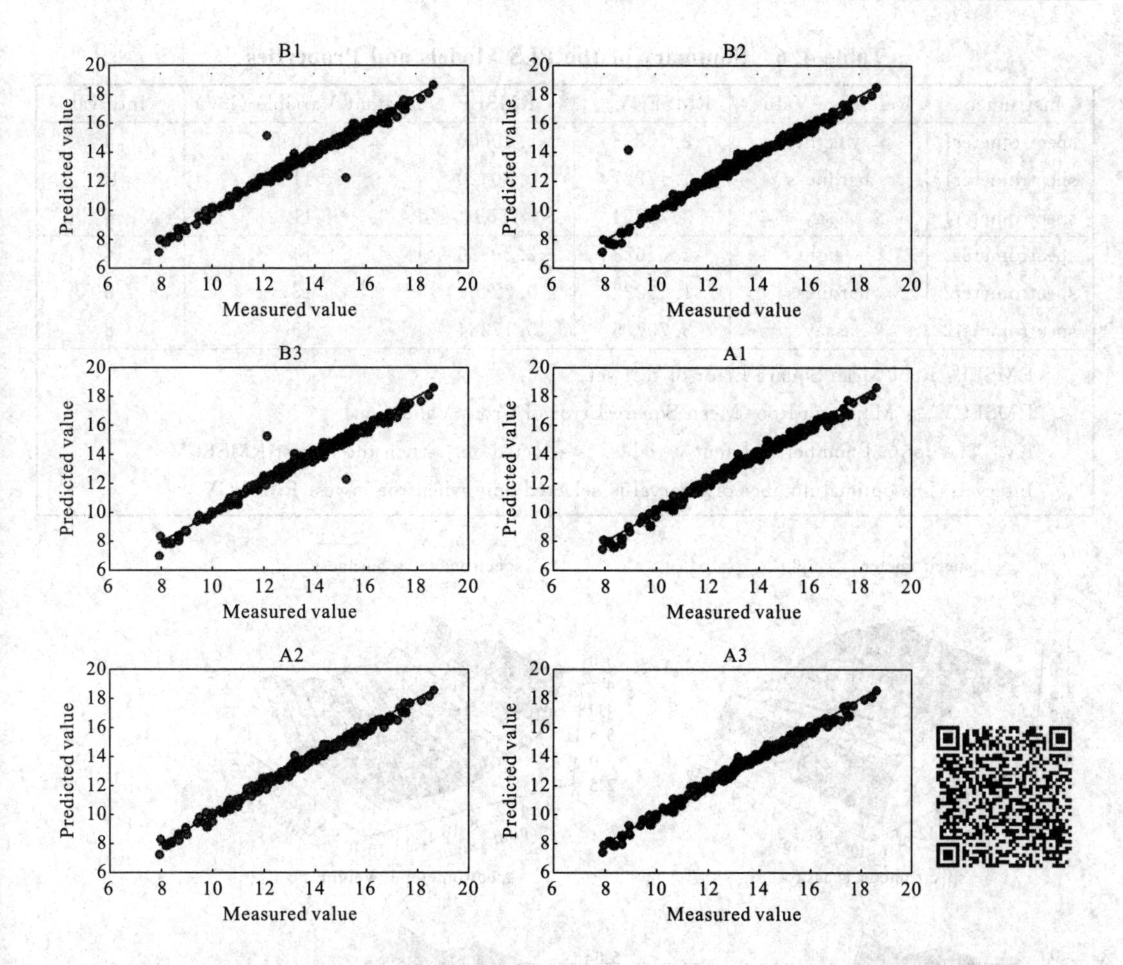

Fig. 4. 28 The actual value compared with the predicted value about the wheat dataset. The blue and red dots represent the results for each sample in the train set and test set, respectively

4. 5. 4. 3 Pharmaceutical Tablet Dataset

This section uses the pharmaceutical data set for experiments. Table 4. 6 lists the relevant attributes of different SPLS models. The number of optimal latent variables and optimal intervals was determined based on the minimum RMSECV criteria. The optimal number of latent variables is 10, 11, 11, 14, 13 and 15, respectively. The optimal number of intervals is 4, 10, 4, 4, 8 and 8, respectively.

Two-way cross-validation is used to select parameters, and find the optimal number of intervals and latent variables. Fig. 4. 29 shows the RMSECV error surface of the SPLS model for the pharmaceutical data set under different instruments. The lowest point of the error surface corresponds to the optimal parameter.

Table 4.6　Summary of the PLS Models and Properties

Instrument	Reference Values	$RMSECV_{min}$	RMSEP	Latent Variable (LV)	Interval
spectrometer1	weight	2.70823	2.70606	10	4
spectrometer1	hardness	1.17227	0.80420	11	10
spectrometer1	assay	3.72854	4.5534	11	4
spectrometer2	weight	2.62623	2.80926	14	4
spectrometer2	hardness	1.19022	0.72491	13	8
spectrometer2	assay	3.79775	5.17384	15	8

RMSEP: Root Mean Square Error of test set

$RMSECV_{min}$: Minimum Root Mean Square Error of Cross-Validation

LV: The optimal number of latent variables is selected only when the lowest RMSECV

Interval: The optimal number of interval is selected only when the lowest RMSECV

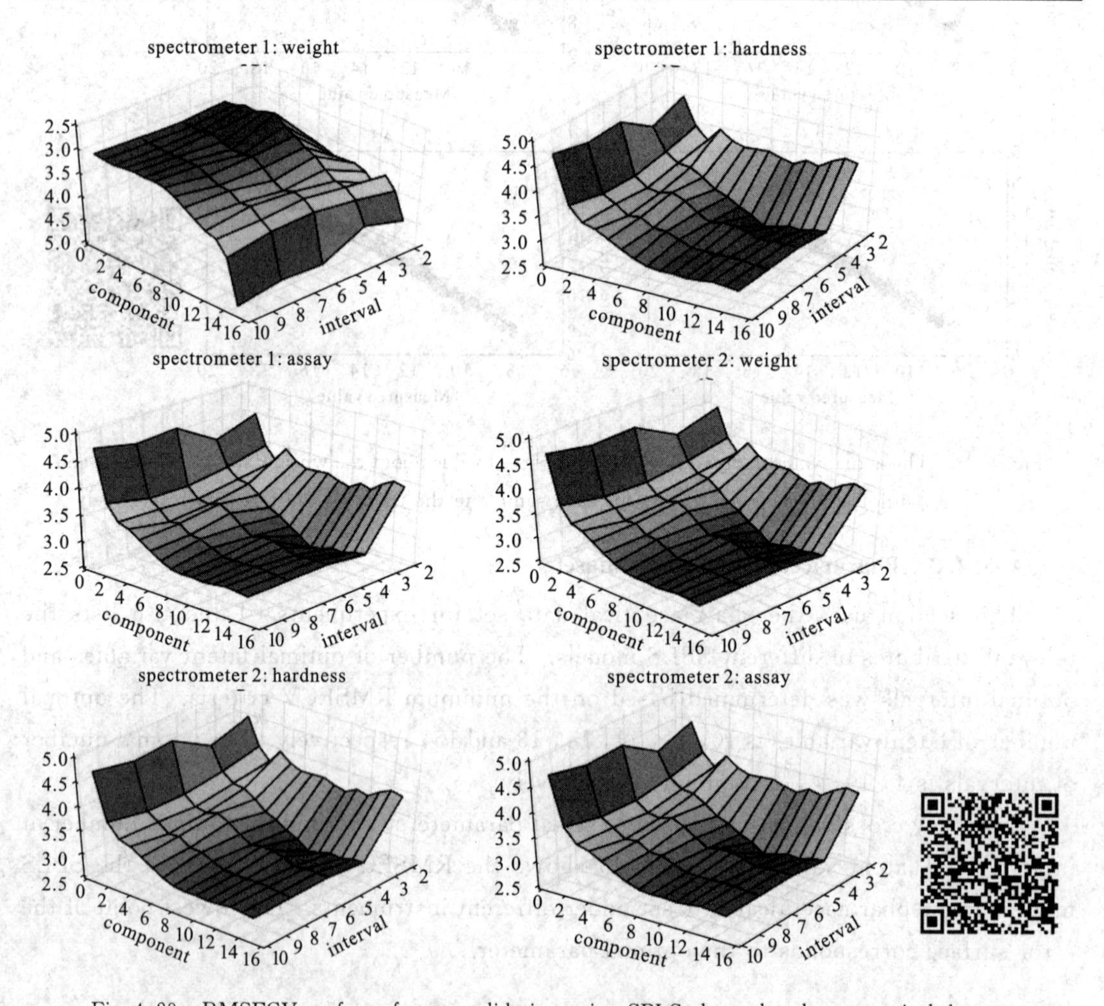

Fig. 4.29　RMSECV surface of cross-validation using SPLS about the pharmaceutical dataset

Fig. 4.30 shows a comparison of expected and predicted concentrations for different SPLS models of the pharmaceutical data set, respectively. It can be seen that a good correlation was found between the expected concentration and the predicted concentration, which confirms the good performance of the SPLS model.

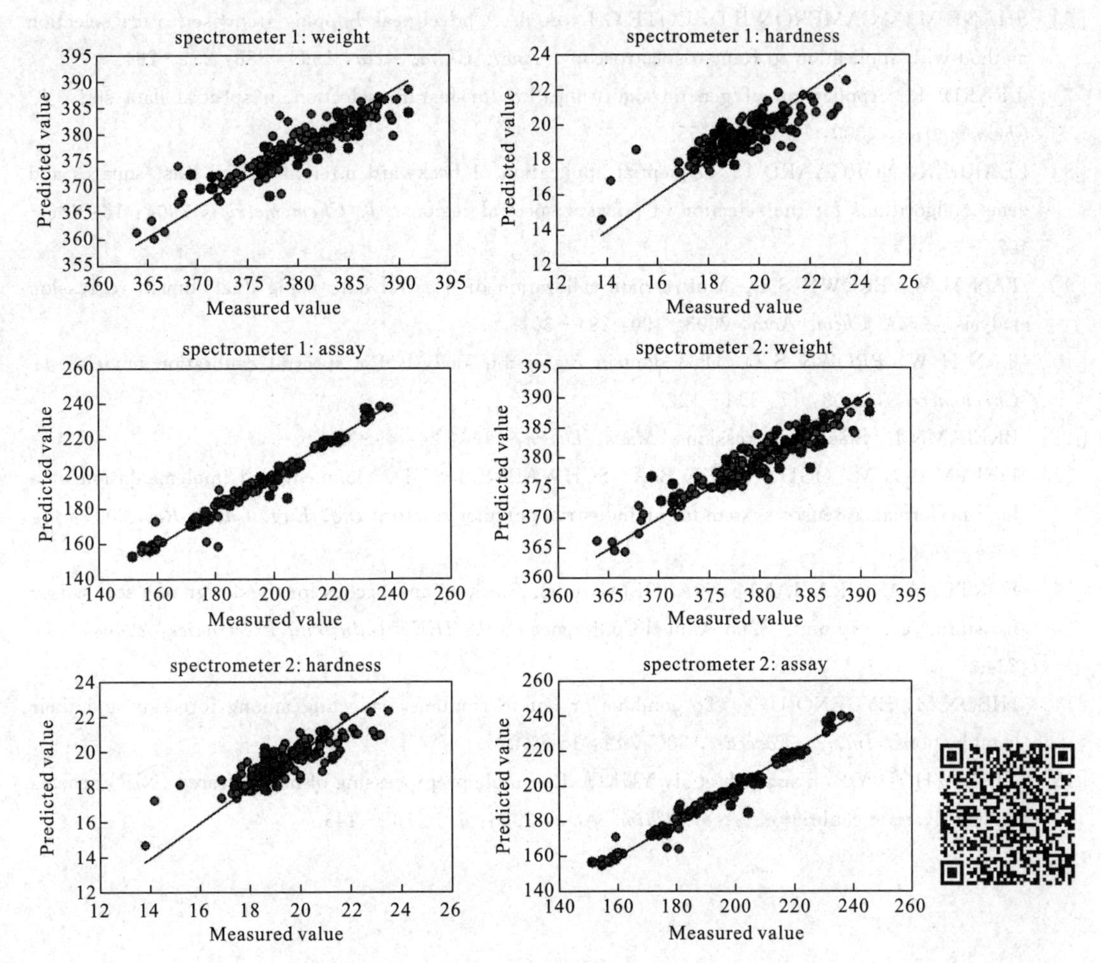

Fig. 4. 30 The actual value compared with the predicted value about the pharmaceutical dataset. The blue and red dots represent the results for each sample in the train set and test set, respectively

References

[1] XU L, SCHECHTER I. Wavelength selection for simultaneous spectroscopic analysis: experimental and theoretical study. Anal. Chem. 1996: 68: 2392 - 2400.

[2] SKULDSSON H A. Variable and subset selection in PLS regression. Chemom. Intell. Lab. Syst., Lab. Inf. Manag, 2001: 55: 23 - 38.

[3] NORGAARD L, SAUDALAND A, WANGNER J, et al. Interval partial least-squares regression (iPLS): a comparative chemometric study with an example from near-infrared spectroscopy. Appl. Spectrosc, 2000: 54: 413 - 419.

[4] JIANG J H, BERRY R J, SIESLER H W, et al. Wavelength interval selection in multicomponent spectral analysis by moving window partial least-squares regression with applications to mid-infrared and near infrared spectroscopic data. *Anal. Chem*, 2002: 74: 3555 - 3565.

[5] CENTNER V, MASSART D L, NOORDOE, et al. Elimination of uninformative variables for multivariate calibration. *Anal. Chem*, 1996: 68: 3851 - 3858.

[6] SHANE M C, CAMERON B D, COTE G L, et al. A novel peak-hopping stepwise feature selection method with application to Raman spectroscopy. *Anal. Chim. Acta*, 1999: 338: 251 - 264.

[7] LEARDI R. Application of genetic algorithm-PLS for feature selection in spectral data sets. *J. Chemometrics*, 2000: 14: 643 - 655.

[8] LEARDI R, NORGAARD L. Sequential application of backward interval partial least squares and genetic algorithms for the selection of relevant spectral regions. *J. Chemometrics*, 2004: 18: 486 - 497.

[9] TAN H W, BROWN S D. Multivariate calibration of spectral data using dual-domain regression analysis. *Anal. Chim. Acta*, 2003: 490: 291 - 301.

[10] TAN H W, BROWN S D. Dual domain regression analysis for spectral calibration models. *J. Chemometrics*, 2003: 17: 111 - 122.

[11] BREIAMN L. Stacked regressions. *Mach. Learn*, 1996: 24: 49 - 64.

[12] POTTMANN M, OGUNNAIKE B A, SCHWABER J S. Development and implementation of a high-performance sensor system for an industrial polymer reactor. *Ind. Eng. Chem. Res*, 2005: 44: 2606 - 2630.

[13] FORTUNA L, GRAZIANI S, NAPOLI G, et al. Stacking approaches for the design of a soft sensor for sulfur recovery unit. 32nd Annual Conference *of the IEEE Industrial Electronics*, 2006: 229 - 234.

[14] HIBON M, EYGENOIU T. To combine or not to combine, selecting among forecasts and their combinations. *Int. J. Forecast*, 2005, 21: 15 - 24.

[15] XU L, ZHOU Y P, Tang L J, et al, YURQ. Ensemble preprocessing of near-infrared (NIR) spectra for multivariate calibration. *Anal. Chim.* Acta, 2008: 616: 138 - 143.

Chapter 5 Regularization

A core problem in machine learning is to design algorithms that not only perform well on training data, but also generalize well on new inputs. In machine learning, many strategies are explicitly designed to reduce test errors (possibly at the expense of increasing training errors). These strategies are collectively referred to as regularization. Machine learners can use many different forms of regularization strategies. In fact, developing a more effective regularization strategy has become one of the main research works in this field.

5.1 Regularization

5.1.1 Classification

One use of regularization is in classification. Empirical learning of classifiers (from a finite data set) is always an underdetermined problem, because it attempts to infer a function of any $\boldsymbol{x}$ given only examples $\boldsymbol{x}_1, \boldsymbol{x}_2, \cdots, \boldsymbol{x}_n$.

A regularization term (or regularizer) $\boldsymbol{R}(f)$ is added to a loss function:

$$\min_f \sum_{i=1}^{n} V(f(x_i), y_i) + \lambda \boldsymbol{R}(f)$$

where V is an underlying loss function that describes the cost of predicting $f(x)$ when the label is $\boldsymbol{y}$, such as the square loss or hinge loss; and $\boldsymbol{\lambda}$ is a parameter which controls the importance of the regularization term. $\boldsymbol{R}(f)$ is typically chosen to impose a penalty on the complexity of f. Concrete notions of complexity used include restrictions for smoothness and bounds on the vector space norm.

A theoretical justification for regularization is that it attempts to impose Occam's razor on the solution. From a Bayesian point of view, many regularization techniques correspond to imposing certain prior distributions on model parameters.

Regularization can serve multiple purposes, including learning simpler models, inducing models to be sparse and introducing group structure (clarification needed) into the learning problem.

The same idea arose in many fields of science. For example, the least-squares method can be viewed as a simple form of regularization, if the regularization term $\lambda \boldsymbol{R}(f)$ is dropped from the optimization problem (i. e. $\lambda = 0$}) and the loss function summation of

$V(f(x_i), y_i)$ is the L_2 norm of the difference between the fitted and actual function values (i. e. $\| \boldsymbol{Y}-f(\boldsymbol{X}) \|_2$).

5.1.2 Tikhonov Regularization

When learning a linear function f, characterized by an unknown vector w such that $f(x)=w \cdot x$, the L_2-norm loss Tikhonov regularization. This is one of the most common forms of regularization. It is also known as ridge regression. It is expressed as:

$$\min_w \sum_{i=1}^{n} V(\hat{x}_i \cdot w, \hat{y}_i)+\lambda \| w \|_2^2$$

In the case of a general function, we take the norm of the function in itsreproducing kernel Hilbert space:

$$\min_w \sum_{i=1}^{n} V(f(\hat{x}_i), \hat{y}_i)+\lambda \| f \|_H^2$$

As the L_2 norm is differentiable, learning problems using Tikhonov regularization can be solved by gradient descent.

5.1.3 Regularizers for Sparsity

5.1.3.1 Proximal Methods

While the L_1 norm does not result in an NP-hard problem, the L_1 norm is convex but is not strictly differentiable due to the kink at $x=0$. Subgradient methods which rely on the subderivative can be used to solve L_1 regularized learning problems. However, faster convergence can be achieved through proximal methods.

For a problem $\min_{w \in H} F(w)+\boldsymbol{R}(w)$ such that F is convex, continuous, differentiable, with Lipschitz continuous gradient (such as the least squares loss function), and $\boldsymbol{R}$ is convex, continuous, and proper, then the proximal method to solve the problem is as follows. First define the proximal operator.

$$\operatorname{prox}_R(v)=\underset{w \in R^D}{\operatorname{argmin}}\left\{\boldsymbol{R}(w)+\frac{1}{2} \| w-v \|^2\right\}$$

and then iterate

$$w_{k+1}=\operatorname{prox}_{\gamma,R}(w_k-\gamma \nabla F(w_k))$$

The proximal method iteratively performs gradient descent and then projects the result back into the space permitted by $\boldsymbol{R}$.

When $\boldsymbol{R}$ is the L_1 regularizer, the proximal operator is equivalent to the soft-thresholding operator,

$$S_\lambda(v) f(n)=\begin{cases} v_i-\lambda, \text{ if } v_i>\lambda \\ 0, \text{ if } v_i \in [-\lambda, \lambda] \\ v_i+\lambda, \text{ if } v_i<-\lambda \end{cases}$$

This allows for efficient computation.

5.1.3.2 Group Sparsity without Overlaps

Groups of features can be regularized by a sparsity constraint, which can be useful for expressing certain prior knowledge into an optimization problem.

In the case of a linear model with non-overlapping known groups, a regularizer can be defined:

$$R(w)=\sum_{g=1}^{G}\|w_g\|_2,\ \text{where } \|w_g\|_2=\sqrt{\sum_{j=1}^{|G_g|}(w_g^j)^2}$$

This can be viewed as inducing a regularizer over the L_2 norm over members of each group followed by an L_1 norm over groups.

This can be solved by the proximal method, where the proximal operator is a block-wise soft-thresholding function:

$$(\text{prox}_{\lambda,R,g}(w_g^j))=\begin{cases}\left(w_g^j-\lambda\dfrac{w_g^j}{\|w_g\|_2}\right), & \text{if } \|w_g\|_2>\lambda\\ 0, & \text{if } \|w_g\|_2\in[-\lambda,\lambda]\\ \left(w_g^j+\lambda\dfrac{w_g^j}{\|w_g\|_2}\right), & \text{if } \|w_g\|_2<-\lambda\end{cases}$$

5.1.3.3 Group Sparsity with Overlaps

The algorithm described for group sparsity without overlaps can be applied to the case where groups do overlap, in certain situations. This will likely result in some groups with all zero elements, and other groups with some non-zero and some zero elements.

If it is desired to preserve the group structure, a new regularizer can be defined:

$$R(w)=\inf\left\{\sum_{g=1}^{G}\|w_g\|_2;w=\sum_{g=1}^{G}\overline{w}_g\right\}$$

For each w_g, $\overline{w}_g$ is defined as the vector such that the restriction of $\overline{w}_g$ to the group g equals w_g and all other entries of $\overline{w}_g$ are zero. The regularizer finds the optimal disintegration of w into parts. It can be viewed as duplicating all elements that exist in multiple groups. Learning problems with this regularizer can also be solved with the proximal method with a complication. The proximal operator cannot be computed in closed form, but can be effectively solved iteratively, inducing an inner iteration within the proximal method iteration.

5.1.4 Other Uses of Regularization in Statistics and Machine Learning

Bayesian learning methods make use of a prior probability that (usually) gives lower probability to more complex models. Well-known model selection techniques include the Akaike information criterion (AIC), minimum description length (MDL), and the Bayesian information criterion (BIC). Alternative methods of controlling overfitting not involving regularization include cross-validation.

Table 5.1 Examples of Applications of Different Methods of Regularization to the Linear Model

Model	Fit Measure	Entropy Measure
AIC(Akaike Information Criterion)/BIC(Bayesian Information Criterion)	$\lVert \boldsymbol{Y}-\boldsymbol{X\beta} \rVert_2$	$\lVert \boldsymbol{\beta} \rVert_0$
Ridge Regression	$\lVert \boldsymbol{Y}-\boldsymbol{X\beta} \rVert_2$	$\lVert \boldsymbol{\beta} \rVert_2$
Lasso	$\lVert \boldsymbol{Y}-\boldsymbol{X\beta} \rVert_2$	$\lVert \boldsymbol{\beta} \rVert_1$
Basis Pursuit Denoising	$\lVert \boldsymbol{Y}-\boldsymbol{X\beta} \rVert_2$	$\lambda \lVert \boldsymbol{\beta} \rVert_1$
Rudin-Osher-Fatemi model	$\lVert \boldsymbol{Y}-\boldsymbol{X\beta} \rVert_2$	$\lambda \lVert \nabla\boldsymbol{\beta} \rVert_1$
Potts Model	$\lVert \boldsymbol{Y}-\boldsymbol{X\beta} \rVert_2$	$\lambda \lVert \nabla\boldsymbol{\beta} \rVert_0$
RLAD(Regularized Least Absolute Deviations Regression)	$\lVert \boldsymbol{Y}-\boldsymbol{X\beta} \rVert_2$	$\lVert \boldsymbol{\beta} \rVert_1$
Dantzig Selector	$\lVert \boldsymbol{X}^{\mathrm{T}}(\boldsymbol{Y}-\boldsymbol{X\beta}) \rVert_2$	$\lVert \boldsymbol{\beta} \rVert_1$

5.2 Ridge Regression: Biased Estimation for Nonorthogonal Problems[1]

Consider the standard model for multiple linear regression, $\boldsymbol{Y}=\boldsymbol{X\beta}+\boldsymbol{\varepsilon}$, where it is assumed that $\boldsymbol{X}$ is $(n\times p)$ and of rank p, $\boldsymbol{\beta}$ is $(p\times 1)$ and unknown, $E[\varepsilon]=0$, and $E[\varepsilon\varepsilon']=\sigma^2\boldsymbol{I}_n$. If an observation on the factors is denoted by $\boldsymbol{x}_v=\{x_{1v}, x_{2v}, \cdots, x_{pv}\}$, the general form $\boldsymbol{X\beta}$ is $\{\sum_{i=1}^{p}\beta_i\theta_i(\boldsymbol{x}_v)\}$ where the θ_i are functions free of unknown parameters.

The usual estimation procedure for the unknown $\boldsymbol{\beta}$ is Gauss-Markov-linear functions of $\boldsymbol{Y}=\{y_v\}$ that are unbiased and have minimum variance. This estimation procedure is a good one if $\boldsymbol{X}'\boldsymbol{X}$, when in the form of a correlation matrix, is nearly a unit matrix. However, if $\boldsymbol{X}'\boldsymbol{X}$ is not nearly a unit matrix, the least squares estimates are sensitive to a number of "errors". The results of these errors are critical when the specification is that $\boldsymbol{X\beta}$ is a true model. Then the least squares estimates often do not make sense when put into the context of the physics, chemistry, and engineering of the process which is generating the data. In such cases, one is forced to treat the estimated predicting function as a black box or to drop factors to destroy the correlation bonds among the $\boldsymbol{X}_i$ used to form $\boldsymbol{X}'\boldsymbol{X}$. Both these alternatives are unsatisfactory if the original intent was to use the estimated predictor for control and optimization. If one treats the result as a black box, he must caution the user of the model not to take partial derivatives (a useless caution in practice), and in the other case, he is left with a set of dangling controllables or observables.

Estimation based on the matrix $[\boldsymbol{X}'\boldsymbol{X}+k\boldsymbol{I}_p]$, $k\geqslant 0$ rather than on $\boldsymbol{X}'\boldsymbol{X}$ has been found to be a procedure that can be used to help circumvent many of the difficulties associated with the usual least squares estimates. In particular, the procedure can be used to portray the sensitivity of the estimates to the particular set of data being used, and it can be used to obtain a point estimate with a smaller mean square error.

5.2.1 Properties of Best Linear Unbiased Estimation

Using unbiased linear estimation with minimum variance or maximum likelihood estimation when the random vector ε, is normal gives

$$\hat{\boldsymbol{\beta}}=(\boldsymbol{X}'\boldsymbol{X})^{-1}\boldsymbol{X}'\boldsymbol{Y} \tag{5.2.1}$$

as an estimate of $\boldsymbol{\beta}$ and this gives the minimum sum of squares of the residuals:

$$\phi(\hat{\boldsymbol{\beta}})=(\boldsymbol{Y}-\boldsymbol{X}\hat{\boldsymbol{\beta}})'(\boldsymbol{Y}-\boldsymbol{X}\hat{\boldsymbol{\beta}}) \tag{5.2.2}$$

The properties of $\hat{\boldsymbol{\beta}}$ are well known (Scott 1966[2]). Here the concern is primarily with cases for which $\boldsymbol{X}'\boldsymbol{X}$ is not nearly a unit matrix (unless specified otherwise, the model is formulated to give an $\boldsymbol{X}'\boldsymbol{X}$ in correlation form). To demonstrate the effects of this condition on the estimation of $\boldsymbol{\beta}$, consider two properties of $\hat{\boldsymbol{\beta}}$—its variance-covariance matrix and its distance from its expected value.

(i) $\mathrm{VAR}(\hat{\boldsymbol{\beta}})=\sigma^2(\boldsymbol{X}'\boldsymbol{X})^{-1}$ (5.2.3)

(ii) $L_1\equiv$ Distance from $\hat{\boldsymbol{\beta}}$ to $\boldsymbol{\beta}$

$$L_1^2=(\hat{\boldsymbol{\beta}}-\boldsymbol{\beta})'(\hat{\boldsymbol{\beta}}-\boldsymbol{\beta}) \tag{5.2.4}$$

$$E[L_2^2]=\sigma^2\,\mathrm{Trace}(\boldsymbol{X}'\boldsymbol{X})^{-1} \tag{5.2.5}$$

or equivalently

$$E[\hat{\boldsymbol{\beta}}'\hat{\boldsymbol{\beta}}]=\boldsymbol{\beta}'\boldsymbol{\beta}+\sigma^2\,\mathrm{Trace}(\boldsymbol{X}'\boldsymbol{X})^{-1} \tag{5.2.5a}$$

When the error ε is normally distributed, then

$$\mathrm{VAR}[L_1^2]=2\sigma^4\,\mathrm{Trace}(\boldsymbol{X}'\boldsymbol{X})^{-2} \tag{5.2.6}$$

These related properties show the uncertainty in $\hat{\boldsymbol{\beta}}$ when $\boldsymbol{X}'\boldsymbol{X}$ moves from a unit matrix to an ill-conditioned one. If the eigenvalues of $\boldsymbol{X}'\boldsymbol{X}$ are denoted by

$$\lambda_{\max}=\lambda_1\geqslant\lambda_2\geqslant\cdots\geqslant\lambda_p=\lambda_{\min}>0 \tag{5.2.7}$$

then the average value of the squared distance from $\hat{\boldsymbol{\beta}}$ to $\boldsymbol{\beta}$ is given by

$$E[L_1^2]=\sigma^2\sum_{i=1}^{p}\frac{1}{\lambda} \tag{5.2.8}$$

and the variance when the error is normal is given by

$$\mathrm{VAR}[L_1^2]=2\sigma^4\sum\left(\frac{1}{\lambda_i}\right)^2 \tag{5.2.9}$$

Lower bounds for the average and variance are $\sigma^2/\lambda_{\min}$ and $2\sigma^4/\lambda_{\min}^2$, respectively. Hence, if the shape of the factor space is such that reasonable data collection results in an $\boldsymbol{X}'\boldsymbol{X}$ with one or more small eigenvalues, the distance from $\hat{\boldsymbol{\beta}}$ to $\boldsymbol{\beta}$ will tend to be large.

Estimated coefficients, $\hat{\boldsymbol{\beta}}$, that are large in absolute value have been observed by all who have tackled live nonorthogonal data problems.

The least squares estimate equation (5.2.1) suffers from the deficiency of mathematical optimization techniques that give point estimates; the estimation procedure does not have built into it a method for portraying the sensitivity of the solution equation (5.2.1) to the optimization criterion equation (5.2.2). The procedures to be discussed in the sections to follow portray the sensitivity of the solutions and utilize nonsensitivity as an aid to analysis.

5.2.2 Ridge Regression

A. E. Hoerl first suggested in 1962 (Hoerl 1962[3]; Hoerl and Kennard 1968[4]) that to control the inflation and general instability associated with the least squares estimates, one can use

$$\hat{\boldsymbol{\beta}}^* = [\boldsymbol{X}'\boldsymbol{X}+k\boldsymbol{I}]^{-1}\boldsymbol{X}'\boldsymbol{Y},\ k\geqslant 0 \quad (5.2.10)$$

$$=\boldsymbol{W}\boldsymbol{X}'\boldsymbol{Y} \quad (5.2.11)$$

The family of estimates given by $k\geqslant 0$ has many mathematical similarities with the portrayal of quadratic response functions (Hoerl 1964[5]). For this reason, estimation and analysis built around equation (5.2.10) has been labeled "ridge regression". The relationship of a ridge estimate to an ordinary estimate is given by the alternative form

$$\hat{\boldsymbol{\beta}}^* = [\boldsymbol{I}_p+k(\boldsymbol{X}'\boldsymbol{X})^{-1}]^{-1}\hat{\boldsymbol{\beta}} \quad (5.2.12)$$

$$=\boldsymbol{Z}\hat{\boldsymbol{\beta}} \quad (5.2.13)$$

This relationship will be explored further in subsequent sections. Some properties of $\hat{\boldsymbol{\beta}}^*$, $\boldsymbol{W}$, and $\boldsymbol{Z}$ that will be used are:

(i) Let $\xi_i(\boldsymbol{W})$ and $\xi_i(\boldsymbol{Z})$ be the eigenvalues of $\boldsymbol{W}$ and $\boldsymbol{Z}$, respectively. Then

$$\xi_i(\boldsymbol{W})=\frac{1}{\lambda_i+k} \quad (5.2.14)$$

$$\xi_i(\boldsymbol{Z})=\frac{\lambda_i}{\lambda_i+k} \quad (5.2.15)$$

where λ_i, are the eigenvalues of $\boldsymbol{X}'\boldsymbol{X}$. The solution of the characteristic equations $|\boldsymbol{W}-\xi\boldsymbol{I}|$ and $|\boldsymbol{Z}-\xi\boldsymbol{I}|=0$.

(ii) $\boldsymbol{Z}=\boldsymbol{I}-k(\boldsymbol{X}'\boldsymbol{X}+k\boldsymbol{I})^{-1}=\boldsymbol{I}-k\boldsymbol{W}$ (5.2.16)

The relationship is readily verified by writing $\boldsymbol{Z}$ in the alternative form $\boldsymbol{Z}=k(\boldsymbol{X}'\boldsymbol{X}+k\boldsymbol{L})^{-1}\boldsymbol{X}'\boldsymbol{X}=\boldsymbol{W}\boldsymbol{X}'\boldsymbol{X}$ and multiplying both sides of equation (5.2.16) on the left by $\boldsymbol{W}^{-1}$.

(iii) $\hat{\boldsymbol{\beta}}^*$ for $k\neq 0$ is shorter than $\hat{\boldsymbol{\beta}}$, i.e.,

$$(\hat{\boldsymbol{\beta}}^*)'(\hat{\boldsymbol{\beta}}^*)<\hat{\boldsymbol{\beta}}'\hat{\boldsymbol{\beta}} \quad (5.2.17)$$

By definition $\hat{\boldsymbol{\beta}}^*=\boldsymbol{Z}\hat{\boldsymbol{\beta}}$. From its definition and the assumptions on $\boldsymbol{X}'\boldsymbol{X}$, $\boldsymbol{Z}$ is clearly symmetric positive definite. Then the following relation holds (Sheffe 1960[6]):

$$(\hat{\boldsymbol{\beta}}^*)'(\hat{\boldsymbol{\beta}}^*) < \xi_{\max}^2(\boldsymbol{Z})\hat{\boldsymbol{\beta}}'\hat{\boldsymbol{\beta}} \tag{5.2.18}$$

But $\xi_{\max}(\boldsymbol{Z}) = \lambda_1/(\lambda_1 + k)$ where λ_1 is the largest eigenvalue of $\boldsymbol{X}'\boldsymbol{X}$ and equation (5.2.18) is established. From equation (5.2.16) and (5.2.15) it is seen that $\boldsymbol{Z}(0)=\boldsymbol{I}$ and that $\boldsymbol{Z}$ approaches $\boldsymbol{0}$ as $k \to \infty$.

For an estimate $\hat{\boldsymbol{\beta}}^*$ the residual sum of squares is

$$\psi^*(k) = (\boldsymbol{Y} - \boldsymbol{X}\hat{\boldsymbol{\beta}}^*)'(\boldsymbol{Y} - \boldsymbol{X}\hat{\boldsymbol{\beta}}^*) \tag{5.2.19}$$

which can be written in the form

$$\psi^*(k) = \boldsymbol{Y}'\boldsymbol{Y} - (\hat{\boldsymbol{\beta}}^*)'\boldsymbol{X}'\boldsymbol{Y} - k(\hat{\boldsymbol{\beta}}^*)'(\hat{\boldsymbol{\beta}}^*) \tag{5.2.20}$$

The expression shows that $\psi^*(k)$ is the total sum of squares less the "regression" sum of squares for $\hat{\boldsymbol{\beta}}^*$ with a modification depending upon the squared length of $\hat{\boldsymbol{\beta}}^*$.

5.2.3 The Ridge Trace

5.2.3.1 Definition of the Ridge Trace

When $\boldsymbol{X}'\boldsymbol{X}$ deviates considerably from a unit matrix, that is, when it has small eigenvalues, and show that the probability can be small that $\hat{\boldsymbol{\beta}}$ will be close to $\boldsymbol{\beta}$. In any except the smallest problems, it is difficult to untangle the relationships among the factors if one is confined to an inspection of the simple correlations that are the elements of $\boldsymbol{X}'\boldsymbol{X}$. That such untangling is a problem is reflected in the "automatic" procedures that have been put forward to reduce the dimensionality of the factor space or to select some "best" subset of the predictors. These automatic procedures include regression using the factors obtained from a coordinate transformation using the principal components of $\boldsymbol{X}'\boldsymbol{X}$, stepwise regression, computation of all 2^p regressions, and some subset of all regressions using fractional factorials or a branch and bound technique (Beale, Kendall, and Mann 1967[7]; Efroyamson 1960[8]; Garside 1965[9]; Gorman and Toman 1966[10]; Hocking and Leslie 1967[11]; Jeffers 1967[12]; Scott 1966[2]). However, with the occasional exception of principal components, these methods don't really give an insight into the structure of the factor space and the sensitivity of the results to the particular set of data at hand. But by computing $\hat{\boldsymbol{\beta}}^*(k)$ and $\phi^*(k)$ for a set of values of k, such insight can be obtained. A detailed study of two nonorthogonal problems and the conclusions that can be drawn from their ridge traces is given in James and Stein (1961[13]).

5.2.3.2 Characterization of the Ridge Trace

Let $\boldsymbol{B}$ be any estimate of the vector $\boldsymbol{\beta}$. Then the residual sums of squares can be written as

$$\begin{aligned}
\phi &= (\boldsymbol{Y} - \boldsymbol{X}\boldsymbol{\beta})'(\boldsymbol{Y} - \boldsymbol{X}\boldsymbol{\beta}) \\
&= (\boldsymbol{Y} - \boldsymbol{X}\hat{\boldsymbol{\beta}})'(\boldsymbol{Y} - \boldsymbol{X}\hat{\boldsymbol{\beta}}) + (\boldsymbol{B} - \hat{\boldsymbol{\beta}})'\boldsymbol{X}'\boldsymbol{X}(\boldsymbol{B} - \hat{\boldsymbol{\beta}}) \\
&= \phi_{\min} + \phi(\boldsymbol{B})
\end{aligned} \tag{5.2.21}$$

Contours of constant ϕ are the surfaces of hyperellipsoids centered at $\hat{\boldsymbol{\beta}}$, the ordinary least squares estimate of $\boldsymbol{\beta}$. The value of ϕ is the minimum value, $\phi_{\min}$, plus the value of the quadratic form in $(\boldsymbol{B}-\hat{\boldsymbol{\beta}})$. There is a continuum of values of $\boldsymbol{B}_0$ that will satisfy the relationship $\phi=\phi_{\min}+\phi_0$ where $\phi_0>0$ is a fixed increment. However, the relationships in Section 5.2.2 show that on the average the distance from $\hat{\boldsymbol{\beta}}$ to $\boldsymbol{\beta}$ will tend to be large if there is a small eigenvalue of $\boldsymbol{X}'\boldsymbol{X}$. In particular, the worse the conditioning of $\boldsymbol{X}'\boldsymbol{X}$, the more $\hat{\boldsymbol{\beta}}$ can be expected to be too long. On the other hand, the worse the conditioning, the further one can move from $\hat{\boldsymbol{\beta}}$ without an appreciable increase in the residual sums of squares. In view of equation (5.2.5a), it seems reasonable that if one moves away from the minimum sum of squares point, the movement should be in a direction which will shorten the length of the regression vector.

The ridge trace can be shown to be following a path through the sums of squares surface so that for a fixed ϕ a single value of $\boldsymbol{B}$ is chosen and that is the one with minimum length. This can be stated precisely as follows:

$$\text{Minimize } \boldsymbol{B}'\boldsymbol{B}$$

$$\text{subject to } (\boldsymbol{B}-\hat{\boldsymbol{\beta}})'\boldsymbol{X}'\boldsymbol{X}(\boldsymbol{B}-\hat{\boldsymbol{\beta}})=\phi_0 \tag{5.2.22}$$

As a Lagrangian problem this isminimize

$$F=\boldsymbol{B}'\boldsymbol{B}+\frac{1}{k}[(\boldsymbol{B}-\hat{\boldsymbol{\beta}})'\boldsymbol{X}'\boldsymbol{X}(\boldsymbol{B}-\hat{\boldsymbol{\beta}})'-\phi_0] \tag{5.2.23}$$

where $(1/k)$ is the multiplier. Then

$$\frac{\partial F}{\partial \boldsymbol{B}}=2\boldsymbol{B}+\frac{1}{k}[2(\boldsymbol{X}'\boldsymbol{X})\boldsymbol{B}-2(\boldsymbol{X}'\boldsymbol{X})\hat{\boldsymbol{\beta}}]=0 \tag{5.2.24}$$

This reduces to

$$\boldsymbol{B}=\hat{\boldsymbol{\beta}}^*=[\boldsymbol{X}'\boldsymbol{X}+kI]^{-1}\boldsymbol{X}'\boldsymbol{Y} \tag{5.2.25}$$

where k is chosen to satisfy the restraint. This is the ridge estimator. Of course, in practice it is easier to choose a $k\geqslant 0$ and then compute ϕ_0. In terms of $\hat{\boldsymbol{\beta}}^*$ the residual sum of squares becomes

$$\phi^*(k)=(\boldsymbol{Y}-\boldsymbol{X}\hat{\boldsymbol{\beta}}^*)'(\boldsymbol{Y}-\boldsymbol{X}\hat{\boldsymbol{\beta}}^*)=\phi_{\min}+k^2\hat{\boldsymbol{\beta}}(\boldsymbol{X}'\boldsymbol{X})^{-1}\hat{\boldsymbol{\beta}}^* \tag{5.2.26}$$

A completely equivalent statement of the path is this: If the squared length of the regression vector $\boldsymbol{B}$ is fixed at R^2, then $\hat{\boldsymbol{\beta}}$ is the value of $\boldsymbol{B}$ that gives a minimum sum of squares. That is, $\hat{\boldsymbol{\beta}}$ is the value of $\boldsymbol{B}$ that minimizes the function

$$F_1=(\boldsymbol{Y}-\boldsymbol{X}\hat{\boldsymbol{\beta}})'(\boldsymbol{Y}-\boldsymbol{X}\hat{\boldsymbol{\beta}}^*)+\frac{1}{k}(\boldsymbol{B}'\boldsymbol{B}-R^2) \tag{5.2.27}$$

5.2.3.3 Likelihood Characterization of the Ridge Trace

Using the assumption that the error vector is Normal $(0,\sigma^2\boldsymbol{I}_n)$ the likelihood function is

$$(2\pi\sigma^2)^{-n/2}\exp\left\{-\frac{1}{2\sigma^2}(\boldsymbol{Y}-\boldsymbol{X\beta})'(\boldsymbol{Y}-\boldsymbol{X\beta})\right\} \quad (5.2.28)$$

The kernel of this function is the quadratic form in the exponential which can be written in the form

$$(\boldsymbol{Y}-\boldsymbol{X\beta})'(\boldsymbol{Y}-\boldsymbol{X\beta})=(\boldsymbol{Y}-\boldsymbol{X}\hat{\boldsymbol{\beta}})'(\boldsymbol{Y}-\boldsymbol{X}\hat{\boldsymbol{\beta}})+(\boldsymbol{\beta}-\hat{\boldsymbol{\beta}})'\boldsymbol{X}'\boldsymbol{X}(\boldsymbol{\beta}-\hat{\boldsymbol{\beta}}) \quad (5.2.29)$$

With equation (5.2.21), this shows that an increase in the residual sum of squares is equivalent to a decrease in the value of the likelihood function. So the contours of equal likelihood also lie on the surface of hyperellipsoids centered at $\boldsymbol{\beta}$.

The ridge trace can thereby be interpreted as a path through the likelihood space, and the question arises as why this particular path can be of special interest. The reasoning is the same as for the sum of squares. Although long vectors give the same likelihood values as shorter vectors, they will not always have equal physical meaning. Implied is a restraint on the possible values of $\hat{\boldsymbol{\beta}}$ that is not made explicit in the formulation of the general linear model given in the Introduction. This implication is discussed further in the sections that follow.

5.2.4 Mean Square Error Properties of Ridge Regression

5.2.4.1 Variance and Bias of a Ridge Estimator

To look at $\hat{\boldsymbol{\beta}}^*$ from the point of view of mean square error it is necessary to obtain an expression for $E[L_1^2(k)]$. Straightforward application of the expectation operator and gives the following:

$$E[L_1^2(k)]=E[(\hat{\boldsymbol{\beta}}^*-\boldsymbol{\beta})'(\hat{\boldsymbol{\beta}}^*-\boldsymbol{\beta})] \quad (5.2.30)$$

$$=E[(\hat{\boldsymbol{\beta}}-\boldsymbol{\beta})'\boldsymbol{Z}'\boldsymbol{Z}(\hat{\boldsymbol{\beta}}-\boldsymbol{\beta})]-(\boldsymbol{Z\beta}-\boldsymbol{\beta})'(\boldsymbol{Z\beta}-\boldsymbol{\beta})$$

$$=\sigma^2\mathrm{Trace}(\boldsymbol{X}'\boldsymbol{X})^{-1}\boldsymbol{Z}'\boldsymbol{Z}+\boldsymbol{\beta}'(\boldsymbol{Z}-\boldsymbol{I})'(\boldsymbol{Z}-\boldsymbol{I})\boldsymbol{\beta} \quad (5.2.31)$$

$$=\sigma^2[\mathrm{Trace}(\boldsymbol{X}'\boldsymbol{X}+k\boldsymbol{I})^{-1}-k\mathrm{Trace}(\boldsymbol{X}'\boldsymbol{X}+k\boldsymbol{I})^{-2}]+k^2\boldsymbol{\beta}'(\boldsymbol{X}'\boldsymbol{X}+k\boldsymbol{I})^{-2}\boldsymbol{\beta} \quad (5.2.32)$$

$$=\sigma^2\sum_{i=1}^{p}\frac{\lambda_i}{(\lambda_i+k)^2}+k^2\boldsymbol{\beta}'(\boldsymbol{X}'\boldsymbol{X}+k\boldsymbol{I})^{-2}\boldsymbol{\beta} \quad (5.2.33)$$

$$=\gamma_1(k)+\gamma_2(k) \quad (5.2.34)$$

The meanings of the two elements of the decomposition, $\gamma_1(k)$ and $\gamma_2(k)$, are readily established. The second element, $\gamma_2(k)$, is the squared distance from $\boldsymbol{Z\beta}$ to $\boldsymbol{\beta}$. It will be zero when $k=0$, since $\boldsymbol{Z}$ is then equal to $\boldsymbol{I}$. Thus, $\gamma_2(k)$ can be considered the square of a bias introduced when $\hat{\boldsymbol{\beta}}^*$ is used rather than $\hat{\boldsymbol{\beta}}$. The first term, $\gamma_1(k)$, can be shown to be the sum of the variances (total variance) of the parameter estimates. In terms of the random variable $\boldsymbol{Y}$,

$$\hat{\boldsymbol{\beta}}^*=\boldsymbol{Z}\hat{\boldsymbol{\beta}}(\boldsymbol{X}'\boldsymbol{X})^{-1}\boldsymbol{X}'\boldsymbol{Y} \quad (5.2.35)$$

Then

$$\mathrm{VAR}[\hat{\boldsymbol{\beta}}^*]=\boldsymbol{Z}(\boldsymbol{X}'\boldsymbol{X})^{-1}\boldsymbol{X}'\mathrm{VAR}[\boldsymbol{Y}]\boldsymbol{X}(\boldsymbol{X}'\boldsymbol{X})^{-1}\boldsymbol{Z}$$

$$=\sigma^2\boldsymbol{Z}(\boldsymbol{X}'\boldsymbol{X})^{-1}\boldsymbol{Z} \quad (5.2.36)$$

The sum of the variances of all the $\hat{\boldsymbol{\beta}}$ is the sum of the diagonal elements of equation (5.2.36).

Fig. 5.1 shows in qualitative form the relationship between the variances, the squared bias, and the parameter k. The total variance decreases as k increases, while the squared bias increases with $\boldsymbol{k}$. As is indicated by the dotted line, which is the sum of $\gamma_1(k)$ and $\gamma_2(k)$ and thus is $E[L_1^2(k)]$, the possibility exists that there are values of k (admissible values) for which the mean square error is less for $\hat{\boldsymbol{\beta}}^*$ than it is for the usual solution $\hat{\boldsymbol{\beta}}$. This possibility is supported by the mathematical properties of $\gamma_1(k)$ and $\gamma_2(k)$. The function $\gamma_1(k)$ is a monotonic decreasing function of k, while $\gamma_2(k)$ is monotonic increasing. However, the most significant feature is the value of the derivative of each function in the neighborhood of the origin. These derivatives are:

$$\lim_{k\to 0^+}\left(\frac{\mathrm{d}\gamma_1}{\mathrm{d}k}\right)=-2\sigma^2\sum\left(\frac{1}{\lambda_i^2}\right) \tag{5.2.37}$$

$$\lim_{k\to 0^+}\left(\frac{\mathrm{d}\gamma_2}{\mathrm{d}k}\right)=0 \tag{5.2.38}$$

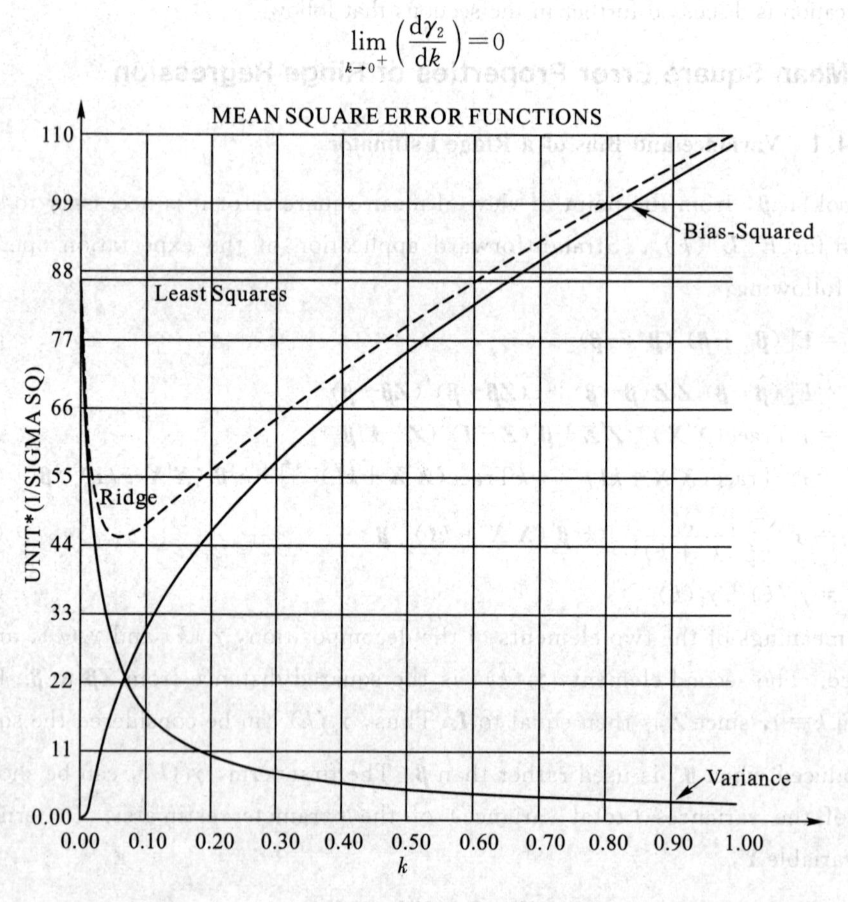

Fig. 5.1 In qualitative form the relationship between the variances, the squared bias, and the parameter k

Thus, $\gamma_1(k)$ has a negative derivative which approaches $-2p\sigma^2$ as $k\to 0^+$ for an orthogonal $\boldsymbol{X}'\boldsymbol{X}$ and approaches $-\infty$ as $\boldsymbol{X}'\boldsymbol{X}$ becomes ill-conditioned and $\lambda_p\to 0$. On the other hand, as $k\to 0^+$, shows that $\gamma_2(k)$ is flat and zero at the origin. These properties lead to the conclusion that it is possible to move to $k>0$, take a little bias, and substantially reduce the variance, thereby improving the mean square error of estimation and prediction. An existence theorem to validate this conclusion is given in Section 5.2.4.2.

5.2.4.2 Theorems on the Mean Square Function

Theorem 1

The total variance $\gamma_1(k)$ is a continuous, monotonically decreasing function of k.

Corollary 1

The first derivative with respect to k of the total variance $\gamma_1'(k)$, approaches $-\infty$ as $k \to 0^+$ and $\lambda_p \to 0$.

Both the theorem and the corollary are readily proved by use of $\gamma_1(k)$ and its derivative expressed in terms of λ_i.

Theorem 2

The squared bias $\gamma_2(k)$ is a continuous, monotonically increasing function of k.

Proof

From $\gamma_2(k)=\sigma^2\sum_{i=1}^{p}\frac{\lambda^i}{(\lambda_i+k)^2}+k^2\boldsymbol{\beta}'(\boldsymbol{X}'\boldsymbol{X}+k\boldsymbol{I})^{-2}\boldsymbol{\beta}$.

Corollary 2

The first derivative of the total variance, $\gamma_1'(k)$, approaches $-\infty$ as $k \to 0^+$ and the matrix $\boldsymbol{X}'\boldsymbol{X}$ becomes singular.

Both the theorem and the corollary are readily proved by use of $\gamma_1(k)$ and its derivative expressed in terms of λ_i

If $\boldsymbol{\Lambda}$ is the matrix of eigenvalues of $\boldsymbol{X}'\boldsymbol{X}$ and $\boldsymbol{P}$ is the orthogonal transformation such that $\boldsymbol{X}'\boldsymbol{X}=\boldsymbol{P}'\boldsymbol{\Lambda}\boldsymbol{P}$, then

$$\gamma_2(k)=\frac{k^2\sum_{i=1}^{p}\alpha_i^2}{(\lambda_i+k)^2} \tag{5.2.39}$$

$$\text{where } \alpha=\boldsymbol{p\beta} \tag{5.2.40}$$

Since λ_i for all i and $k>0$, each element (λ_i+k) is positive and there are no singularities in the sum. Clearly, $\gamma_2(0)=0$. Then $\gamma_2(k)$ is a continuous function for $k \geqslant 0$. For $k>0$ can be written as

$$\gamma_2(k)=\frac{\sum_{i=1}^{p}\alpha_i^2}{[1+(\lambda_i+k)]^2} \tag{5.2.41}$$

Since $\lambda_i>0$ for all i, the functions λ_i/k are clearly monotone decreasing for increasing k and each term of $\gamma_2(k)$ is monotone increasing. So $\gamma_2(k)$ is monotone increasing.

Corollary 3

The squared bias $\gamma_2(k)$ approaches $\boldsymbol{\beta}'\boldsymbol{\beta}$ as an upper limit.

Proof

From $\lim\limits_{k\to\infty}\gamma_2(k)=\sum\limits_{i=1}^{p}\alpha_i^2=\alpha'\alpha=\boldsymbol{\beta}'\boldsymbol{P}'\boldsymbol{P}\boldsymbol{\beta}=\boldsymbol{\beta}'\boldsymbol{\beta}$.

Corollary 4

The derivative $\gamma_2'(k)$ approaches zero as $k\to 0^+$.

Proof

From equation (5.2.39) it is readily established that

$$\frac{\mathrm{d}\gamma_2(k)}{\mathrm{d}k}=\frac{2k\sum\limits_{i=1}^{p}\lambda\alpha_i^2}{(\lambda_i+k)^3} \tag{5.2.42}$$

Each term in the sum $2k\sum\limits_{i=1}^{p}\lambda\alpha_i^2/(\lambda_i+k)^3$ is a continuous function. And the limit of each term as $k\to 0^+$ is zero.

Theorem 3

(Existence Theorem) There always exists a $k>0$ such that $E[L_1^2(k)]<E[L_1^2(0)]=\sigma^2\sum\limits_{i=1}^{p}(1+\lambda_i)$.

Proof

From equation (5.2.34), (5.2.37), and (5.2.42)

$$\frac{\mathrm{d}E[L_1^2(k)]}{\mathrm{d}k}<\frac{\mathrm{d}\gamma_1(k)}{\mathrm{d}k}+\frac{\mathrm{d}\gamma_2(k)}{\mathrm{d}k}\ \frac{-2\sigma^2\sum\limits_{i=1}^{p}\lambda_i}{(\lambda_i+k)^3}+\frac{2k\sum\limits_{i=1}^{p}\lambda\sigma_i^2}{(\lambda_i+k)^3} \tag{5.2.43}$$

First note that $\gamma_1(0)=\sigma^2\sum\limits_{1}^{p}(1+\lambda_i)$ and $\gamma_2(0)=0$. In Theorems 1 and 2, it was established that $\gamma_1(k)$ and $\gamma_2(k)$ are monotonically decreasing and in creasing, respectively.

Their first derivatives are always non-positive and non- negative, respectively. Thus, to prove the theorem, it is only necessary to show that there always exists a $k>0$ such that $\mathrm{d}E[L_1^2(k)]/\mathrm{d}k<0$. The condition for this is shown by equation (5.2.43) to be:

$$k<\frac{\sigma^2}{\alpha_{\max}^2\ k}<\frac{\sigma^2}{\alpha_{\max}^2} \tag{5.2.44}$$

5.2.4.3 Some Comments on the Mean Square Error Function

The properties of $E[L_1^2(k)]=\gamma_1(k)+\gamma_2(k)$ show that it will go through a minimum. And since $\gamma_2(k)$ approaches $\boldsymbol{\beta}'\boldsymbol{\beta}$ as a limit as $k\to\infty$, this minimum will move toward $k=0$ as the magnitude of $\boldsymbol{\beta}'\boldsymbol{\beta}$ increases. Since $\boldsymbol{\beta}'\boldsymbol{\beta}$ is the squared length of the unknown regression vector, it would appear to be impossible to choose a value of $k\neq 0$ and thus achieve a smaller mean square error without being able to assign an upper bound to $\boldsymbol{\beta}'\boldsymbol{\beta}$. On the other hand, it is clear that $\boldsymbol{\beta}'\boldsymbol{\beta}$ does not become infinite in practice, and one should be able to find a value or values for k that will put $\hat{\boldsymbol{\beta}}^*$ closer to $\boldsymbol{\beta}$ than is $\hat{\boldsymbol{\beta}}$. In other words, unboundedness, in the strict mathematical sense, and practical unboundedness are two different things. In Section 5.2.7 some recommendations for choosing a $k > 0$ are given, and the implicit assumptions of boundedness are explored further.

5.2.5 A General Form of Ridge Regression

It is always possible to reduce the general linear regression problem as defined in the Introduction to a canonical form in which the $\boldsymbol{X}'\boldsymbol{X}$ matrix is diagonal. In particular there exists an orthogonal transformation $\boldsymbol{P}$ such that $\boldsymbol{X}'\boldsymbol{X}=\boldsymbol{P}'\boldsymbol{\Lambda}\boldsymbol{P}$ where $\boldsymbol{\Lambda}=(\delta_{ij}\lambda_i)$ is the matrix of eigenvalues of $\boldsymbol{X}'\boldsymbol{X}$. Let

$$\boldsymbol{X}=\boldsymbol{X}'\boldsymbol{P} \tag{5.2.45}$$

and

$$\boldsymbol{Y}=\boldsymbol{X}'\boldsymbol{\alpha}+\boldsymbol{\varepsilon} \tag{5.2.46}$$

where

$$\boldsymbol{\alpha}=\boldsymbol{P}\boldsymbol{\beta},\quad (\boldsymbol{X}^*)'(\boldsymbol{X}^*)=\boldsymbol{\Lambda},\quad \text{and}\quad \boldsymbol{\alpha}'\boldsymbol{\alpha}=\boldsymbol{\beta}'\boldsymbol{\beta} \tag{5.2.47}$$

Then the general ridge estimation procedure is defined from

$$\boldsymbol{\alpha}^*=[(\boldsymbol{X}^*)'(\boldsymbol{X}^*)+\boldsymbol{K}]^{-1}(\boldsymbol{X}^*)'\boldsymbol{Y} \tag{5.2.48}$$

where

$$\boldsymbol{K}=(\delta_{ij}\lambda_i),\quad k_{ij}\geqslant 0$$

All the basic results given in Section 5.2.4 can be shown to hold for this more general formulation. Most important is that there is an equivalent to the existence theorem, Theorem 3. In the general form; one seeks a k_i for each canonical variate defined by $\boldsymbol{X}^*$. By defining $(L_1^*)^2=(\hat{\boldsymbol{\alpha}}^*-\boldsymbol{\alpha})'(\hat{\boldsymbol{\alpha}}^*-\boldsymbol{\alpha})$ it can be shown that the optimal values for the k_i will be $k_i=\sigma^2/\alpha_i^2$. There is no graphical equivalent to the RIDGE TRACE but an iterative procedure initiated at $\hat{k}_i=\hat{\sigma}^2/\hat{\alpha}_i^2$ can be used.

5.2.6 Relation to Other Work in Regression

Ridge regression has points of contact with other approaches to regression analysis and to work with the same objective. Three should be mentioned.

Stein (1960[14], 1962[15]), and James and Stein (1961)[13] investigated the improvement in mean square error by a transformation on $\hat{\boldsymbol{\beta}}$ of the form $C\hat{\boldsymbol{\beta}}$, $0\leqslant C<1$, which is a shortening of the vector \$. They show that such a $C>0$ can always be found and indicate how it might be computed.

A Bayesian approach to regression can be found in Jeffreys(1961)[16] and Raiffa and Schlaifer (1961[16]). Viewed in this context, each ridge estimate can be considered as the posterior mean based on giving the regression coefficients, $\boldsymbol{\beta}$, a prior normal distribution with mean zero and variance-covariance matrix $\boldsymbol{\Sigma}=(\delta_{ij}\delta^2/k)$. For those that do not like the philosophical implications of assuming $\boldsymbol{\beta}$ to be a random variable, all this is equivalent to constrained estimation by a nonuniform weighting on the values of $\boldsymbol{\beta}$.

Constrained estimation in a context related to regression can be found in 1961[18]. For the model in the present paper, let $\boldsymbol{\beta}$ be constrained to be in a closed, bounded convex set C, and, in particular, let C be a hypersphere of radius R. Let the estimation criterion be minimum residual sum of squares $\phi=(\boldsymbol{Y}-\boldsymbol{XB})'(\boldsymbol{Y}-\boldsymbol{XB})$ where $\boldsymbol{B}$ is the value giving the minimum. Under the constraint, if $\hat{\boldsymbol{\beta}}'\hat{\boldsymbol{\beta}}\leqslant R^2$, than $\boldsymbol{B}$ is chosen to be $\hat{\boldsymbol{\beta}}$; otherwise $\boldsymbol{B}$ is chosen to be $\hat{\boldsymbol{\beta}}^*$ where k is chosen so that $(\hat{\boldsymbol{\beta}}^*)'(\hat{\boldsymbol{\beta}}^*)\leqslant R^2$.

5.2.7 Selecting a Better Estimate of $\boldsymbol{\beta}$

In Section 5.2.2 and in the example of Section 5.2.3, it has been demonstrated that the ordinary least squares estimate of the regression vector $\boldsymbol{\beta}$ suffers from a number of deficiencies when $\boldsymbol{X}'\boldsymbol{X}$ does not have a uniform eigenvalue spectrum. A class of biased estimators $\hat{\boldsymbol{\beta}}^*$, obtained by augmenting the diagonal of $\boldsymbol{X}'\boldsymbol{X}$ with small positive quantities, has been introduced both to portray the sensitivity of the solution to $\boldsymbol{X}'\boldsymbol{X}$ and to form the basis for obtaining an estimate of $\boldsymbol{\beta}$ with a smaller mean square error. In examining the properties of $\hat{\boldsymbol{\beta}}^*$, it can be shown that its use is equivalent to making certain boundedness assumptions regarding either the individual coordinates of $\boldsymbol{\beta}$ or its squared length, $\boldsymbol{\beta}'\boldsymbol{\beta}$. As Barnard (1963[18]) has recently pointed out, an alternative to unbiasedness in the logic of the least squares estimator $\hat{\boldsymbol{\beta}}$ is the prior assurance of bounded mean square error with no boundedness assumption on $\boldsymbol{\beta}$. If it is possible to make specific mathematical assumptions about $\boldsymbol{\beta}$, then it is possible to constrain the estimation procedure to reflect these assumptions.

The inherent boundedness assumptions in using $\hat{\boldsymbol{\beta}}^*$ make it clear that it will not be possible to construct a clear-cut, automatic estimation procedure to produce a point estimate (a single value of k or a specific value for each k_i) as can be constructed to produce $\hat{\boldsymbol{\beta}}$. However, this is no drawback to its use because with any given set of data it is not difficult to select a $\hat{\boldsymbol{\beta}}^*$ that is better than $\hat{\boldsymbol{\beta}}$. In fact, put in context, any set of data which is a candidate for analysis using linear regression has implicit in it restrictions on the possible values of the estimates that can be consistent with known properties of the data generator. Yet it is difficult to be explicit about these restrictions; it is especially difficult to be mathematically explicit. In a recent paper (Clutton-Brock 1965[19]) it has been shown that for the problem of estimating the mean μ of a distribution, a set of data has in it implicit restrictions on the values of σ that can be logical contenders as generators. Of course, in linear regression the problem is much more difficult; the number of possibilities is so large. First, there is the number of parameters involved. To have ten to twenty regression coefficients is not uncommon. And their signs have to be considered. Then there is $\boldsymbol{X}'\boldsymbol{X}$ and the $\binom{p}{2}$ different factor correlations and the ways in which they can be related. Yet in the final analysis these many different influences can be integrated to make an assessment as to whether the estimated values are consistent with the data and the properties of the data generator. Guiding one along the way, of course, is the objective of the study. In Hoerl and Kennard (1970[21]) it is shown for two problems how such an assessment can be made.

Based on experience, the best method for achieving a better estimate $\hat{\boldsymbol{\beta}}^*$ is to use $k_i = k$ for all i and use the Ridge Trace to select a single value of k and a unique $\hat{\boldsymbol{\beta}}^*$. These kinds of things can be used to guide one to a choice.

1) At a certain value of k the system will stabilize and have the general characteristics of an orthogonal system.

2) Coefficients will not have unreasonable absolute values with respect to the factors for which they represent rates of change.

3) Coefficients with apparently incorrect signs at $k=0$ will have changed to have the proper sign.

4) The residual sum of squares will not have been inflated to an unreasonable value. It will not be large relative to the minimum residual sum of squares or large relative to what would be a reasonable variance for the process generating the data.

Another approach is to use estimates of the optimum values of k_i developed in Section 5.2.5. A typical approach here would be as follows:

5) Reduce the system to canonical by the transformations $\boldsymbol{X}=\boldsymbol{X}^*\boldsymbol{P}$ and $\boldsymbol{\alpha}=\boldsymbol{P}\boldsymbol{\beta}$.

6) Determine estimates of the optimum k_i's using $\hat{k}_{i0}=\hat{\sigma}^2/\hat{\alpha}_i^2$. Use the $\hat{k}_{i0}$ to obtain $\hat{\boldsymbol{\beta}}^*$.

7) The $\hat{k}_{i0}$ will tend to be too small because of the tendency to overestimate $\boldsymbol{\alpha}'\boldsymbol{\alpha}$. Since use of the $\hat{k}_{i0}$ will shorten the length of the estimated regression vector, $\hat{k}_{i0}$ can be re-estimated using the $\hat{\alpha}_i^*$. This re-estimation can be continued until there is a stability achieved in $(\hat{\alpha}_i^*)'(\hat{\alpha}_i^*)$ and $\hat{k}_{i0}=\hat{\sigma}^2/(\hat{\alpha}_i^*)^2$.

References

[1] ARTHUR E KENNARD R W. Ridge Regression: Biased Estimation for Nonorthogonal Problems. Technometrics,1970:55 - 67.

[2] SCOTT J T. Factor Analysis and Regression, Econometrica, 1966, 34: 552 - 562.

[3] HOERL A E. Application of Ridge Analysis to Regression Problems, Chemical Engineering Progress, 1962: 58, 54 - 59.

[4] HOERL A E, KENNARD R W. On Regression Analysis and Biased Estimation. Technometrics, 1968: 10, 422 - 423.

[5] HOERL A E. Ridge Analysis. Chemical Engineering Progress Symposium Series, 1964, 60: 67 - 77.

[6] SCHEFFE H. The Analysis of Variance. New York: Wiley, 1960.

[7] BEALE E M L, KENDALL M G, MANN D W. The of Variables in Multivariate Analysis, Biometrika, 1967, 54: 356 - 366.

[8] EFROYAMSON M A. Multiple Regression Analysis in Mathematical Methods for Digital Computers, eds. A. Ralston and H. S. Wilf, New York: Wiley, chap. 17, 1960.

[9] GARSIDE M J. The Best Subset in Multiple Regression Analysis. Applied Statistics, 1965:14.
[10] GORMAN J W, TOMAN R J. Selection of Variables for Fitting Equations to Data. Technometrics, 1966: 8, 27 - 51.
[11] HOCKING R R, LESLIE R N. Selection of the Best Subset in Regression Analysis. Technometrics, 1967: 9, 531 - 540.
[12] JEFFERS J N R. Two Case Studies in the Application of Principal Component Analysis. Applied Statistics, 1967: 16, 225 - 236.
[13] JAMES W, STEIN C M. Estimation With Quadratic Loss, Proceedings of the 4th Berkeley Symposium, 1961, 1: 361 - 379.
[14] STEIN C M. Multiple Regression in Essays in Honor of Harold Hotelling, Stanford. CA: Stanford University Press, chap. 37, 1960.
[15] STEIN C M. Confidence Sets for the Mean of a Multivariate Normal Distribution, Journal of the Royal Statistical Society. Ser. B, 1962, 24: 265 - 296.
[16] JEFFREY S H. Theory of Probability (3rd ed.), London: Oxford University Press, 1961.
[17] RAIFFA H, SCHLAIFER R. Applied Statistical Decision Theory, Boston: Harvard University Press, 1961.
[18] BARNARD G A. Thelogicofleast squares. Journal of the Royal Statistical Society, Series B,1963, 25: 124 - 127.
[19] CLUTTON-BROCK M. Using the observations to estimate prior distribution. Journal of the Royal Statistical Society, Series B,1965, 27: 17 - 27.
[20] DISCRADIN C B M. Using the Observations to Estimate Prior Distribution, Journal of the Royal Statistical Society, Series B, 1965, 27: 17 - 27.
[21] HOERL A E, KENNARD R W. Ridge Regression. Applications to Non-orthogonal Problems. Technometrics, 12. 1970.

5.3 Lasso

The "lasso" minimizes the residual sum of squares subject to the sum of the absolute value of the coefficients being less than a constant. Because of the nature of this constraint it tends to produce some coefficients that are exactly 0 and hence gives interpretable models. There is also an interesting relationship with recent work in adaptive function estimation by Donoho and Johnstone. The lasso idea is quite general and can be applied in a variety of statistical models: extensions to generalized regression models and tree-based models are briefly describe.

5.3.1 Introduction

Consider the usual regression situation: we have data $(\boldsymbol{x}^i, \boldsymbol{y}_i)$, $i = 1, 2, \cdots, N$, where $\boldsymbol{x}^i = (x_{i1}, x_{i2}, \cdots, x_{iP})^{\mathrm{T}}$ and y_i are the regressors and response for the i-th observation. The ordinary least squares (OLS) estimates are obtained by minimizing the residual squared error. There are two reasons why the data analyst is often not satisfied with the OLS estimates. The first is prediction accuracy: the OLS estimates often have low bias but large variance; prediction accuracy can sometimes be improved by shrinking or setting to 0 some coefficients. By doing so we sacrifice a little bias to reduce the variance of the predicted values and hence may improve the overall prediction accuracy. The second reason is interpretation. With a large number of predictors, we often would like to determine a smaller subset that exhibits the strongest effects.

The two standard techniques for improving the OLS estimates, subset selection and ridge regression, both have drawbacks. Subset selection provides interpretable models but can be extremely variable because it is a discrete process—regressors are either retained or dropped from the model. Small changes in the data can result in very different models being selected and this can reduce its prediction accuracy. Ridge regression is a continuous process that shrinks coefficients and hence is more stable: however, it does not set any coefficients to 0 and hence does not give an easily interpretable model.

A new technique was proposed, called the lasso, for "least absolute shrinkage and selection operator". It shrinks some coefficients and sets others to 0, and hence tries to retain the good features of both subset selection and ridge regression.

5.3.2 Theory of the Lasso

5.3.2.1 Definition

Suppose that we have data $(\boldsymbol{x}_i, \boldsymbol{y}_i)$, $i = 1, 2, \cdots, N$, where $\boldsymbol{x}^i = (x_{i1}, x_{i2}, \cdots, x_{iP})^{\mathrm{T}}$ are the predictor variables and $\boldsymbol{y}_i$ are the responses. As in the usual regression set-up, we assume either that the observations are independent or that the $\boldsymbol{y}_{\mathrm{is}}$ are conditionally independent given the $\boldsymbol{x}_{ij}$ s. We assume that the $\boldsymbol{x}_{ij}$ are standardized so that $\sum_i \frac{\boldsymbol{x}_{ij}}{N} = 0$, $\sum_i \frac{\boldsymbol{x}_{ij}^2}{N} = 1$.

Letting $\beta = (\hat{\beta}_1, \hat{\beta}_2, \cdots, \hat{\beta}_p)^{\mathrm{T}}$, the lasso estimate $(\hat{\alpha}, \hat{\beta})$ is defined by

$$(\hat{\alpha}, \hat{\beta}) = \operatorname{argmin}\left\{ \sum_{i=1}^{N} (\boldsymbol{y}_i - \alpha - \sum_j \beta_j x_{ij})^2 \right\} \text{ subject to } \sum_j |\beta_j| \leqslant t \qquad (5.3.1)$$

Here $t \geqslant 0$ is a tuning parameter. Now, for all t, the solution for α is $\hat{\alpha} = \overline{y}$. We can assume without loss of generality that $\overline{y}$ and hence omit α.

Computation of the solution to equation (5.3.1) is a quadratic programming problem with linear inequality constraints.

The parameter $t \geqslant 0$ controls the amount of shrinkage that is applied to the estimates. Let $\hat{\beta}_j^o$ be the full least squares estimates and let $t_0 = \sum |\hat{\beta}_j^o|$. Values of $t < t_0$ will cause shrinkage of the solutions towards 0, and some coefficients may be exactly equal to 0. For example, if $t = t_0/2$, the effect will be roughly similar to finding the best subset of size $p/2$. Note also that the design matrix need not be of full rank.

The motivation for the lasso came from an interesting proposal of Breiman (1993)[1]. Breiman's non-negative garotte minimizes

$$\sum_{i=1}^{N}(y_i-\alpha-\sum_j c_j\hat{\beta}_j^0 x_{ij})^2 \qquad \text{subject to } c_j\geqslant 0,\ \sum c_j\leqslant t \tag{5.3.2}$$

The garotte starts with the OLS estimates and shrinks them by non-negative factors whose sum is constrained. In extensive simulation studies, Breiman showed that the garotte has consistently lower prediction error than subset selection and is competitive with ridge regression except when the true model has many small non-zero coefficients.

A drawback of the garotte is that its solution depends on both the sign and the magnitude of the OLS estimates. In overfit or highly correlated settings where the OLS estimates behave poorly, the garotte may suffer as a result. In contrast, the lasso avoids the explicit use of the OLS estimates.

Frank and Friedman (1993)[5] proposed using a bound on the L^q-norm of the parameters, where q is some number greater than or equal to 0; the lasso corresponds to $q=1$.

5.3.2.2 Orthonormal Design Case

Insight about the nature of the shrinkage can be gleaned from the orthonormal design case. Let $\boldsymbol{X}$ be the $n\times p$ design matrix with ij-th entry x_{ij}, and suppose that $\boldsymbol{X}^{\mathrm{T}}\boldsymbol{X}=\boldsymbol{I}$, the identity matrix.

The solutions to equation (5.3.1) are easily shown to be

$$\hat{\beta}_j=\operatorname{sign}(\hat{\beta}_j^0)(|\hat{\beta}_j^0|-\gamma)^+ \tag{5.3.3}$$

where γ is determined by the condition $\sum|\hat{\beta}_j|=t$. Interestingly, this has exactly the same form as the soft shrinkage proposals of Donoho and Johnstone (1994)[2] and Donoho et al. (1995)[4], applied to wavelet coefficients in the context of function estimation. The connection between soft shrinkage and a minimum L_1-norm penalty was also pointed out by Donoho et al. (1992)[3] for non-negative parameters in the context of signal or image recovery.

In the orthonormal design case, best subset selection of size k reduces to choosing the k largest coefficients in absolute value and setting the rest to 0. For some choice of λ this is equivalent to setting $\hat{\beta}_j=\hat{\beta}_j^0$ if $|\hat{\beta}_j^0|>\lambda$ and to 0 otherwise. Ridge regression minimizes

$$\sum_{i=1}^{N}(y_j-\sum_j\beta_j x_{ij})^2+\lambda\sum_j\beta_j^2$$

or, equivalently, minimizes

$$\sum_{i=1}^{N}(y_j-\sum_j\beta_j x_{ij})^2 \qquad \text{subject to } \sum_j\beta_j^2\leqslant t \tag{5.3.4}$$

The ridge regression solutions are

$$\frac{1}{1+\gamma}\hat{\beta}_j^0$$

Where γ depends on λ or t. The garotte estimates are

$$\left(1-\frac{\gamma}{\hat{\beta}_j^{02}}\right)^+ \hat{\beta}_j^0$$

Fig. 5. 2 shows the form of these functions. Ridge regression scales the coefficients by a constant factor, whereas the lasso translates by a constant factor, truncating at 0. The garotte function is very similar to the lasso, with less shrinkage for larger coefficients. As our simulations will show, the differences between the lasso and garotte can be large when the design is not orthogonal.

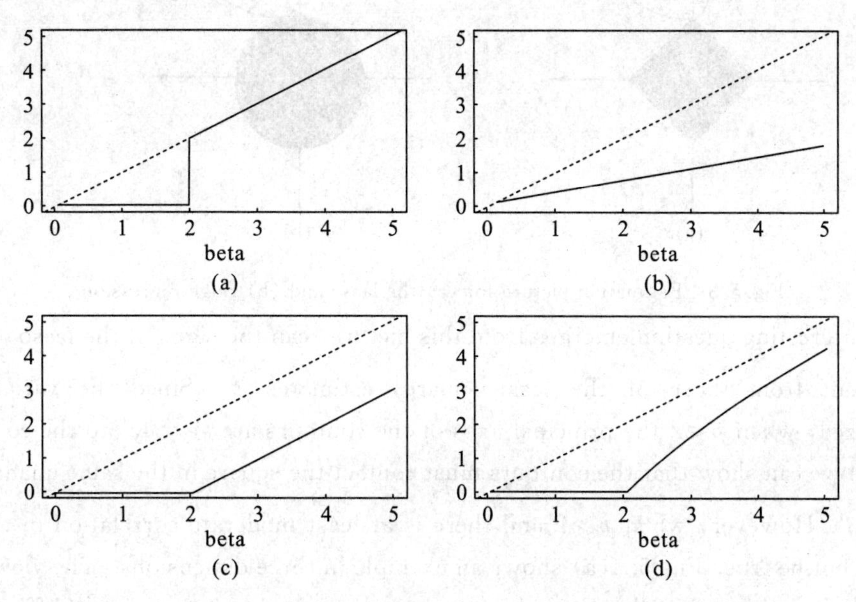

Fig. 5. 2 (a) Subset regression, (b) ridge regression, (c) the lasso and (d) the garotte. ------: form of coefficient shrinkage in the orthonormal design case ——: 45°-line for reference

5. 3. 2. 3 Geometry of Lasso

It is clear from Fig. 5. 2 why the lasso will often produce coefficients that are exactly 0. Why does this happen in the general (non-orthogonal) setting? And why does it not occur with ridge regression, which uses the constraint $\sum_j \beta_j^2 \leqslant t$ rather than $\sum |\hat{\beta}_j| \leqslant t$? Fig. 5. 3 provides some insight for the case $p = 2$.

The criterion $\sum_{i=1}^{N}(y_j - \sum_j \beta_j x_{ij})^2$ equals the quadratic function

$$(\boldsymbol{\beta}-\hat{\boldsymbol{\beta}}^0)^{\mathrm{T}}\boldsymbol{X}^{\mathrm{T}}\boldsymbol{X}(\boldsymbol{\beta}-\hat{\boldsymbol{\beta}}^0)$$

(plus a constant). The elliptical contours of this function are shown by the full curves in Fig. 5. 3(a); they are centered at the OLS estimates; the constraint region is the rotated square. The lasso solution is the first place that the contours touch the square, and this will sometimes occur at a corner, corresponding to a zero coefficient. The picture for ridge regression is shown in Fig. 5. 3(b): there are no corners for the contours to hit and hence zero solutions will rarely result.

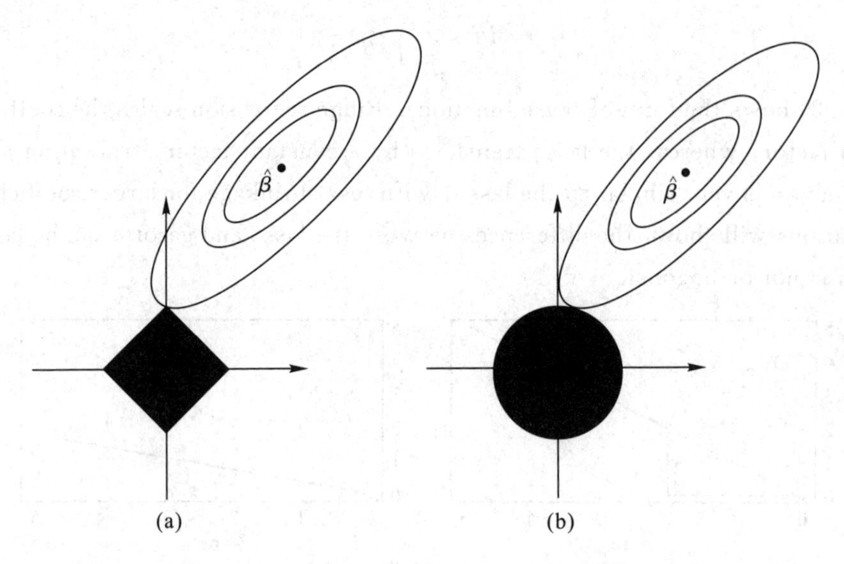

Fig. 5. 3　Estimation picture for (a) the lasso and (b) ridge regression

An interesting question emerges from this picture: can the signs of the lasso estimates be different from those of the least squares estimates $\hat{\beta}_j^0$? Since the variables are standardized, when $p=2$ the principal axes of the contours are at $\pm 45°$ to the co-ordinate axes, and we can show that the contours must contact the square in the same quadrant that contains $\hat{\beta}^0$. However, when $p>2$ and there is at least moderate correlation in the data, this need not be true. Fig. 5. 4(a) shows an example in three dimensions. The view in Fig. 5. 4(b) confirms that the ellipse touches the constraint region in an octant different from the octant in which its center lies.

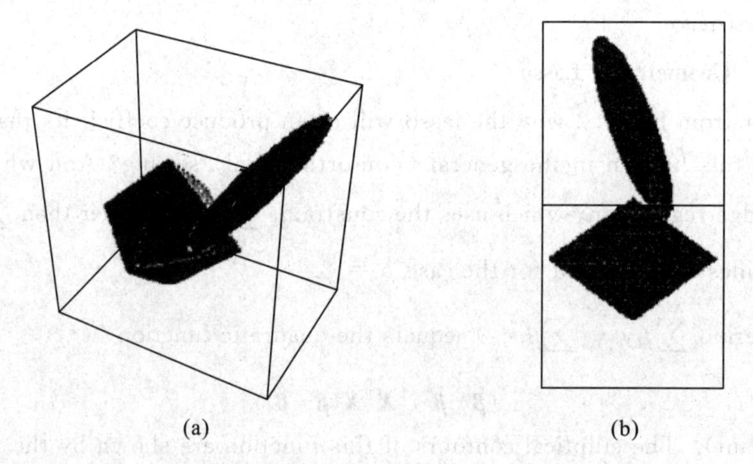

Fig. 5. 4　(a) Example in which the lasso estimate falls in an octant different from the overall least squares estimate; (b) overhead view

Whereas the garotte retains the sign of each $\hat{\beta}_j^0$, the lasso can change signs. Even in cases where the lasso estimate has the same sign vector as the garotte, the presence of the

OLS estimates in the garotte can make it behave differently. The model $\sum c_j \hat{\beta}_j^0 x_{ij}$ with constraint $\sum c_j \leqslant t$ can be written as $\sum \hat{\beta}_j x_{ij}$ with constraint $\sum \hat{\beta}_j / \hat{\beta}_j^0 \leqslant t$. If for example $p=2$ and $\hat{\beta}_1^0 > \hat{\beta}_2^0 > 0$ then the effect would be to stretch the square in Fig. 5. 3(a) horizontally. As a result, larger values of β_1 and smaller values of β_2 will be favored by the garotte.

5.3.2.4 More on Two-predictor Case

Suppose that $p=2$, and assume without loss of generality that the least squares estimates $\hat{\beta}_j^0$ are both positive. Then we can show that the lasso estimates are

$$\hat{\beta} = (\hat{\beta}_j^0 - \gamma)^+$$

where γ is chosen so that $\hat{\beta}_1 + \hat{\beta}_2 = t$. This formula holds for $t \leqslant \hat{\beta}_1^0 + \hat{\beta}_2^0$ and is valid even if the predictors are correlated. Solving for γ yields

$$\hat{\beta}_1 = \left(\frac{t}{2} + \frac{\hat{\beta}_1^0 - \hat{\beta}_2^0}{2}\right)^+$$

$$\hat{\beta}_2 = \left(\frac{t}{2} - \frac{\hat{\beta}_1^0 - \hat{\beta}_2^0}{2}\right)^+$$

In contrast, the form of ridge regression shrinkage depends on the correlation of the predictors. Fig. 5. 5 shows an example. We generated 100 data points from the model $y = 6x_1 + 3x_2$ with no noise. Here x_1 and x_2 are standard normal variates with correlation ρ. The curves in Fig. 5. 5 show the ridge and lasso estimates as the bounds on $\hat{\beta}_1^2 + \hat{\beta}_2^2$ and $|\hat{\beta}_1| + |\hat{\beta}_2|$ are varied. For all values of ρ the lasso estimates follow the full curve. The ridge estimates (broken curves) depend on ρ. When $\rho = 0$ ridge regression does proportional shrinkage. However, for larger values of ρ the ridge estimates are shrunken differentially and can even increase a little as the bound is decreased. As pointed out by Jerome Friedman, this is due to the tendency of ridge regression to try to make the coefficients equal to minimize their squared norm.

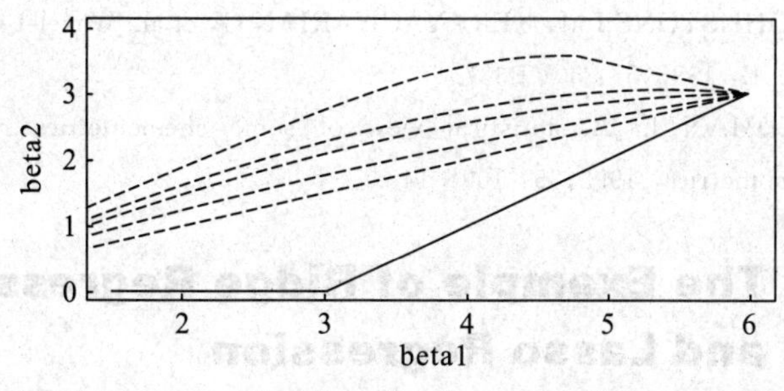

Fig. 5. 5 Lasso (—) and ridge regression (----) for the two-predictor example: the curves show the (β_1, β_2) pairs as the bound on the lasso or ridge parameters is varied; starting with the bottom broken curve and moving upwards, the correlation ρ is 0, 0. 23, 0. 45, 0. 68 and 0. 90

5.3.2.5 Standard Errors

Since the lasso estimate is a non-linear and non-differentiable function of the response values even for a fixed value of t, it is difficult to obtain an accurate estimate of its standard error. One approach is via the bootstrap: either t can be fixed or we may optimize over t for each bootstrap sample. Fixing t is analogous to selecting a best subset, and then using the least squares standard error for that subset.

An approximate closed form estimate may be derived by writing the penalty $\sum |\beta_j|$ as $\sum \beta_j^2 / |\beta_j|$. Hence, at the lasso estimate β, we may approximate the solution by a ridge regression of the form $\boldsymbol{\beta}^* = (\boldsymbol{X}^{\mathrm{T}}\boldsymbol{X} + \lambda \boldsymbol{W})^{-1}\boldsymbol{X}^{\mathrm{T}}\boldsymbol{y}$ where $\boldsymbol{W}$ is a diagonal matrix with diagonal elements $|\hat{\beta}_j|$, $\boldsymbol{W}$ denotes the generalized inverse of $\boldsymbol{W}$ and λ is chosen so that $\sum |\boldsymbol{\beta}_j|^* = t$. The covariance matrix of the estimates may then be approximated by

$$(\boldsymbol{X}^{\mathrm{T}}\boldsymbol{X} + \lambda \boldsymbol{W})^{-1}\boldsymbol{X}^{\mathrm{T}}\boldsymbol{X}(\boldsymbol{X}^{\mathrm{T}}\boldsymbol{X} + \lambda \boldsymbol{X}^{-})^{-1}\hat{\sigma}^2$$

Where $\hat{\sigma}^2$ is an estimate of the error variance. A difficulty with this formula is that it gives an estimated variance of 0 for predictors with $\hat{\beta}_j = 0$.

This approximation also suggests an iterated ridge regression algorithm for computing the lasso estimate itself, but this turns out to be quite inefficient. However, it is useful for selection of the lasso parameter t.

References

[1] BREIMAN L. Better subset selection using the non-negative garotte. Technical Report. University of California, Berkeley, 1993.

[2] DONOHO D, JOHNSTONE I. Ideal spatial adaptation by wavelet shrinkage. Biometrika, 1994, 81: 425 - 455.

[3] DONOHO D, JOHNSTONE I M, HOCH J C, et al. Maximum entropy and the nearly black object (with discussion). J. R. Statist. Soc. B, 1992: 54, 41 - 81.

[4] DONOHO D L, JOHNSTONE I M, KERKYACHARIAN G, et al. Wavelet shrinkage: asymptopia? J. R. Statist. Soc. B, 1995, 57: 301 - 337.

[5] FRANK I, FRIEDMAN J. A statistical view of some chemometrics regression tools (with discussion). Technometrics, 1993, 35: 109 - 148.

5.4 The Example of Ridge Regression and Lasso Regression

5.4.1 Example

In this section, we compile code to show the fitting curves corresponding to different regularization coefficients in Ridge and Lasso. Where x is an array of evenly spaced intervals from 0 to 9, the size of which is (20×1). y equals x plus a random number from the standard normal distribution.

```
numpy.random.seed(seed=None)
    Seed the generator.
numpy.random.randn(d0, d1, …, dn):
    Return a sample (or samples) from the "standard normal" distribution.
numpy.linspace(start, stop, num=50, endpoint=True,
               retstep=False, dtype=None, axis=0)
    Return evenly spaced numbers over a specified interval.
    Returns num evenly spaced samples, calculated over the interval [start, stop].
    The endpoint of the interval can optionally be excluded.
        numpy.logspace(start, stop, num=50, endpoint=True,
                       base=10.0, dtype=None, axis=0):
    Return numbers spaced evenly on a log scale.
```

```
example.py
#-*-coding: utf-8-*-

from sklearn.linear_model import Lasso
from sklearn.linear_model import Ridge
import numpy as np
import matplotlib.pyplot as plt

np.random.seed(0) # Set seed for reproducibility
N =20
x = np.linspace(0, 9, N)
x = np.sort(x)
y = x + np.random.randn(N)
x.shape = -1, 1
y.shape = -1, 1

# print "x : ", x, np.shape(x)
# print "y : ", y, np.shape(y)

alphas=np.logspace(-3, 2, 10)
print "alphas : ", alphas

for alpha_ in alphas:

    clf_Ridge = Ridge(alpha=alpha_)
    clf_Ridge.fit(x, y)
```

```
    y_Ridge = clf_Ridge.predict(x)
#       print(clf_Ridge.coef_)
#       print(clf_Ridge.intercept_)

    clf_Lasso = Lasso(alpha=alpha_)
    clf_Lasso.fit(x, y)
    y_Lasso = clf_Lasso.predict(x)
#       print(clf_Lasso.coef_)
#       print(clf_Lasso.intercept_)

    plt.figure(1)
    plt.plot(x, y, 'ro')
    plt.plot(x, y_Ridge, '-')
    plt.xlabel('x')
    plt.ylabel('y')
    plt.title("Ridge")

    plt.figure(2)
    plt.plot(x, y, 'ro')
    plt.plot(x, y_Lasso, '-')
    plt.xlabel('x')
    plt.ylabel('y')
    plt.title("Lasso")

plt.show()
```

In these two figures, different colors correspond to different regularization coefficients. Fig. 5.6 shows the fitting curves of different regularization coefficients in Ridge. Fig. 5.7 shows the fitting curves of different regularization coefficients in Lasso.

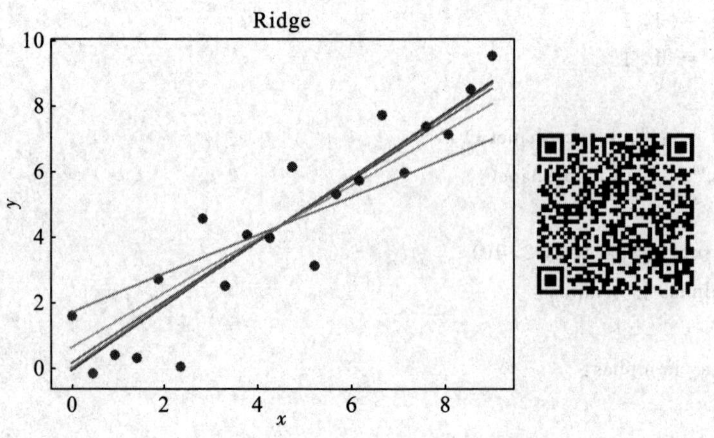

Fig. 5.6 Ridge regression fitting curves

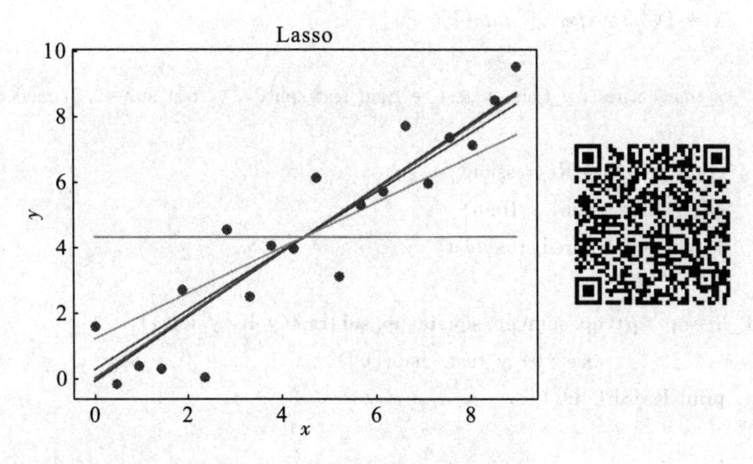

Fig. 5.7 Lasso regression fitting curves

5.4.2 Practical Example

In this section, we show the practical application of Ridge and Lasso on the corn dataset. We use the cross validation in LassoCV and RidgeCV to select the regularization coefficient. Simultaneous compared with linear regression. Finally, the regularization coefficient and the predicted root mean square error (RMSEP) of the four components (moisture, oil, protein, starch) corresponding to the three instruments in the corn data set are listed.

```
example. py
    #-*-coding: utf-8-*-

    import numpy as np
    import matplotlib. pyplot as plt
    from sklearn. linear_model import LinearRegression, LassoCV, RidgeCV
    from scipy. io import loadmat, savemat
    from sklearn. cross_validation import train_test_split

    if __name__ == "__main__":

        D = loadmat('NIRCorn. mat')
        print D. keys()

        plt. figure(figsize=(9, 7))

        for i in range(4):
            y = D['cornprop'][:, [i]]
    #           x = D['m5spec']['data'][0][0]
    #           x = D['mp5spec']['data'][0][0]
```

```
        x = D['mp6spec']['data'][0][0]

        x_train, x_test, y_train, y_test = train_test_split(x, y, test_size=0.2, random_state=0)

        clf_lr = LinearRegression()
        clf_lr.fit(x_train, y_train)
        y_lr = clf_lr.predict(x_test)

RMSEP_lr=np.sqrt(np.sum(np.square(np.subtract(y_lr, y_test)),
                axis=0)/y_test.shape[0])
        print RMSEP_lr

        clf_Ridge = RidgeCV()
        clf_Ridge.fit(x_train, y_train)
        y_Ridge = clf_Ridge.predict(x_test)

RMSEP_Ridge=np.sqrt(np.sum(np.square(np.subtract(y_Ridge, y_test)),
                axis=0)/y_test.shape[0])
print clf_Ridge.alpha_, RMSEP_Ridge[0]

        clf_Lasso = LassoCV()
        clf_Lasso.fit(x_train, y_train.ravel())
        y_Lasso = clf_Lasso.predict(x_test).reshape(-1, 1)

RMSEP_Lasso=np.sqrt(np.sum(np.square(np.subtract(y_Lasso, y_test)),
        axis=0)/y_test.shape[0])
        print clf_Lasso.alpha_, RMSEP_Lasso

        plt.subplot(2, 2, i+1)
        plt.plot(y_test, y_lr, 'bo', label='LinearRegression')
        plt.plot(y_test, y_Ridge, 'ro', label='Ridge')
        plt.plot(y_test, y_Lasso, 'go', label='Lasso')
        plt.plot([min(y_test), max(y_test)], [min(y_test), max(y_test)], color='black')

        plt.xlabel('Measured values')
        plt.ylabel('Predicted values')
        plt.legend(loc='upper left')
        plt.title(i)

    plt.subplots_adjust(hspace =0.3, wspace = 0.3)
    plt.show()
```

Fig. 5. 8 shows the comparison of measured values versus predicted values of LinearRegression, Ridge, and Lasso for the four components of M5 instrument. From these four plots, we can find that LinearRegression exhibits the best performance in the four components.

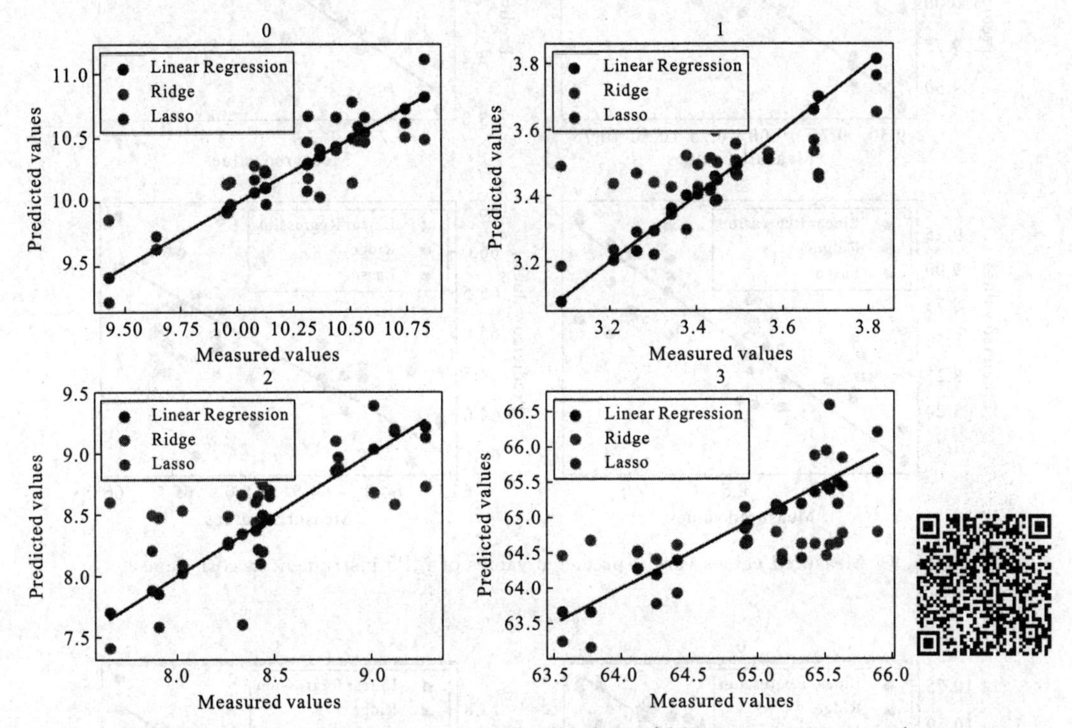

Fig. 5. 8 Measured values versus predicted values of M5 instrument in corn datase

Fig. 5. 9 shows the comparison of measured values versus predicted values of LinearRegression, Ridge, and Lasso for the four components of MP5 instrument. In the moisture and oil, Lasso has the best performance. However, in the protein and starch, LinearRegression exhibits the best performance.

Fig. 5. 10 displays the comparison of measured values versus predicted values of LinearRegression, Ridge, and Lasso for the four components of MP6 instrument. Lasso gains the best performance only in the moisture. Among the other three components, LinearRegression has the best performance.

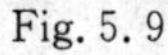

Fig. 5. 9 Fig. 5. 10

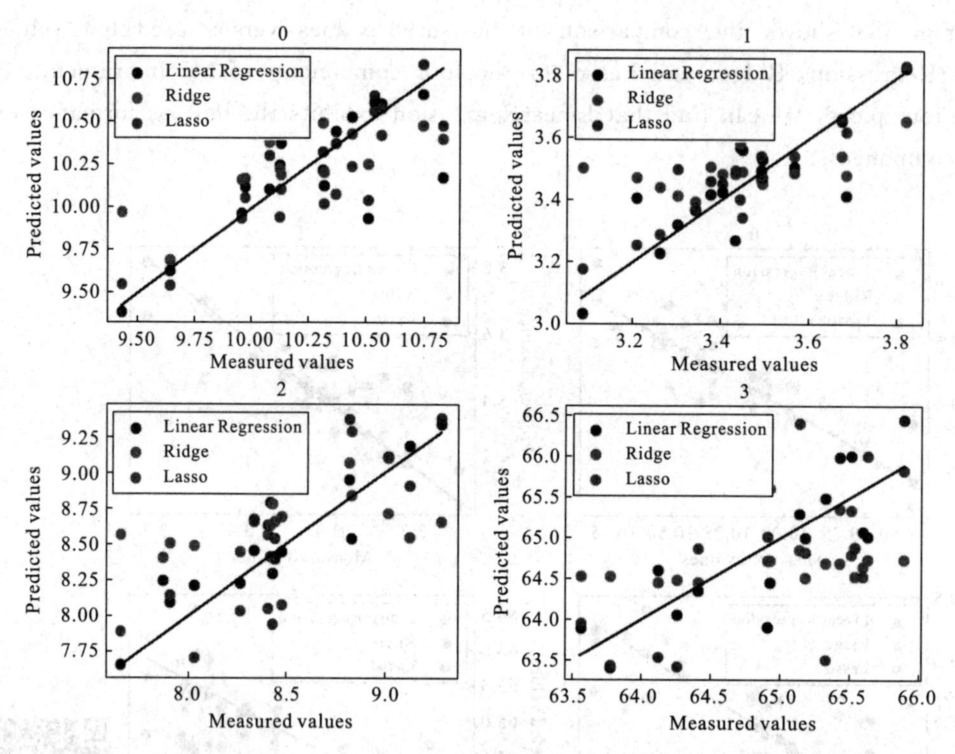

Fig. 5.9　Measured values versus predicted values of MP5 instrument in corn dataset

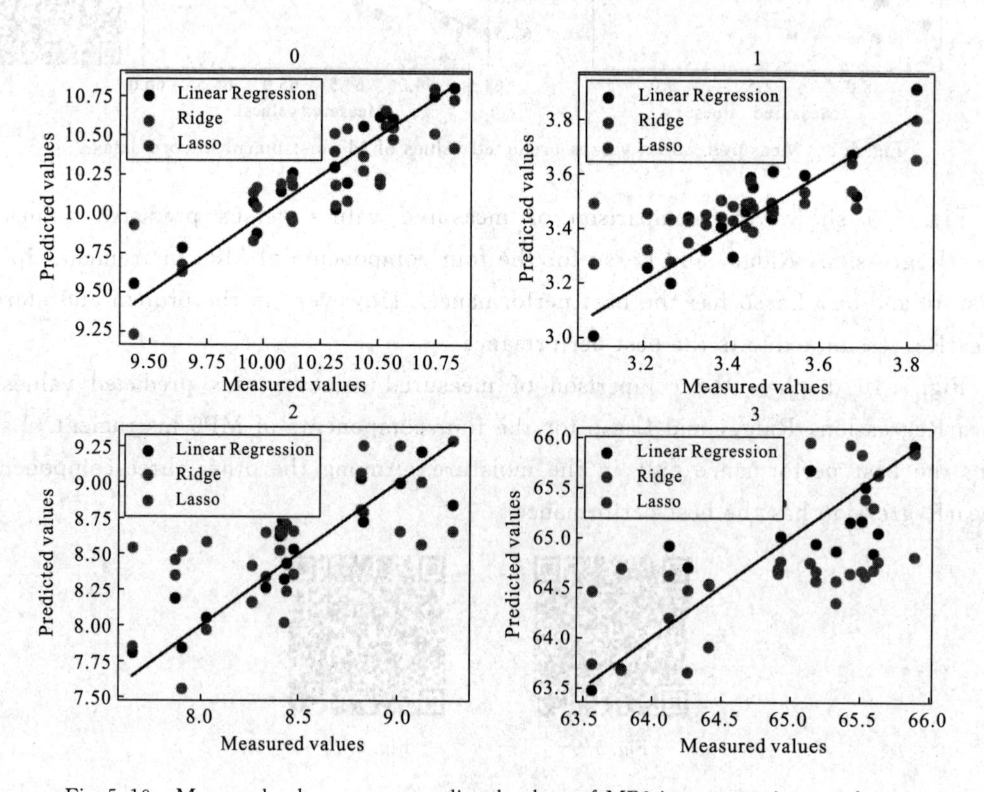

Fig. 5.10　Measured values versus predicted values of MP6 instrument in corn dataset

Table 5. 2 lists the regularization coefficient (alpha) and RMSEP of the four components corresponding to the three instruments in the corn data set. From table 5. 2, we can note that Lasso has the best performance only in the moisture and oil of MP5 instrument and the moisture of MP6 instrument. In other cases, LinearRegression works best, and obtains the smallest RMSEP.

Table 5. 2 Summary of Results

Instrument	Component	Method	Alpha	RMSEP
m5spec	moisture	LinearRegression		0. 004776
		Ridge	0. 1	0. 222966
		Lasso	1. 28E-05	0. 183542
	oil	LinearRegression		0. 016041
		Ridge	0. 1	0. 167543
		Lasso	1. 30E-05	0. 081364
	protein	LinearRegression		0. 042262
		Ridge	0. 1	0. 451078
		Lasso	8. 84E-06	0. 282023
	starch	LinearRegression		0. 107264
		Ridge	0. 1	0. 721352
		Lasso	5. 44E-06	0. 52641
mp5spec	moisture	LinearRegression		0. 238383
		Ridge	0. 1	0. 242359
		Lasso	1. 22E-05	0. 195496
	oil	LinearRegression		0. 1058
		Ridge	0. 1	0. 1642
		Lasso	9. 17E-06	0. 075685
	protein	LinearRegression		0. 189644
		Ridge	0. 1	0. 435509
		Lasso	8. 57E-06	0. 412035
	starch	LinearRegression		0. 487533
		Ridge	0. 1	0. 693999
		Lasso	4. 41E-06	0. 742767

continue

Instrument	Component	Method	Alpha	RMSEP
mp6spec	moisture	LinearRegression		0.12908
		Ridge	0.1	0.233897
		Lasso	1.25E-05	0.152965
	oil	LinearRegression		0.091652
		Ridge	0.1	0.161801
		Lasso	6.97E-06	0.087941
	protein	LinearRegression		0.189663
		Ridge	0.1	0.445122
		Lasso	8.74E-06	0.25082
	starch	LinearRegression		0.40006
		Ridge	0.1	0.707962
		Lasso	4.54E-06	0.450186

5.5 Sparse PCA

Principal component analysis (PCA) is widely used in data processing and dimensionality reduction. However, PCA suffers from the fact that each principal component is a linear combination of all the original variables, thus it is often difficult to interpret the results. We introduce a new method called sparse principal component analysis (SPCA) using the lasso (elastic net) to produce modified principal components with sparse loadings. We show that PCA can be formulated as a regression-type optimization problem, then sparse loadings are obtained by imposing the lasso (elastic net) constraint on the regression coefficients. Efficient algorithms are proposed to realize SPCA for both regular multivariate data and gene expression arrays. We also give a new formula to compute the total variance of modified principal components. As illustrations, SPCA is applied to real and simulated data, and the results are encouraging.

5.5.1 Introduction

Principal component analysis (PCA)(Jolliffe 1986)[4] is a popular data processing and dimension reduction technique. As an un-supervised learning method, PCA has numerous applications such as handwritten zip code classification (Hastie et al. 2001)[3] and human face recognition (Hancock et al. 1996). Recently PCA has been used in gene expression data analysis (Misra et al. 2002)[7]. Hastie et al. (2000)[2] propose the so-called Gene Shaving techniques using PCA to cluster high variable and coherent genes in microarray data.

PCA seeks the linear combinations of the original variables such that the derived variables capture maximal variance. PCA can be done via the singular value decomposition (SVD) of the data matrix. In detail, let the data $\boldsymbol{X}$ be a $n \times p$ matrix, where n and p are the number of observations and the number of variables, respectively. Without loss of generality, assume the column means of $\boldsymbol{X}$ are all 0. Suppose we have the SVD of $\boldsymbol{X}$ as

$$\boldsymbol{X} = \boldsymbol{U}\boldsymbol{D}\boldsymbol{V}^{\mathrm{T}} \tag{5.5.1}$$

where T means transpose. $\boldsymbol{U}$ are the principal components (PCs) of unit length, and the columns of $\boldsymbol{V}$ are the corresponding loadings of the principal components. The variance of the i-th PC is $D_{i,i}^2$. In gene expression data the PCs $\boldsymbol{U}$ are called the eigen-arrays and $\boldsymbol{V}$ are the eigen-genes (Alter et al. 2000). Usually the first q ($q \ll p$) PCs are chosen to represent the data, thus a great dimensionality reduction is achieved.

The success of PCA is due to the following two important optimal properties:

1. Principal components sequentially capture the maximum variability among $\boldsymbol{X}$, thus guaranteeing minimal information loss;

2. Principal components are uncorrelated, so we can talk about one principal component without referring to others.

However, PCA also has an obvious drawback, i. e., each PC is a linear combination of all p variables and the loadings are typically nonzero. This makes it often difficult to interpret the derived PCs. Rotation techniques are commonly used to help practitioners to interpret principal components (Jolliffe 1995).[5] Vines (2000)[1] considered simple principal components by restricting the loadings to take values from a small set of allowable integers such as 0, 1 and -1.

We feel it is desirable not only to achieve the dimensionality reduction but also to reduce the size of explicitly used variables. An ad hoc way is to artificially set the loadings with absolute values smaller than a threshold to zero. This informal thresholding approach is frequently used in practice but can be potentially misleading in various respects [1](Cadima & Jolliffe 1995). McCabe (1984)][7] presented an alternative to PCA which found a subset of principal variables. Jolliffe & Uddin (2003)[6] introduced SCoTLASS to get modified principal components with possible zero loadings.

Recall the same interpretation issue arising in multiple linear regression, where the response is predicted by a linear combination of the predictors. Interpretable models are obtained via variable selection. The lasso (Tibshirani 1996)[9] is a promising variable selection technique, simultaneously producing accurate and sparse models. Zou & Hastie (2003)[12] propose the elastic net, a generalization of the lasso, to further improve upon the lasso. In this paper we introduce a new approach to get modified PCs with sparse loadings, which we call sparse principal component analysis (SPCA). SPCA is built on the fact that PCA can be written as a regression-type optimization problem, thus the lasso (elastic net) can be directly integrated into the regression criterion such that the resulting modified PCA produces sparse loadings.

5.5.2 Motivation and Method Details

In both lasso and elastic net, the sparse coefficients are a direct consequence of the L_1 penalty, not depending on the squared error loss function. Jolliffe & Uddin (2003)[6] proposed SCoTLASS by directly putting the L_1 constraint in PCA to get sparse loadings. SCoTLASS successively maximizes the variance

$$\boldsymbol{\alpha}_k^{\mathrm{T}}(\boldsymbol{X}^{\mathrm{T}}\boldsymbol{X})\boldsymbol{\alpha}_k \tag{5.5.2}$$

subject to

$$\boldsymbol{\alpha}_k^{\mathrm{T}}\boldsymbol{\alpha}_k=1 \quad \text{and (for } k\geqslant 2) \quad \boldsymbol{\alpha}_h^{\mathrm{T}}\boldsymbol{\alpha}_k=0,\ h<k \tag{5.5.3}$$

and the extra constraints

$$\sum_{j=1}^{p} |\boldsymbol{\alpha}_{\mathrm{k,j}}| \leqslant \mathrm{t} \tag{5.5.4}$$

for some tuning parameter t. Although sufficiently small t yields some exact zero loadings, SCoTLASS seems to lack of a guidance to choose an appropriate t value. One might try several t values, but the high computational cost of SCoTLASS makes it an impractical solution. The high computational cost is due to the fact that SCoTLASS is not a convex optimization problem. Moreover, the examples in Jolliffe & Uddin (2003)[6] show that the obtained loadings by SCoTLASS are not sparse enough when requiring a high percentage of explained variance.

We consider a different approach to modify PCA, which can more directly make good use of the lasso. In light of the success of the lasso (elastic net) in regression, we state our strategy. We seek a regression optimization framework in which PCA is done exactly. In addition, the regression framework should allow a direct modification by using the lasso (elastic net) penalty such that the derived loadings are sparse.

5.5.2.1 Direct Sparse Approximations

We first discuss a simple regression approach to PCA. Observe that each PC is a linear combination of the p variables, thus its loadings can be recovered by regressing the PC on the p variables

Theorem 1

$\forall i$, denote $\boldsymbol{Y}_i=\boldsymbol{U}_i\boldsymbol{D}_i$. $\boldsymbol{Y}_i$ is the i-th principal component. $\forall \lambda>0$, suppose $\hat{\beta}_{\mathrm{ridge}}$ is the ridge estimates given by

$$\hat{\boldsymbol{\beta}}_{\mathrm{ridge}}=\arg\min_{\beta}|\boldsymbol{Y}_i-\boldsymbol{X}\boldsymbol{\beta}|^2+\lambda|\boldsymbol{\beta}|^2 \tag{5.5.5}$$

Let $\hat{\nu}=\dfrac{\hat{\beta}_{\mathrm{ridge}}}{|\hat{\beta}_{\mathrm{ridge}}|}$, then $\hat{\nu}=\boldsymbol{V}_i$.

The theme of this simple theorem is to show the connection between PCA and a regression method is possible. Regressing PCs on variables was discussed in Cadima & Jolliffe (1995)[2], where they focused on approximating PCs by a subset of k variables. We extend it to a more general ridge regression in order to handle all kinds of data, especially the gene expression data. Obviously when $n>p$ and $\boldsymbol{X}$ is a full rank matrix, the theorem does not require a positive λ. Note that if $p>n$ and $\lambda=0$, ordinary multiple regression has no unique solution that is exactly $\boldsymbol{V}_i$. The same story happens when $n>p$ and $\boldsymbol{X}$ is not a full rank matrix. However, PCA always gives a unique solution in all situations. As shown in theorem 1, this discrepancy is eliminated by the positive ridge penalty ($\lambda|\boldsymbol{\beta}|^2$). Note that after normalization the coefficients are independent of λ, therefore the ridge penalty is not used to penalize the regression coefficients but to ensure the reconstruction of principal components. Hence we keep the ridge penalty term throughout this section.

Now let us add the L_1 penalty to equation (5.5.5) and consider the following optimization problem

$$\hat{\boldsymbol{\beta}}=\arg\min_{\beta}|\boldsymbol{Y}-\boldsymbol{X\beta}|^2+\lambda|\boldsymbol{\beta}|^2+\lambda_1|\boldsymbol{\beta}|_1 \tag{5.5.6}$$

We call $\hat{\boldsymbol{V}}=\frac{\hat{\beta}}{\hat{\beta}}$ an approximation to $\boldsymbol{V}_i$, and $\boldsymbol{X}\hat{\boldsymbol{V}}_i$ the i-th approximated principal component. Equation 5.5.6 is called naive elastic net (Zou & Hastie 2003)[12] which differs from the elastic net by a scaling factor $(1+\lambda)$. Since we are using the normalized fitted coefficients, the scaling factor does not affect $\hat{\boldsymbol{V}}_i$. Clearly, large enough λ_1 gives a sparse $\hat{\boldsymbol{\beta}}$, hence a sparse $\hat{\boldsymbol{V}}_i$. Given a fixed λ, equation 5.5.6 is efficiently solved for all λ_1 by using the LARS-EN algorithm (Zou & Hastie 2003)[12]. Thus we can flexibly choose a sparse approximation to the i-th principal component.

5.5.2.2 Sparse Principal Components Based on SPCA Criterion

Theorem 1 depends on the results of PCA, so it is not a genuine alternative. However, it can be used in a two-stage exploratory analysis: first perform PCA, then use equation (5.5.6) to find suitable sparse approximations.

We now present a "self-contained" regression-type criterion to derive PCs. We first consider the leading principal component.

Theorem 2 let $\boldsymbol{X}_i$ denote the i-th row vector of the matrix $\boldsymbol{X}$. For any $\lambda>0$, let

$$(\hat{\boldsymbol{\alpha}},\hat{\boldsymbol{\beta}})=\arg\min_{\alpha,\beta}\sum_{i=1}^{n}|\boldsymbol{X}_i-\boldsymbol{\alpha\beta}^{\mathrm{T}}\boldsymbol{X}_i|^2+\lambda|\boldsymbol{\beta}|^2$$
$$\text{subject to } |\alpha|^2=1 \tag{5.5.7}$$

The next theorem extends theorem 2 to derive the whole sequence of PCs.

Theorem 3 suppose we are considering the first k principal components. Let $\boldsymbol{\alpha}$ and $\boldsymbol{\beta}$ be $p\times k$ matrices. $\boldsymbol{X}_i$ denote the i-th row vector of the matrix $\boldsymbol{X}$. For any $\lambda>0$, let

$$(\hat{\boldsymbol{\alpha}},\hat{\boldsymbol{\beta}})=\arg\min_{\alpha,\beta}\sum_{i=1}^{n}|\boldsymbol{X}_i-\boldsymbol{\alpha\beta}^{\mathrm{T}}\boldsymbol{X}_i|^2+\lambda\sum_{j=1}^{n}|\boldsymbol{\beta}_j|^2$$
$$\text{subject to } \boldsymbol{\alpha\alpha}^{\mathrm{T}}=\boldsymbol{I}_k \tag{5.5.8}$$

Theorem 3 effectively transforms the PCA problem to a regression-type problem. The critical element is the object function $\sum_{i=1}^{n} | \boldsymbol{X}_i - \boldsymbol{\alpha\beta}^{\mathrm{T}} \boldsymbol{X}_i |^2$. If we restrict $\boldsymbol{\beta} = \boldsymbol{\alpha}$, then $\sum_{i=1}^{n} | \boldsymbol{X}_i - \boldsymbol{\alpha\beta}^{\mathrm{T}} \boldsymbol{X}_i |^2 = \sum_{i=1}^{n} | \boldsymbol{X}_i - \boldsymbol{\alpha\alpha}^{\mathrm{T}} \boldsymbol{X}_i |^2$, whose minimizer under the orthonormal constraint on $\boldsymbol{\alpha}$ is exactly the first k loading vectors of ordinary PCA. This is actually an alternative derivation of PCA other than the maximizing variance approach, e. g. Hastie et al. (2001)[3]. Theorem 6 shows that we can still have exact PCA while relaxing the restriction $\boldsymbol{\beta} = \boldsymbol{\alpha}$ and adding the ridge penalty term. As can be seen later, these generalizations enable us to flexibly modify PCA.

To obtain sparse loadings, we add the lasso penalty into the criterion (5. 5. 8) and consider the following optimization problem

$$(\hat{\boldsymbol{\alpha}}, \hat{\boldsymbol{\beta}}) = \arg\min_{\alpha,\beta} \sum_{i=1}^{n} |\boldsymbol{X}_i - \boldsymbol{\alpha\beta}^{\mathrm{T}} \boldsymbol{X}_i|^2 + \lambda \sum_{j=1}^{k} |\boldsymbol{\beta}_j|^2 + \lambda_{1,j} \sum_{j=1}^{k} |\boldsymbol{\beta}_j|_1$$

$$\text{subject to } \boldsymbol{\alpha\alpha}^{\mathrm{T}} = \boldsymbol{I}_k \tag{5. 5. 9}$$

Whereas the same λ is used for all k components, different $\lambda_{1,j}$ are allowed for penalizing the loadings of different principal components. Again, if $p > n$, a positive λ is required in order to get exact PCA when the sparsity constraint (the lasso penalty) vanishes ($\lambda_{1,j} = 0$). Equation (5. 5. 9) is called the SPCA criterion hereafter.

5. 5. 2. 3 Numerical Solution

We propose an alternatively minimization algorithm to minimize the SPCA criterion. From the proof of theorem 3 (see appendix for details) we get

$$\sum_{i=1}^{n} |\boldsymbol{X}_i - \boldsymbol{\alpha\beta}^{\mathrm{T}} \boldsymbol{X}_i|^2 + \lambda \sum_{j=1}^{k} |\boldsymbol{\beta}_j|^2 + \lambda_{1,j} \sum_{j=1}^{k} |\boldsymbol{\beta}_j|_1$$

$$= \mathrm{Tr}\boldsymbol{X}^{\mathrm{T}}\boldsymbol{X} + \sum_{j=1}^{k} (\boldsymbol{\beta}_j^{\mathrm{T}} (\boldsymbol{X}^{\mathrm{T}}\boldsymbol{X} + \lambda)\boldsymbol{\beta}_j - 2\boldsymbol{\alpha}_j^{\mathrm{T}} \boldsymbol{X}^{\mathrm{T}} \boldsymbol{X\alpha}_j + \lambda_{1,j} |\boldsymbol{\beta}_j|_1) \tag{5. 5. 10}$$

Hence if given $\boldsymbol{\alpha}$, it amounts to solve k independent elastic net problems to get $\hat{\boldsymbol{\beta}}_j$ for $j = 1, 2, \cdots, k$.

On the other hand, we also have (details in appendix)

$$\sum_{i=1}^{n} |\boldsymbol{X}_i - \boldsymbol{\alpha\beta}^{\mathrm{T}} \boldsymbol{X}_i|^2 + \lambda \sum_{j=1}^{k} |\boldsymbol{\beta}_j|^2 + \sum_{j=1}^{k} \lambda_{1,j} |\boldsymbol{\beta}_j|_1 \tag{5. 5. 11}$$

$$= \mathrm{Tr}\boldsymbol{X}^{\mathrm{T}}\boldsymbol{X} + 2\mathrm{Tr}\boldsymbol{\alpha}^{\mathrm{T}}\boldsymbol{X}^{\mathrm{T}}\boldsymbol{X\beta} + \mathrm{Tr}\boldsymbol{\beta}^{\mathrm{T}}(\boldsymbol{X}^{\mathrm{T}}\boldsymbol{X} + \lambda)\boldsymbol{\beta} + \sum_{j=1}^{k} \lambda_{1,j} |\boldsymbol{\beta}_j|_1$$

Thus if $\boldsymbol{\beta}$ is fixed, we should maximize $\mathrm{Tr}\boldsymbol{\alpha}^{\mathrm{T}} (\boldsymbol{X}^{\mathrm{T}} \boldsymbol{X})\boldsymbol{\beta}$ subject to $\boldsymbol{\alpha}^{\mathrm{T}} \boldsymbol{\alpha} = \boldsymbol{I}_k$, whose solution is given by the following theorem.

Theorem 4 let $\boldsymbol{\alpha}$ and $\boldsymbol{\beta}$ be $m \times k$ matrices and $\boldsymbol{\beta}$ has rank k. Consider the constrained maximization problem

$$\hat{\boldsymbol{\alpha}} = \arg\max_{\alpha} \mathrm{Tr}(\boldsymbol{\alpha}^{\mathrm{T}}\boldsymbol{\beta}) \quad \text{Subject to } \boldsymbol{\alpha\alpha}^{\mathrm{T}} = \boldsymbol{I}_k \tag{5. 5. 12}$$

Suppose the SVD of $\boldsymbol{\beta}$ is $\boldsymbol{\beta} = \boldsymbol{UDV}^{\mathrm{T}}$, then $\hat{\boldsymbol{\alpha}}=\boldsymbol{UV}^{\mathrm{T}}$.

Here are the steps of our numerical algorithm to derive the first k sparse PCs.

General SPCA Algorithm

1) Let $\boldsymbol{\alpha}$ start at $\mathbf{V}[1: k]$, the loadings of first k ordinary principal components.

2) Given fixed $\boldsymbol{\alpha}$, solve the following naive elastic net problem for $j=1, 2, \cdots, k$.

$$\boldsymbol{\beta}_j=\arg\min_{\beta^*}\boldsymbol{\beta}^{*\mathrm{T}}(\boldsymbol{X}^{\mathrm{T}}\boldsymbol{X}+\lambda)\boldsymbol{\beta}^* -2\boldsymbol{\alpha}_j^{\mathrm{T}}\boldsymbol{X}^{\mathrm{T}}\boldsymbol{X}\boldsymbol{\beta}^* +\lambda_{1,j}|\beta^*|_1 \tag{5.5.13}$$

3) For each fixed $\boldsymbol{\beta}$, do the SVD of $\boldsymbol{X}^{\mathrm{T}}\boldsymbol{X}\boldsymbol{\beta} = \boldsymbol{UDV}^{\mathrm{T}}$, then update $\boldsymbol{\alpha}= \boldsymbol{UV}^{\mathrm{T}}$.

4) Repeat steps 2)- 3), until $\boldsymbol{\beta}$ converges.

5) Normalization: $\hat{\mathbf{V}}_j=\dfrac{\beta}{|\hat{\beta}|}$, $j=1, 2, \cdots, k$

5.5.2.4 Adjusted Total Variance

The ordinary principal components are uncorrelated and their loadings are orthogonal. Let $\hat{\Sigma}= \boldsymbol{X}^{\mathrm{T}}\boldsymbol{X}$, then $\mathbf{V}^{\mathrm{T}}\mathbf{V} = \boldsymbol{I}_k$ and $\mathbf{V}^{\mathrm{T}}\hat{\Sigma}\mathbf{V}$ is diagonal. It is easy to check that only the loadings of ordinary principal components can satisfy both conditions. In Jolliffe & Uddin (2003)[6] the loadings were forced to be orthogonal, so the uncorrelated property was sacrificed. SPCA does not explicitly impose the uncorrelated components condition too.

Let $\hat{\boldsymbol{U}}$ be the modified PCs. Usually the total variance explained by $\hat{\boldsymbol{U}}$ is calculated by trace$(\hat{\boldsymbol{U}}^{\mathrm{T}}\hat{\boldsymbol{U}})$. This is unquestionable when $\hat{\boldsymbol{U}}$ are uncorrelated. However, if they are correlated, the computed total variance is too optimistic. Here we propose a new formula to compute the total variance explained by $\hat{\boldsymbol{U}}$, which takes into account the correlations among $\hat{\boldsymbol{U}}$.

Suppose $(\hat{\boldsymbol{U}}, i=1, 2, \cdots, k)$ are the first k modified PCs by any method. Denote $(\hat{\boldsymbol{U}}_{j,1}, \cdots, {}_{j-1})$ the reminder of $\hat{\boldsymbol{U}}_j$ after adjusting the effects of $\hat{\boldsymbol{U}}_1, \cdots, \hat{\boldsymbol{U}}_{j-1}$, that is

$$\hat{\boldsymbol{U}}_{j,1}, \cdots, {}_{j-1}=\hat{\boldsymbol{U}}_j-H_1, \cdots, {}_{j-1}\hat{\boldsymbol{U}}_j \tag{5.5.14}$$

where $H_{1,\ldots,j-1}$ is the projection matrix on $\hat{\boldsymbol{U}}_j$, $i=1, 2, \cdots, j-1$. Then the adjusted variance of $\hat{\boldsymbol{U}}_j$ is $|\hat{\boldsymbol{U}}_{j=1}, \cdots, {}_{j-1}|^2$, and the total explained variance is given by $\sum_{j=1}^{k}|\hat{\boldsymbol{U}}_{j=1}, \cdots, {}_{j-1}|^2$. When the modified PCs $\hat{\boldsymbol{U}}$ are uncorrelated, then the new formula agrees with trace$(\hat{\boldsymbol{U}}^{\mathrm{T}}\hat{\boldsymbol{U}})$. Note that the above computations depend on the order of $\hat{\boldsymbol{U}}_i$. However, since we have a natural order in PCA, ordering is not an issue here.

Using the QR decomposition, we can easily compute the adjusted variance. Suppose $\hat{\boldsymbol{U}}=\boldsymbol{QR}$, where $\boldsymbol{Q}$ is orthonormal and $\boldsymbol{R}$ is upper triangular. Then it is straightforward to see that

$$|\hat{\boldsymbol{U}}_{j=1}, \cdots, {}_{j-1}|^2=\boldsymbol{R}_{j,j}^2 \tag{5.5.15}$$

Hence the explained total variance is equal to $\sum_{j=1}^{k} \boldsymbol{R}_{j,j}^2$.

5.5.2.5 Computation Complexity

PCA is computationally efficient for both $n > p$ or $p \gg n$ data. We separately discuss the computational cost of the general SPCA algorithm for $n > p$ and $p \gg n$.

1) $n>p$

Traditional multivariate data fit in this category. Note that although the SPCA criterion is defined using $\boldsymbol{X}$, it only depends on $\boldsymbol{X}$ via $\boldsymbol{X}^{\mathrm{T}}\boldsymbol{X}$. A trick is to first compute the $p\times p$ matrix $\hat{\boldsymbol{\Sigma}}=\boldsymbol{X}^{\mathrm{T}}\boldsymbol{X}$ once for all, which requires np^2 operations. Then the same $\hat{\boldsymbol{\Sigma}}$ is used at each step within the loop. Computing $\boldsymbol{X}^{\mathrm{T}}\boldsymbol{X}\boldsymbol{\beta}$ costs $\boldsymbol{p}^2\boldsymbol{k}$ and the SVD of $\boldsymbol{X}^{\mathrm{T}}\boldsymbol{X}\boldsymbol{\beta}$ is of order $O(pk^2)$. Each elastic net solution requires at most $O(p^3)$ operations. Since $k\leqslant p$, the total computation cost is at most $np^2+mO(p^3)$, where m is the number of iterations before convergence. Therefore the SPCA algorithm is able to efficiently handle data with huge n, as long as p is small (say $p < 100$).

2) $p\gg n$

Gene expression arrays are typical examples of this $p\gg n$ category. The trick of $\hat{\boldsymbol{\Sigma}}$ is no longer applicable, because $\hat{\boldsymbol{\Sigma}}$ is a huge matrix ($p\times p$) in this case. The most consuming step is solving each elastic net, whose cost is of order $O(pJ^2)$ for a positive finite λ, where J is the number of nonzero coefficients. Generally speaking the total cost is of order $mO(pJ^2k)$, which is expensive for a large J.

5.5.3 SPCA for $p \geqslant n$ and Gene Expression Arrays

Gene expression arrays are a new type of data where the number of variables (genes) are much bigger than the number of samples. Our general SPCA algorithm still fits this situation using a positive λ. However the computation cost is expensive when requiring a large number of nonzero loadings. It is desirable to simplify the general SPCA algorithm to boost the computation.

Observe that theorem 3 is valid for all $\lambda > 0$, so in principle we can use any positive λ. It turns out that a thrifty solution emerges if $\lambda\to\infty$. Precisely, we have the following theorem.

Theorem 5

Let $\hat{\boldsymbol{V}}_i(\lambda)=\dfrac{\hat{\beta}_i}{|\hat{\beta}_i|}$ be the loadings derived from criterion (5.5.9). Define($\hat{\boldsymbol{\alpha}}^*$, $\hat{\boldsymbol{\beta}}^*$) as the solution of the optimization problem

$$(\hat{\boldsymbol{\alpha}}^*, \hat{\boldsymbol{\beta}}^*)=\arg\min_{\alpha,\beta} -2\mathrm{Tr}\boldsymbol{\alpha}^{\mathrm{T}}\boldsymbol{X}^{\mathrm{T}}\boldsymbol{X}\boldsymbol{\beta}+\sum_{j=1}^{k}\beta_j^2+\sum_{j=1}^{k}\lambda_{i,j}|\boldsymbol{\beta}_j|_1$$

$$\text{subject to } \boldsymbol{\alpha}\boldsymbol{\alpha}^{\mathrm{T}}=\boldsymbol{I}_k \tag{5.5.16}$$

When $\lambda \to \infty$, $\hat{\mathbf{V}}_i(\lambda) \to \frac{\hat{\beta}_i}{|\hat{\beta}_i|}$.

By the same statements in Section 5. 5. 2. 3, criterion (5. 5. 16) is solved by the following algorithm, which is a special case of the general SPCA algorithm with $\lambda \to \infty$.

Gene Expression Arrays SPCA Algorithm

Replacing step 2 in the general SPCA algorithm with

Step 2 * : Given fixed α, for $j = 1, 2, \cdots, k$

$$\beta = (|\boldsymbol{\alpha}_j^T \mathbf{X}^T \mathbf{X}|)_+ Sign(\boldsymbol{\alpha}_j^T \mathbf{X}^T \mathbf{X}) \tag{5. 5. 17}$$

The operation in equation (5. 5. 17) is called soft-thresholding. Fig. 5. 11 gives an illustration of how the soft-thresholding rule operates. Recently soft-thresholding has become increasingly popular in the literature. For example, nearest shrunken centroids (Tibshirani et al. 2002)[10] adopts the soft-thresholding rule to simultaneously classify samples and select important genes in microarrays.

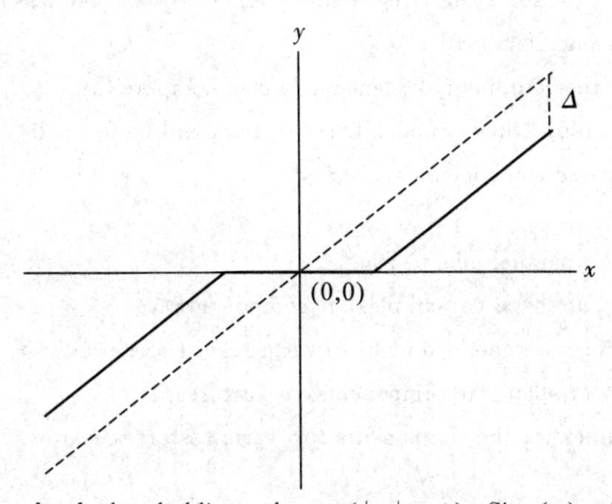

Fig. 5. 11 An illstration of soft-thresholding rule $y = (|x| - \Delta)_+ \text{Sign}(x)$ with $\Delta = 1$. the literature. For example, nearest shrunken centroids (Tibshirani et al. 2002) adopts the soft-thresholding rule to simultaneously classify samples and select important genes in microarrays

5.5.4 Demo of SPCA

Software version python 2. 7 and a Microsoft Windows 7 operating system. Sparse PCA from the sklearn package. The Sparse PCA is mainly composed of two functions fit and transform.

```
class SparsePCA(BaseEstimator, TransformerMixin):
    """Sparse Principal Components Analysis (SparsePCA)
    Finds the set of sparse components that can optimally reconstruct
    the data. The amount of sparseness is controllable by the coefficient
    of the L1 penalty, given by the parameter alpha.
    Parameters
```

```
    ----------
    n_components : int,
        Number of sparse atoms to extract.
    alpha : float,
Sparsity controlling parameter. Higher values lead to sparser
        components.
    ridge_alpha : float,
        Amount of ridge shrinkage to apply in order to improve
        conditioning when calling the transform method.
    max_iter : int,
        Maximum number of iterations to perform.
tol : float,
        Tolerance for the stopping condition.
    method : {'lars', 'cd'}
        lars: uses the least angle regression method to solve the lasso problem
        (linear_model. lars_path)
        cd: uses the coordinate descent method to compute the
        Lasso solution (linear_model. Lasso). Lars will be faster if
        the estimated components are sparse.
    n_jobs : int,
        Number of parallel jobs to run.
    U_init : array of shape (n_samples, n_components),
        Initial values for the loadings for warm restart scenarios.
    V_init : array of shape (n_components, n_features),
        Initial values for the components for warm restart scenarios.
    verbose :
        Degree of verbosity of the printed output.
    random_state : int or RandomState
        Pseudo number generator state used for random sampling.
    Attributes
    ----------
    components_ : array, [n_components, n_features]
        Sparse components extracted from the data.
    error_ : array
        Vector of errors at each iteration.
    n_iter_ : int
        Number of iterations run.
    """
def __init__(self, n_components=None, alpha=1, ridge_alpha=0.01,
                 max_iter=1000, tol=1e-8, method='lars', n_jobs=1, U_init=None,
                 V_init=None, verbose=False, random_state=None):
        self.n_components = n_components
        self.alpha = alpha
```

```
        self.ridge_alpha = ridge_alpha
        self.max_iter = max_iter
        self.tol = tol
        self.method = method
        self.n_jobs = n_jobs
        self.U_init = U_init
        self.V_init = V_init
        self.verbose = verbose
        self.random_state = random_state

    def fit(self, X, y=None):
        """Fit the model from data in X.
        Parameters
        ----------
        X: array-like, shape (n_samples, n_features)
            Training vector, where n_samples in the number of samples
            and n_features is the number of features.
        Returns
        -------
        self : object
            Returns the instance itself.
        """
        random_state = check_random_state(self.random_state)
        X = check_array(X)
        if self.n_components is None:
            n_components = X.shape[1]
        else:
            n_components =self.n_components
        code_init =self.V_init.T if self.V_init is not None else None
        dict_init =self.U_init.T if self.U_init is not None else None
        Vt, _, E, self.n_iter_ = dict_learning(X.T, n_components, self.alpha,
                                               tol=self.tol,
                                               max_iter=self.max_iter,
                                               method=self.method,
                                               n_jobs=self.n_jobs,
                                               verbose=self.verbose,
                                               random_state=random_state,
                                               code_init=code_init,
                                               dict_init=dict_init,
                                               return_n_iter=True
                                               )

        self.components_ = Vt.T
        self.error_ = E
```

```
        return self

    def transform(self, X, ridge_alpha=None):
        """Least Squares projection of the data onto the sparse components.
        To avoid instability issues in case the system is under-determined,
        regularization can be applied (Ridge regression) via the
        'ridge_alpha' parameter.
        Note that Sparse PCA components orthogonality is not enforced as in PCA
        hence one cannot use a simple linear projection.
        Parameters
        ----------
        X: array of shape (n_samples, n_features)
            Test data to be transformed, must have the same number of
            features as the data used to train the model.
        ridge_alpha: float, default: 0.01
            Amount of ridge shrinkage to apply in order to improve
            conditioning.
        Returns
        -------
        X_new array, shape (n_samples, n_components)
            Transformed data.
        """
        check_is_fitted(self, 'components_')

        X = check_array(X)
        ridge_alpha = self.ridge_alpha if ridge_alpha is None else ridge_alpha
        U = ridge_regression(self.components_.T, X.T, ridge_alpha,
                                    solver='cholesky')
        self.u = U
        s = np.sqrt((U ** 2).sum(axis=0))
        s[s == 0] = 1
        U /= s
        return U
>>>import numpy as np
>>>from sklearn.datasets import make_friedman1
>>>from sklearn.decomposition import SparsePCA
>>>X, _ = make_friedman1(n_samples=200, n_features=30, random_state=0)
>>>transformer = SparsePCA(n_components=5, normalize_components=True,
...         random_state=0)
>>>transformer.fit(X)
SparsePCA(...)
>>>X_transformed = transformer.transform(X)
>>>X_transformed.shape
(200, 5)
```

References

[1] CADIMA J, JOLLIFFFE I. Loadings and correlations in the interpretation of principal components , Journal of Applied Statistics, 1995(22):203 - 214.

[2] HASTINE T, TIBASHIRANI R, EISEN M, et al. Gene shaving as a method for identifying distinct sets of genes with similar expression patterns. Genome Biology, 2000(1):1 - 21.

[3] HASTIE T, TIBSHIRANI R, RIEDMAN J. The Elements of Statistical Learning; Datamining, Inference and Prediction, Springer Verlag, New York. 2001.

[4] JOLIFFFFE I. Principal component analysis. Springer Verlag, New York. 1986.

[5] JOLIFFFFE I. Rotation of principal components: choice of normalization constraints. Journal of Applied Statistics, 1995(22): 29 - 35.

[6] JOLIFFFFE I T, UDDIN M. A modifified principal component technique based on the lasso. Journal of Computational and Graphical Statistics, 2003(12):531 - 547.

[7] MCCABE G. Principal variables. Technometrics, 1984(26): 137 - 144.

[8] MISRA J, SCHMITT W, HWANG D, et al. Interactive exploration of microarray gene expression patterns in a reduced dimensional space. Genome Research, 2002(12): 1112 - 1120.

[9] TIBSHIRANI R. Regression shrinkage and selection via the lasso. Journal of the Royal Statistical Society. Series B, 1996, 58: 267 - 288.

[10] TIBSHIRANI R, HASTIE T, NARASIMHAN B, et al. Diagnosis of multiple cancer types by shrunken centroids of gene, Proceedings of the National Academy of Sciences, 2002(99): 6567 - 6572.

[11] VINES S. Simple principal components , Applied Statistics, 2000, 49: 441 - 451.

[12] ZOU H, HASTIE T. Regression shrinkage and selection via the elastic net, with applications to microarrays. Technical report. Department of Statistics, Stanford University, 2003.

Chapter 6 Transfer Method

This chapter will introduce transfer methods which extend the linear regression model established in the one domain to the others. Through the phenomenon that the spectral signals measured by different infrared spectroscopes are different, the partial least squares model of infrared spectroscopy is not suitable for other instruments. This is calibration transfer learning problem which is introduced firstly and then two PLS feature-based calibration transfer methods and their Python path are given, so does as the experimental results and data analysis process.

6.1 Calibration Transfer of Spectral Models[1]

6.1.1 Introduction

Within the past decades, the combination of spectroscopic measurements and multivariate calibration techniques has become increasingly applied and a widely acknowledged approach for the extraction of (bio-) chemical information in various applications fields. In this chapter we will focus on near-infrared (NIR) spectral data, as commonly used for on-line/in-line monitoring of (bio-) chemical processes, quality control of pharmaceutical, petroleum, agriculture and other products, environmental analysis, medical diagnostics and academic research[2-7]. For the extraction of quantitative information from the rather featureless NIR spectra, multivariate calibration techniques such as principal component regression (PCR)[8-10] and partial least squares (PLS)[11, 12] have proven successful[13-16].

In general, the process of data collection, model calibration and model optimization (e. g. with respect to outlier detection, data preprocessing, variable selection and the determination of a reasonable model complexity) is a time-consuming and cost-intensive one. As such, one naturally hopes that the obtained model remains adequate for the intended purpose, reliable and accurate for an extended time period. Different approaches for the development of such a robust calibration model have been proposed and are e. g. discussed in [17, 18, 14, 19]. Unfortunately, changes in the environmental conditions, an adaptation of the measurement setup, as e. g. arising of required maintenance operations, or an intended or unintended modification of the measured substance itself can all affect the spectral measurements and result in a calibration model that is no longer adequate for the intended purpose[14, 20, 2, 3]. In such a case, there are two possibilities to overcome this problem: (i) the calibration of a completely new model or (ii) the application of methods enabling the incorporation of information on the original calibration model and/or data.

Techniques within the second option are preferable, as they promise to be more effective, and are therefore widely studied in literature. They are referred to as calibration transfer or instrument standardization in chemometrics, while being known as transfer learning, domain adaptation or multi-task learning tools in the machine learning community. In this contribution, we propose novel calibration transfer techniques that do not require the availability of (reall) transfer standards, i. e. a set of samples measured under both the old and new measurement/environmental/sample condition, and can be applied using no or only few reference measurements (of the desired response variable) in the new setting. These calibration transfer approaches draw on methods from machine learning (e. g. transfer component analysis (TCA) [21, 22]) and sophisticated techniques from the area of chemometrics and are evaluated on two different data sets (one of which is publicly accessible) in the Section.

6.1.2 Calibration Transfer Setting

6.1.2.1 Notation and Assumptions

In the following we denote information (data, model) originating from the primary measurement/environmental/sample condition as master information and information corresponding to the new setting as slave or secondary information: $\boldsymbol{X}_m$ and $\boldsymbol{X}_S$ denote the $n_{M\times p}$ and $n_{S\times p}$ matrices of spectral measurements in the master (M) and slave (S) setting, respectively, where $\boldsymbol{x}_i$ denotes the i-th row of $\boldsymbol{X}$ and corresponds to a single spectrum. The reference measurements of the response variable in the master and slave setting, are denoted as $\boldsymbol{y}_M$ and $\boldsymbol{y}_S$ respectively. Predictions are generally denoted by $\hat{\boldsymbol{y}}$. If there are no values of the response variable available (i. e. no y value) for a set of spectra (that is no reference measurement has been conducted), then these spectra are referred to as 'without-response' and are denoted as $\boldsymbol{X}_S$, while spectra coming with reference value are simply written as $\boldsymbol{X}_S$ or ($\boldsymbol{X}_S$) and called "with-response".

In this section, we assume the following data (Fig. 6. 1) to be available at the time of transfer: Spectra $\boldsymbol{X}_M$ have been used together with their reference measurements $\boldsymbol{y}_M$ to build the master calibration model. Few spectra $\boldsymbol{X}_S$ together with the responses $\boldsymbol{y}_S$ and a set of "without-response" spectra, $\boldsymbol{X}_S^o$ are available in the slave setting to perform the calibration transfer. We generally find us in the setting where there are (much) more master data than slave data ($n_M \gg n_S$) and there are (much) more spectra without responses than with responses ($n_S^o \geqslant n_S$). This is due to the fact that spectra do basically come for free in an in-line application while corresponding reference measurements for getting the response $\boldsymbol{y}_S$ are costly. Hence, any method which is capable to exploit information contained in X_S^o is beneficial.

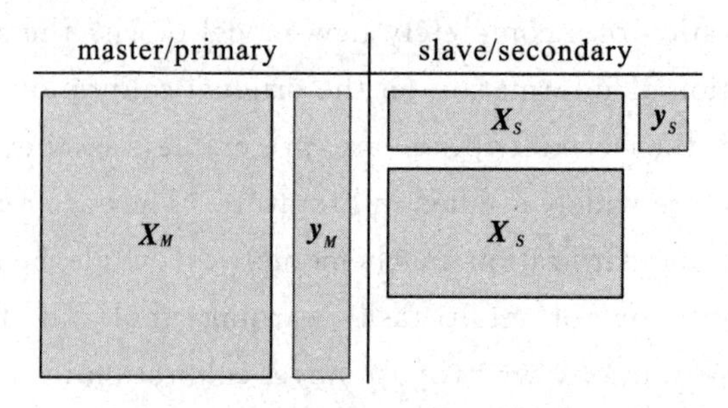

Fig. 6.1 Assumed Data Setting. We assume the following data to be available at the time of transfer: The master/primary data (left) consisting of spectral measurements X_M and the corresponding response y_M obtained by reference measurements. In the slave/secondary setting there are also spectral measurements X_S and the corresponding response y_S (although typically less: $n_S < n_M$) and in addition there are 'without-response' spectra X_S^o

6.1.2.2 Transfer Standards

In chemometric literature, transfer methods are generally based on the analysis of differences between the master and slave setting. Towards this end, the vast majority of approaches requires a set of samples, referred to as transfer standards, to be measured under both condition. A comparison of the corresponding spectral responses, Z_M and Z_S, provides information on the difference between the settings and can then be used to develop a way to correct them.

While the number of transfer standards does not have to be excessively large, in general, it has been shown that they need to be very well chosen in order to guarantee a successful transfer[14, 17, 23, 24]. Specifically, they should exhibit two important properties: representativity and stability[17]. It is intuitively clear that transfer standards should be representative for the secondary setting, as otherwise the computed correction may only cover a part of the changes occurring between master and slave. Chemical and physical stability is important, if transfer standards have to be stored for some time. In general, a lack of stability or representativity can be mitigated by the choice of an appropriate selection approach.

There are mainly three approaches to differentiate:

(a) The selection of a representative subset of samples among the master data.

(b) the selection of a representative subset of samples among the slave data.

(c) the use of an independent set of samples.

In order to ensure the selection of a representative sample in (a) and (b), different techniques have been developed. These include the selection of samples with high leverage, as proposed in related paper[25], and the application of the Kennard-Stone algorithm[26], as suggested in [18]. Assuming that a representative sample can be found in (a) and (b), there are still the following issues to consider:

Regarding (a) it is clear that the chosen set of samples needs to be measured in the slave setting. This is only possible, if the transfer samples are stable enough to still be available in unmodified condition at that time.

Approach (b) uses samples from the slave setting as transfer standards; thus, stability should not be much of an issue here. For the samples to be re-measured under the master condition, however, the master setting has to still exist at the time of slave measuring. In case of maintenance operations, as the required change of a spectral light source covered in the considered new industrial application, the master measurement setup is replaced by the slave measurement setup and therefore no longer available.

For approach (c) to be effective, the chosen independent / generic transfer standards need to vary in the areas of the spectrum upon which the calibration depends[20]. In addition to this restriction, they certainly need to be able and permitted to be measured in the master and slave setting. Regrettably, this is not the case for the NIR in-line installation of the investigated application.

As such, we believe that calibration transfer methods, which do not require explicit transfer standards are highly beneficial in industrial inline settings as the considered melamine resin production. This leaves the question how else to deal with such a situation. We believe that there are only two options, namely the one of applying methods not requiring transfer standards at all or the option of developing artificial transfer standards enabling the use of common standard-based transfer techniques. Both options will be studied in this contribution.

6.1.3 Related Work

While there are several ways to divide the different transfer approaches into categories, we decided for the separation into methods not requiring transfer standards and ones requiring transfer standards, where the second group becomes relevant for this contribution in the sense that we propose a way to replace real transfer standards by artificial ones in this section.

6.1.3.1 Methods Not Requiring Transfer Standards

Model updating. In settings, where the direct application of the master model is no option, a natural approach is to incorporate the new variance of the slave data and to calibrate a new model on the augmented calibration set[14]. This technique, referred to as $M+S$ in the following, is very common in literature and practice. In cases where $n_M \gg n_S$, however, the master data may simply outweigh the incorporated slave information and result in rather inaccurate predictions[27, 24, 14]. Thus, either more slave data need to be collected or an up-weighting of the slave information may be attempted. In addition to these straightforward approaches, automatic model updating methods have been studied[20]. As noted in [20], though, it is hard to imagine that these methods would be allowed in aregulated environment. A potential disadvantage of model updating techniques is that they require the training, and usually also optimization, of a new calibration model.

Prediction correction. A common approach to obtain predictions for the slave case in practice consists of the idea to perform a univariate slope and bias correction (USBC) of the predictions from the master calibration model. This correction relates the predictions from the master model via a linear model to the slave domain. The prediction $\hat{y}_i$ for a new spectrum $\boldsymbol{x}_i$ from the slave setting is obtained as $\hat{y}_i = k \cdot \hat{y}_{M,i} + d$, with , $\hat{y}_{M,i} = \boldsymbol{x}_i \boldsymbol{\beta} + \boldsymbol{\beta}_0$ where $\boldsymbol{\beta}$ and $\boldsymbol{\beta}_0$ are the regression coefficients of the master model. k and d are found as the parameters for the (least squares) regression line through the points $\{(y_i, \hat{y}_i)\}$, $i=1, 2, \cdots, n_S$. A major advantage of this method is that no new calibration model needs to be trained and optimized and theoretically only a few reference samples in the slave domain are needed. Due to its simple univariate nature, though, USBC is generally only able to cope with rather simple master to slave changes.

6.1.3.2 Correction of Master and Slave Spectra

If no transfer standards are available, a natural approach to guarantee a better matching of spectral responses and thereby a more robust model is the application of common signal pre-processing methods for both master and slave spectra[14]. The most commonly used pre-processing techniques can be classified in methods applying simple mathematical operations and signal centering or scaling, ones performing noise, offset and baseline filtering, as done via baseline correction, smoothing and derivative calculation, and techniques applying sample normalization, e. g. via multiplicative scatter/signal correction (MSC) and standard normal variate (SNV) normalization[14, 28, 2, 29].

More advanced pre-processing approaches include: finite impulse response (FIR) filtering[13], a modified FIR approach introduced in[30], generalized moving window MSC (MW-MSC)[2] and a preprocessing via wavelets[14, 31].

Above these techniques, multivariate filtering via orthogonal signal correction (OSC) has been investigated. OSC[32, 33] aims at the removal of information in the spectra that is orthogonal to the response variable. It may be used as a common pre-processing method or be applied in a slightly modified version for calibration transfer. In the latter case, a small set of data ($\boldsymbol{X}_S$ and $\boldsymbol{y}_S$) in the slave setting and a possibly different small set of data ($\boldsymbol{X}_M$ and $\boldsymbol{y}_M$) in the master domain are combined and used together in the common OSC algorithm. Details on how this approach is used/modified for our contribution are found in Section 6.2.

All of these pre-processing methods have in common that a new calibration model needs to be calibrated. Additionally, it has been noted that simple pre-processing techniques may not be able to cope with more complicated changes, while the more advanced ones tend to reduce net analyte signal as they remove variation not common to both master and slave spectra[20]. In this chapter, we only include results for our proposed version of transfer OSC.

Correction of slave spectra. Typical approaches to correct slave spectra in a way to achieve data that are similar to corresponding data obtained under the master setting are usually based on transfer standards. Two basic standard-free techniques, however, are listed under Section 6. 2. Methods in this class bear the advantage that no new calibration model needs to be trained and optimized. However, similar to prediction correction, only relatively simple master to slave changes are usually covered well by these methods.

6. 1. 3. 3 Robust Models via Projection

The idea to project master and slave data to a common feature space has for instance been followed in a machine learning technique referred to as transfer component analysis (TCA)[21]. The basic concept behind TCA is the intention to make the master and slave data distributions as similar as possible while preserving crucial (geometric and statistical) properties of the original data. The question is the notion of similarity. In principle any similarity measures of probability distributions can be employed; starting from simple ones like requiring similar means and covariances (which is necessary but not sufficient) up to more complex measures like Kullback-Leibler divergence or Bhattacharyya distance, Pan et al. implement this idea via the learning of a shared subspace between the master and slave data that is able to deal with nonlinearities as well as complex changes in the data employing the maximum mean discrepancy in the common feature space. for details see [21, 22]. Once the subspace is found, standard machine learning methods can be used to train classifiers or regression models across domains. Further techniques belonging to this group include kernel principal component analysis (KPCA)[34-36], domain generalization via invariant feature representation (DICA)[37], covariance operator inverse regression (COIR)[38] as well as approaches to uncover a joint (latent) feature subspace in multi-task learning (MTL) settings[39, 40].

An advantage of TCA and related methods consists of the possible use of different kernels, thereby allowing to model complex changes also. On the negative side, they generally require a new model to be calibrated and computations involved within the process to determine the projected space are rather complex and expensive in terms of time and computational resources.

6. 1. 3. 4 Methods Requiring Transfer Standards

As has been noted in Section 6. 1. 2. 2, we study two options to deal with the lack of transfer standards in the investigated industrial application: One is the application of methods not requiring transfer standards at all and the other one is the option to develop artificial transfer standards enabling the use of common standard-based transfer techniques. Approaches falling in the former category have been presented in Section 6. 1. 3. 1, while standard-based techniques and the corresponding artificial transfer standards are covered here and in Section 6. 2, respectively.

Correction of slave spectra. As methods in this category require the availability of transfer standards, they are widely known as standardization methods in chemometrics.

The majority of transfer approaches proposed in chemometric literature belongs to this group[17, 14]. They compare the spectral responses of transfer standards, $\boldsymbol{Z}_M$ and $\boldsymbol{Z}_S$, measured under both, the master and slave, settings and estimate a matrix $\boldsymbol{F}$ transferring the slave spectra $\boldsymbol{Z}_S$ as closely as possible to the corresponding master spectra $\boldsymbol{Z}_M = \boldsymbol{Z}_S \cdot \boldsymbol{F}$. If applied to new slave spectra, $\boldsymbol{X}_S^*$, the matrix $\boldsymbol{F}$ is assumed to achieve $\boldsymbol{X}_M^* \approx \boldsymbol{X}_S^* \boldsymbol{F}$ where $\boldsymbol{X}_M^*$ denotes the spectral responses that would have been obtained if the new samples had been measured under the master condition. Hence, any methods for computing $\boldsymbol{F}$ enables to use the original master model for the transferred slave spectra. Several methods exist to compute a transfer matrix $\boldsymbol{F}$.

For one of the simplest methods, introduced by Shank and Westerhaus[41] the transfer matrix contains non-zero entries only along the diagonal, i. e. constitutes a simple slope and bias correction for each wavelength (which in the most simple case may be the same for each wavelength).

Arguably the most prominent and widely used methods are direct standardization (DS) and piecewise direct standardization (PDS)[25, 42, 14].

For DS, the transfer matrix $\boldsymbol{F}^{DS}$ is computed as the pseudo-inverse of $\boldsymbol{Z}_S$ via singular value decomposition[29, 43]. A way to overcome the commonly occurring overfitting[44,14] is to reduce the number of wavelengths involved in the mapping process. This represents the idea underlying the PDS approach.

In PDS, each wavelength of the master setting is related to a window of wavelengths in the slave setting resulting in a banded transfer matrix, $\boldsymbol{F}^{PDS}$. This assumption is in accordance with the perception that for many transfer applications the spectral correlations between master and slave are limited to smaller regions[14]. A drawback of the PDS procedure is the need for an optimization of p PLS or PCR models. The determination of a reasonable number of latent variables to be used in these models and the choice of a suitable window size represent a crucial task and may lead to artifacts and/or discontinuities in the PDS transferred slave spectra, if not performed carefully[42, 3]. Different approaches for a possible solution or reduction of this problem have been proposed in others literature[42, 45, 46]. PDS is typically employed as a reference for other novel techniques[14] , as its local character and multivariate nature cover a wide variety of transfer settings and, if not affected by artifacts, do generally yield reasonably accurate results for small sets of transfer samples already[20].

An alternative approach in this group is referred to as spectral space transformation (SST)[3].

Correction of master and slave spectra. A standard-based method correcting both master and slave spectra before the development of a new slave calibration model is generalized least squares weighting (GLSW)[47, 48]. GLSW addresses the objective of removing variation not common in the master and slave setting. A graphical illustration and corresponding explanation on the development of the GLS weighting matrix has been provided in [48]. There is only one parameter to be optimized in GLSW. An appropriate value can usually be determined via the minimization of some form of error criterion. All drawbacks listed in the corresponding category of standard-free methods do also apply to GLSW. Naturally, the requirement of transfer standards represents another restriction.

Robust models via projection. A number of advanced standard-based transfer projection techniques has been published in literature. Among these are: Canonical correlation analysis (CCA)[49], spectral regression (SR)[44], transfer by orthogonal projection (TOP)[50-53] and error removal by orthogonal subtraction (EROS)[54, 51, 53].

6.1.4 New or Adapted Methods

6.1.4.1 Correction of Slave Spectra

The following two basic standard-free techniques were recently introduced by the authors in [55] and are simple instantiations of the more general idea to bring the probability distribution of the slave observations closer to the probability distribution in the master domain. They are motivated by the assumption that a linear filter may be sufficient to model a change in the measurement setting. Such a linear filter is supposed to be approximated as an element-wise multiplication, more suitable for intensity spectra, or element-wise addition, suitable for absorption spectra, in the wavelength domain. Hence, an additive mean correction (AMC), performs a subtraction of the mean slave spectrum and the addition of the mean master spectrum, thereby forcing the mean of the corrected slave spectra to equal the mean of the master spectra. For a spectrum, $\boldsymbol{x}_S$ in the slave space the correction is performed as:

$$\boldsymbol{x}_M^{\mathrm{AMC}} = \boldsymbol{x}_S - \boldsymbol{\mu}_S + \boldsymbol{\mu}_M \tag{6.1.1}$$

where $\boldsymbol{\mu}_M$ denotes the mean master spectrum and $\boldsymbol{\mu}_S$ the mean slave spectrum. In this contribution, all available slave spectra (with and without corresponding responses) are used to compute $\boldsymbol{\mu}_S$.

6.1.4.2 Correction of Master and Slave Spectra

In this chapter, we apply the transfer OSC approach described in [28] by combining the same number of master and slave spectra we then mean center the resulting matrix, accompany it with the corresponding autoscaled vector of combined reference response values and hand these data over to the OSC algorithm. We fix the parameters tolerance of initial iterations and the tolerance of reconstruction[56] while the most crucial parameter, the number of latent OSC variables to remove from the data, is optimized. This is performed utilizing the "without-response" slave spectra in a way to minimize the Euclidean distance between the mean OSC processed master spectrum and the mean OSC processed "without-response" slave spectrum:

$$k^{\mathrm{OSC}} = \arg\min_{k\in 1,2,\cdots,K} \| \boldsymbol{\mu}_M^{\mathrm{OSC}} - \boldsymbol{\mu}_S^{\mathrm{OSC}}(k) \|_2 \tag{6.1.2}$$

where $\boldsymbol{\mu}_M^{\mathrm{OSC}}(k)$ denotes the mean of the OSC transferred master spectra when k latent OSC variables are used and $\boldsymbol{\mu}_S^{OSC}(k)$ denotes the corresponding vector for the "without-response" slave data. The idea to perform parameter optimization in this way is novel and enables access to information contained in the additionally available "without response" slave spectra. Note that the use of the Euclidean distance between mean spectra is used for simplicity here. Any other measure of spectral similarity measure may also be utilized.

6.1.4.3 Robust Models via Projection

This chapter combines TCA with ordinary least squares (OLS) regression, denoted as TCR in the following. While the original TCA approach incorporates only spectral information, we developed a modified version that enables the use of all available data in some form. The underlying experimental protocol is depicted in Fig. 6.2 and summarized as follows: A certain fraction (e.g. 0.5) of the available master data is drawn randomly, combined with all "without-response" spectra $\boldsymbol{X}_S^o$ from the slave domain and used to learn the shared TCA subspace[21, 22]. Afterwards, concatenate the remaining master samples with the data from the slave domain and use this information to optimize TCR's parameters. The use of the spectral data with its response variable from both master and slave domains to find the optimal setting of parameters, has a considerable positive effect on TCR's performance in practice. A related approach is to replace TCA by PCA in the above mentioned protocol. We refer to this method as PCRT. While the original TCA/PCA formulation uses only "without-response" data for the determination of the latent space, the proposed novel version of TCA/PCA utilizes all available data in some form. In this way, we expect more accurate models to be obtained.

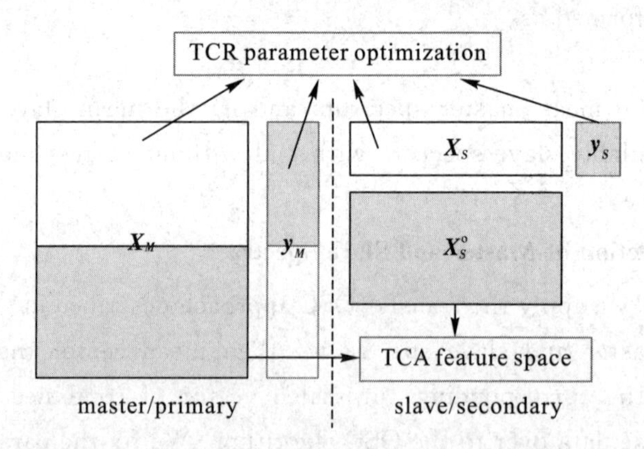

Fig. 6.2 Data used to learn the common TCA feature space and data used for the TCR parameter optimization

6.1.5 Standard-free Alternatives to Methods Requiring Transfer Standards

In industrial applications where no transfer standards can be obtained, a procedure similar to the one investigated in dynamic orthogonal projection (DOP)[50, 51] may be utilized to determine artificial/virtual transfer standards. A practical procedure proposed here is based on a small set of data with responses which is available for the slave setting and performs the following steps:

• For each sample in the slave space with a response value y, look for all master samples($\boldsymbol{x}_i$, $\boldsymbol{y}_i$) with reference measurements y_i being close to $\boldsymbol{y}$($|\boldsymbol{y}-\boldsymbol{y}_i|<\varepsilon$ and ε chosen depending on the response range). If additional information differentiating the samples is available, then this may be additionally incorporated in such a selection step. In the new industrial application considered in this work, e. g., the knowledge of recipe numbers (in certain cases, a common master model is built for products based on several related recipes) is available and can be used to select a smaller set of master samples assumed to be very similar to the considered slave sample.

• We then compute the mean over these selected master spectra and use this mean spectrum in place of a real transfer standard measurement.

Via this procedure, all methods requiring transfer standards become available for our and similar industrial applications. Results will be presented for PDS and GLSW, where the use of artificial transfer standards will be indicated by a prefix A-.

To exploit the available 'without-response' slave data, parameter optimization for A-PDS and A-GLSW is performed in a similar way to the one outlined for OSC transformed spectra, i. e. in case of A-PDS the term

$$\| \boldsymbol{\mu}_M - \boldsymbol{\mu}_S^{\mathrm{PDS}}(w) \|_2 \tag{6.1.3}$$

is minimized over different window widths w, and for A-GLSW we minimize

$$\| \boldsymbol{\mu}_M^{\mathrm{GLSW}}(\alpha) - \boldsymbol{\mu}_S^{\mathrm{GLSW}}(\alpha) \|_2 \tag{6.1.4}$$

with respect to the parameter α, where $\boldsymbol{\mu}_M$ denotes the mean over the (non-transferred) master spectra, $\boldsymbol{\mu}_S^{\mathrm{PDS}}(w)$ and $\boldsymbol{\mu}_S^{\mathrm{GLSW}}(\alpha)$ are the means over the A-PDS and A-GLSW transferred "without-response" slave spectra $\boldsymbol{X}_S^o$ based on the corresponding parameters, respectively, and $\boldsymbol{\mu}_M^{\mathrm{GLSW}}(\alpha)$ denotes the mean over the A-GLSW transferred master data using the parameter value α in GLSW. As already noted, the Euclidean distance between mean spectra is used for simplicity and may be replaced by any other measure of spectral similarity.

References

[1] MALLI, B, BIRLUTIU A, NATSCHLÄGER T. Standard-free calibration transfer—An evaluation of different techniques. Chemom. Intell. Lab. Syst. 2017,161: 49 - 60.

[2] KRAMER K E, MORRIS R E, ROSEPSEHRSSON S L. Comparison of two multiplicative signal correction strategies for calibration transfer without standards, Chemom. Intell. Lab. Syst. ,2008,92 (1):33 - 43.

[3] DU W. Maintaining the predictive abilities of multivariate calibration models by spectral space transformation, Anal. Chim. Acta, 2011, 690 (1): 64 - 70.

[4] HUANG H B. Near infrared spectroscopy for on/in-line monitoring of quality in foods and beverages: a review, J. Food Eng. , 2008, 87(3): 303 - 313.

[5] LOPES J A, COSTA P F, ALVES T P, et al. Chemometrics in bioprocess engineering: process analytical technology (PAT) applications, Chemom. Intell. Lab. Syst. ,2004,74(2):269 - 275.

[6] ROGGO Y, CHALUS P, LEMAMARTINEZ L C, et al. A review of near infrared spectroscopy and chemometrics in pharmaceutical technologies, J. Pharm. Biomed. Anal. ,2007,44(3):683 - 700.

[7] CEN H Y, HE Y. Theory and application of near infrared reflectance spectroscopy in determination of food quality, Trends Food Sci. Technol,2007,18(2):72 - 83.

[8] NAES T, MARTENS H. Principal component regression in NIR analysis: view-points, background details and selection of components, J. Chemom. 1988,2(2):155 -167.

[9] GELADI P, ESBENSEN K. Regression on multivariate images: principal component regression for modeling, prediction and visual diagnostic tools, J. Chemom. ,1991,5(2):97 - 111.

[10] JOLLIFFE I T. A note on the use of principal components in regression, J. R. Stat. Soc. Ser. C (Appl. Stat.),1982,31(3):300 - 303

[11] WOLD H. Estimation of Principal Components and Related Models by Iterative Least squares. Academic Press, New York, 1966.

[12] WOLD S, SJOSROM M, ERIKSSON L. PLS-regression: a basic tool of chemometrics, Chemom. Intell. Lab. Syst. ,2001,58(2):109 - 130.

[13] BLANK T B, SUM S T, BROWN S D, et al. Transfer of near-infrared multivariate calibrations without standards, Anal. Chem. ,1996,68(17):2987 - 2995.

[14] FEUDALE R N, WOODY N A, TAN H, et al. Transfer of multivariate calibration models: a review, Chemom. Intell. Lab. Syst. ,2002,64(2):181 - 192.

[15] BUCHANAN B, HONIGS D. Trends in near-infrared analysis, TrAC Trends Anal. Chem. ,1986, 5(6):154 - 157.

[16] WIESNER K, FUCHS K, GIGLER A M, et al. Trends in near infrared spectroscopy and multivariate data analysis from an industrial perspective, Procedia Eng. ,2014,87:867 - 870.

[17] BOUVERESSE E, MASSART D. Standardisation of near-infrared spectrometric instruments: a review, Vib. Spectrosc,1996,11(1):3 - 15.

[18] NOORD O E D. Multivariate calibration standardization, Chemom. Intell. Lab. Syst. ,1994,25(2): 85 - 97.

[19] WANG D N, BROWN S D, MAN R. Stacked PLS for calibration transfer without standards, J. Chemom,2011,25(3):130 - 137.

[20] WISE B M, ROGINSKI R T. A calibration model maintenance roadmap, IFAC-Pap. Online,2015, 48(8):260 - 265.

[21] PAN S J, TSANG I, KWORK J, et al. Domain adaptation via transfer component analysis, Neural Netw. IEEE Trans. ,2011,22(2):199 - 210.

[22] PAN S J, TSANG I, KWORK J, et al. Transfer learning via dimensionality reduction, in: Proceedings of the 23rd National Conference on Artificial Intelligence, Volume 2, AAAI'08, AAAI Press, 2008:677 - 682.

[23] SIANO G G, GOICOECHEA H C. Representative subset selection and standardizationtechniques. A comparative study using NIR and a simulated fermentative processUV data, Chemom. Intell. Lab. Syst. ,2007,88(2):204 - 212.

[24] CAPRON X, WALCZAK B, NOORDO D, et al. Selection and weighting of samplesin multivariate regression model updating, Chemom. Intell. Lab. Syst. 2005,76(2):205 - 214.

[25] WANG Y D, VELTKAMP D J, KOWALSKI B R. Multivariate instrument standardization, Anal. Chem. ,1991,63(23):2750 - 2756.

[26] KENNARD R W, STONEL A. Computer aided design of experiments, Technometrics, 1969, 11 (1):137 - 148.

[27] STORK C L, KOWALSKI B R, Weighting schemes for updating regression models - atheoretical approach, Chemom. Intell. Lab. Syst,1999,48(2):151 - 166.

[28] WOODY N A, FEUDALE R N, MYLES A J, et al. Transfer of multivariate calibrations between four near-infrared spectrometers using orthogonal signal correction, Anal. Chem. ,2004,76(9):2595 - 2600.

[29] WISE B, GALLAGHE BRO N R, SHAVER J, et al. Chemometrics tutorial for PLS_tool box and Solo, Eigenvector Research, Inc., 3905.

[30] TAN H, SUMS T, BROWN S D, Improvement of a standard-free method for near infrared calibration transfer, Appl. Spectrosc,2002,56(8):1098 - 1106.

[31] GELADI P, BARRING H, DABAKK E, et al. Calibration transfers for predicting lake-water pH from near infrared spectra of lake sediments, J. Infrared Spectrosc,1999,7(4):251 - 264.

[32] WOLD S, ANTTI H, LINDGRE F N, et al. Orthogonal signal correction of near infrared spectra, Chemom. Intell. Lab. Syst. ,1998,44 (1 - 2):175 - 185.

[33] SJOBLOM J, SVENSSON O, JOSEFSON M, et al. An evaluation of orthogonal signal correction applied to calibration transfer of near infrared spectra, Chemom. Intell. Lab. Syst. ,1998,44(1): 229 - 244.

[34] SCHOLKOPF B, SMOLA A, MULLER K R. Kernel principal component analysis, in: Artificial Neural Networks ICANN'97, Springer, 1997: 583 - 588.

[35] SCHILKOPF B, SMOLA A, MULLER K R. Nonlinear component analysis as a kernel eigen value problem, Neural Comput,1998,10:1299 - 1319.

[36] FUKUMIZU K, BACH FR, JORDAN MI. Dimensionality reduction for supervisedlearning with reproducing kernel hilbert spaces, J. Mach. Learn. Res. ,2004,5:73 - 99.

[37] MUANDET K, BALDUZZI D, SCHOLKOPF B. Domain generalization via invariant feature representation, in: Proceedings of the 30th International Conference on Machine Learning (ICML - 13), 2013:10 - 18.

[38] KIM M, PAVLOVIC V. Central subspace dimensionality reduction using covariance operators, IEEE Trans. Pattern Anal. Mach. Intell. ,2011,33(4):657 - 670.

[39] ARGYRION A, EVGENIOU T, PONTIL M. Multi-task feature learning, in: Proceedings of the Advances in Neural Information Processing Systems 19 of the Twentieth Annual Conference on Neural Information Processing Systems, Vancouver, British Columbia, Canada, December 4 - 7, 2006,200(6):41 - 48. URL ? http://papers. nips. cc/paper/3143 - multi - task - feature - learning?

[40] BIRLUTIU A, GROOT P, HESKES T. Multi-task preference learning with an application to hearing aid personalization, Neurocomputing,2010,73(7 - 9):1177 - 1185.

[41] SHENK J S, WESTERHAUS M O. US Patent 4866644,1989.

[42] BOUVERESSE E, MASSART D. Improvement of the piecewise direct standardisation procedure for the transfer of NIR spectra for multivariate calibration, Chemom. Intell. Lab. Syst. ,1996,32(2): 201 - 213.

[43] CLINE A K, DHILLON I S. Computation of the singular value decomposition, in: L. Hogben (Ed.)Handbook of Linear AlgebraSecond Edition, Chapman and Hall/CRC, 2013.

[44] PENG J, PENG S, JIANG A, et al. Near-infrared calibration transfer based onspectral regression, Spectrochim. Acta Part A: Mol. Biomol. Spectrosc,2011,78(4):1315 - 1320.

[45] TAN H W, BROWN S D. Wavelet hybrid direct standardization of near-infrared multivariate calibrations, J. Chemom,2001,15(8):647 - 663.

[46] WANG Y D, LYSAGHT M J, KOWALSKI B R. Improvement of multivariate calibration through instrument standardization, Anal. Chemom,1992,64(5):562 - 564.

[47] WISE B M, MARTARTENS H, HOY M, et al. Calibration transfer by generalized least squares, 2001.

[48] WISE B M, MARTENS H, HOY M. URL Generalized least squares for calibration transfer[Online (accessed22 - October - 2015)].

[49] FAN W. Calibration model transfer for near-infraredspectra based on canonical correlation analysis, Anal. Chim. Acta,2008,623(1):22 - 29.

[50] BOULET J C, ROGER J M. Pretreatments by means of orthogonal projections, Chemom. Intell. Lab. Syst. ,2012,117:61 - 69.

[51] IGEN B, ROGER J M, ROUSSEL S, et al. Improving the transfer of near infrared prediction models by orthogonal methods, Chemom. Intell. Lab. Syst. ,2009,99(1):57 - 65.

[52] SALGUERO L. CHAPARRO B, PALAGOS F, et al. Calibration transfer of intact olive NIR spectra between a pre-dispersive instrument and a portable spectrometer, Comput. Electron. Agric, 2013,96:202 - 208.

[53] IGNE B. URL Intra and inter-brand calibration transfer for near infrared spectrometers, Graduate Theses and Dissertations. Paper 10294.

[54] ZHU Y, FEARN T, SAMUEL D, et al. Error removal by orthogonal subtraction (EROS): a customised pre-treatment for spectroscopic data, J. Chemom,2008,22(2):130 - 134.

[55] MALLI B, NASTSCHLAGER T, PAWLICZEK M, et al. Application-oriented standard-free methods for calibration transfer, Lenzing. Berichte,2015,92:33 - 46.

[56] INC E R. URL Eigenvector Wiki -OSC, [Online;(accessed 16 - October - 2015) (Mar. 2012).

6.2 PLS Subspace Based Calibration Transfer for NIR Quantitative Analysis[1]

In this section, a novel projection method is proposed, which is a feature transfer model based on PLS subspace (PLSCT). PLSCT establishes the PLS model of the calibration set of the master instrument firstly, constructing a low-dimensional PLS subspace, which is a feature space constructed by the spectral feature vectors. The PLS model is then used to extract the predicted features of the master spectra and the pseudo predicted features of the slave spectra, that is, to project all spectra of the master instrument and slave instrument into this PLS subspace. Then, the ordinary least squares method is used to explore the relationship between the two features in the identical PLS subspace, the relationship will then be resorted to construct a feature transfer relationship model.

Notice that the pseudo predicted feature of the slave spectra is acquired by the PLS model established by the master instrument rather than the PLS model of the slave instrument. And PLSCT does not need the response variable corresponding to the standard set. In addition, compared with PDS, PLSCT corrects the feature of the spectra rather than the spectra. In contrast to CCACT, PLSCT uses PLS to find the covariance between the spectra and the response variable, instead of using CCA to find the correlation between the master spectra and the slave spectra.

6.2.1 Calibration Transfer Method

6.2.1.1 Notation

In this section, we defined the spectral matrix as $\boldsymbol{X}$, $n \times p$ represents the size of the matrix, n represents the number of samples, p represents the number of variables, and $\boldsymbol{x}_i$ represents the spectral variables corresponding to the i-th sample of the matrix. The response variables are defined as $\boldsymbol{y}$ and the predicted values are defined as $\hat{\boldsymbol{y}}$. In order to distinguish the spectra collected on the two instruments, we added a superscript to the back of the matrix, such as defining the spectra from the master instrument as $\boldsymbol{X}^m$, defining the spectra from the slave instrument as $\boldsymbol{X}^s$, the predicted feature matrix of the master spectra obtained by the master instrument calibration model is $\hat{\boldsymbol{T}}^m$, the pseudo predicted feature matrix of the slave spectra obtained by the master instrument calibration model is $\hat{\boldsymbol{T}}^{s\text{-}m}$. At the same time, a subscript was added to the back of the matrix to distinguish different data sets. For instance, $\boldsymbol{X}^m_{\text{cal}}$, $\boldsymbol{X}^m_{\text{std}}$, and $\boldsymbol{X}^m_{\text{test}}$ represent the calibration set, standard set and test set of the master instrument, respectively. $\boldsymbol{X}^s_{\text{cal}}$, $\boldsymbol{X}^s_{\text{std}}$, $\boldsymbol{X}^s_{\text{test}}$ and represent the calibration set, standard set and test set of the slave instrument, respectively.

6.2.1.2 Proposed PLSCT Method

In the PLSCT, the PLS model was built on the calibration set of the master instrument to construct the PLS subspace, which is also the feature space constructed by the feature vectors of the spectra of the master instrument calibration set. The number of latent variables (LVs) in the PLS model is determined by cross-validation.

$$\boldsymbol{\beta}^m = \boldsymbol{W}^m((\boldsymbol{P}^m)^{\mathrm{T}}\boldsymbol{W}^m)^{-1}(\boldsymbol{q}^m)^{\mathrm{T}} \tag{6.2.1}$$

On the basis of this PLS model, the predicted feature matrix of standard set in the master instrument $\boldsymbol{X}^m_{\text{std}}$ can be calculated via it, that is, the spectra of the master instrument can be projected into the PLS subspace:

$$\hat{\boldsymbol{T}}^m_{\text{std}} = \boldsymbol{X}^m_{\text{std}}\boldsymbol{W}^m((\boldsymbol{p}^m)^{\mathrm{T}}\boldsymbol{W}^m)^{-1} \tag{6.2.2}$$

Similarly, the pseudo predicted feature matrix of standard set in the slave instrument $\boldsymbol{X}^s_{\text{std}}$ can be calculated via this PLS model as well as $\boldsymbol{X}^m_{\text{std}}$, in other words, the spectra of the slave instrument can be projected into this PLS subspace:

$$\boldsymbol{T}_{\mathrm{std}}^{s\_m}=\boldsymbol{X}_{\mathrm{std}}^{s}\boldsymbol{W}^{m}((\boldsymbol{P}^{m})^{\mathrm{T}}\boldsymbol{W}^{m})^{-1} \tag{6.2.3}$$

The two predicted feature matrices obtained are derived from the same PLS model of the master instrument, that is to say, all spectra are projected into the identical PLS subspace constructed by the master instrument. In the identical PLS subspace, there should be a linear relationship between the two feature matrices. So $\widetilde{\boldsymbol{T}}_{\mathrm{std}}^{s\_m}$ and $\hat{\boldsymbol{T}}_{\mathrm{std}}^{m}$ can be built as:

$$\widetilde{\boldsymbol{T}}_{\mathrm{std}}^{s\_m}\xi=\hat{\boldsymbol{T}}_{\mathrm{std}}^{m} \tag{6.2.4}$$

The linear relationship between the two feature matrices can be solved through the ordinary least squares method, by the following equation:

$$\xi=((\widetilde{\boldsymbol{T}}_{\mathrm{std}}^{s\_m})^{\mathrm{T}}\widetilde{\boldsymbol{T}}_{\mathrm{std}}^{s\_m})^{-1}(\widetilde{\boldsymbol{T}}_{\mathrm{std}}^{s\_m})^{\mathrm{T}}\hat{\boldsymbol{T}}_{\mathrm{std}}^{m} \tag{6.2.5}$$

Once ξ is computed, for the test set from the slave instrument $\boldsymbol{X}_{\mathrm{test}}^{s}$, applying equation (6.2.6) to calculate the predicted values corresponding to the spectra:

$$\hat{\boldsymbol{y}}_{\mathrm{test}}=\boldsymbol{X}_{\mathrm{test}}^{s}\boldsymbol{W}^{m}((\boldsymbol{P}^{m})^{\mathrm{T}}\boldsymbol{W}^{m})^{-1}\xi(\boldsymbol{q}^{m})^{\mathrm{T}} \tag{6.2.6}$$

6.2.2 Experimental

6.2.2.1 Dataset Description

The corn, wheat and pharmaceutical tablet datasets used in this section are described in detail in Section 3.6.3 and are therefore not described here.

6.2.2.2 Determination of the Optimal Parameters

The number of latent variables used in the PLS model was selected by a 10-fold cross-validation. In order to avoid over-fitting caused by the inclusion of redundant latent variables, the optimal number of latent variables was achieved based on the statistical F-test ($\alpha=0.05$).

The predicted feature from the standard set of slave instrument is a pseudo predicted feature $\widetilde{\boldsymbol{T}}_{\mathrm{std}}^{s\_m}$ constructed by the PLS model of the master instrument. Compared with the predicted feature $\widetilde{\boldsymbol{T}}_{\mathrm{std}}^{s}$ constructed by the PLS model of the slave instrument, the $\widetilde{\boldsymbol{T}}_{\mathrm{std}}^{s\_m}$ may contain some noise, which has a great influence on the solution of the transfer matrix $\boldsymbol{\xi}$, further affecting the performance of the PLSCT model. In order to optimize the model, we used leave-one-out cross-validation to select the best number of factors in the standard set based on the minimum root mean square error of cross-validation (RMSECV) criterion. The response variable of the standard set used in cross-validation was the predicted value of the master instrument standard set obtained by the PLS model of the master instrument.

For the PDS method, its window sizes were set to 3, 5, and 7, respectively.

6.2.2.3 Model Performance Evaluation

In order to verify the prediction performance of different calibration models, we calculated the root mean square error of prediction (RMSEP). The calculation of RMSEP is as follows:

$$\text{RMSEP}=\sqrt{\frac{\sum_{i=1}^{n}(y_i-\hat{y}_i)}{n}} \tag{6.2.7}$$

where y_i represents the measured value associated to the i-th test sample, $\hat{y}_i$ is its final predicted value, while n is the number of samples in the test set.

In order to compare the prediction performance difference between the proposed model and other models more directly, equation (6.2.8) was used to calculate the RMSEP improvement of the PLSCT method compared with other methods:

$$h=\left(1-\frac{\text{RMSEP}_{\text{PLSCT}}}{\text{RMSEP}_{\text{other}}}\right)\times 100\% \tag{6.2.8}$$

Where $\text{RMSEP}_{\text{PLSCT}}$ represents the prediction error of the PLSCT method, $\text{RMSEP}_{\text{other}}$ represents the prediction error of other comparison methods.

In addition, by comparing prediction error of the different models, the Wilcoxon signed rank test at the 95% confidence level was utilized to point out whether there was a significant difference between PLSCT and other methods. In python, we used the wilcoxon function in the scipy package to directly calculate the p-value between the two prediction errors. If $p>0.05$, there is no significant difference between the two methods. Otherwise, there is significant difference.

6.2.3 Results and Discussion

6.2.3.1 The Analysis of the Corn Dataset

First of all, Table 6.1 lists the latent variables (LVs) and the root mean square error of prediction (RMSEP) of Calibration, Direct transfer and Recalibration. The RMSEP was 0.010156 when using the calibration model of the master instrument to predict the spectra of the test set measured on the master instrument. However, when directly using the calibration model of the master instrument to predict the spectra of the test set measured on the slave instrument, the RMSEP was 1.41931, which indicates that if the model of the master instrument is directly applied to the slave instrument, a large prediction error will be generated.

The number of the factors for constructing the pseudo predicted feature matrix from the standard set of the slave spectra ($\tilde{\boldsymbol{T}}_{\text{std}}^{s\_m}$) and the predicted feature matrix from the standard set of the master spectra ($\hat{\boldsymbol{T}}_{\text{std}}^{m}$), which is a key parameter in the PLSCT model, was determined by leave-one-out cross-validation. Fig 6.3(a),(b) illustrates the effects of selecting the number of factors used to build $\tilde{\boldsymbol{T}}_{\text{std}}^{s\_m}$ and $\hat{\boldsymbol{T}}_{\text{std}}^{m}$ on the cross-validation error when the number of the samples in the standard set is set to 25 and 30. From the results in Fig. 6.3(a),(b), inferring that when the number of the samples in the standard set is set to 25 and 30, the number of factors should be set to 3. At this time, the root mean square error of cross-validation (RMSECV) reached the minimum and PLSCT achieves the best performance.

Table 6.1 Root Mean Square Error of Prediction (RMSEP) Obtained by Calibration, Direct Transfer, and Recalibration on Three Spectra Datasets

Instrument	Methods	LVs	RMSEP
Corn	Calibration[1]	13	0.010156
	Direct transfer[2]		1.41931
	Recalibration[3]	5	0.208522
Wheat	Calibration[1]	12	0.258014
	Direct transfer[2]		0.85131
	Recalibration[3]	8	0.530799
Pharmaceutical tablet	Calibration[1]	7	3.123115
	Direct transfer[2]		4.514284
	Recalibration[3]	2	3.31598

[1] Calibration: the calibration model of the calibration set of the master instrument;

[2] Direct transfer: the calibration model of master instrument is used on the slave instrument without modification;

[3] Recalibration: the calibration model of the calibration set of the slave instrument.

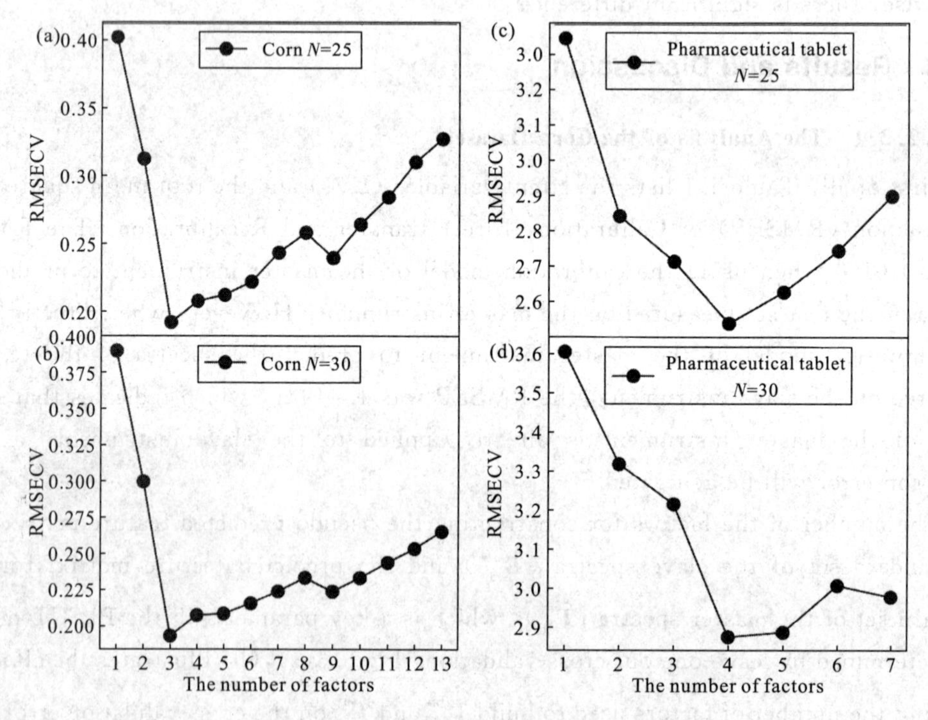

Fig. 6.3 The effect of selecting the number of factors when building $\tilde{T}^{s\_m}_{std}$ and T^{m}_{std} on the cross-validation error. (a) Corn dataset and the number of the samples in the standard set is 25, (b) Corn dataset and the number of the samples in the standard set is 30, (c) Pharmaceutical tablet dataset and the number of the samples in the standard set is 25, (d) Pharmaceutical tablet dataset and the number of the samples in the standard set is 30

In addition, the measured values of the moisture content of the corn dataset obtained from different models are compared with the predicted values when the number of the samples in the standard set is set to 30 are shown in Fig. 6. 4. In this case, the slope of the line was equal to 1. A point on the line indicates that the predicted value was equal to the measured value. As shown in Fig. 6. 4, PLSCT exhibited the smallest differences between the measured values and predicted values. This is attributed to the implementation of the feature transfer in the PLS subspace. The detailed description is shown in Fig. 6. 5.

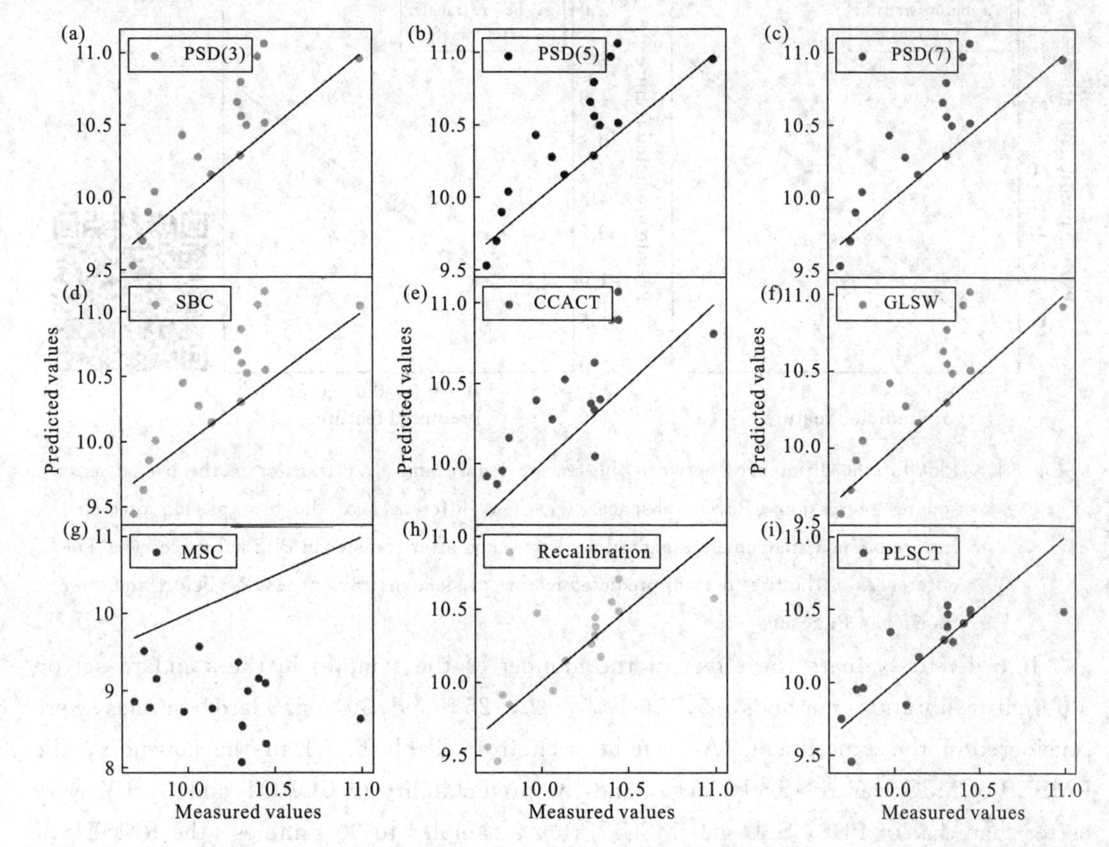

Fig. 6. 4 Measured values versus predicted values of moisture content for corn dataset as determined by (a) piecewise direct standardization with a window size of 3 (PDS(3)), (b) piecewise direct standardization with a window size of 5 (PDS(5)), (c) piecewise direct standardization with a window size of 7 (PDS(7)), (d) slope and bias correction (SBC), (e) calibration transfer method based on canonical correlation analysis (CCACT), (f) generalized least squares (GLSW), (g) multiplicative signal correction (MSC), (h) Recalibration and (i) partial least squares regression subspace based calibration transfer (PLSCT)

For comparison, the differences between the feature before and after transfer in the PLS subspace, the relationship between the first pseudo predicted feature of the slave instrument and the first predicted feature of the master instrument is displayed in Fig. 6. 5. In these two plots, the blue dots represent the feature before transfer, and the red dots represent the feature after transfer. The closer the dots are to a straight line, the smaller the differences between the pseudo predicted feature of the slave instrument and the predicted feature of the master instrument. Fig. 6. 5 (a), (b) depicts the differences between features in the standard set and the test set, respectively. Obviously, after transfer, the differences between the first pseudo predicted feature of the slave instrument and the first predicted feature of the master instrument was significantly reduced, not only in the standard set, but also in the test set.

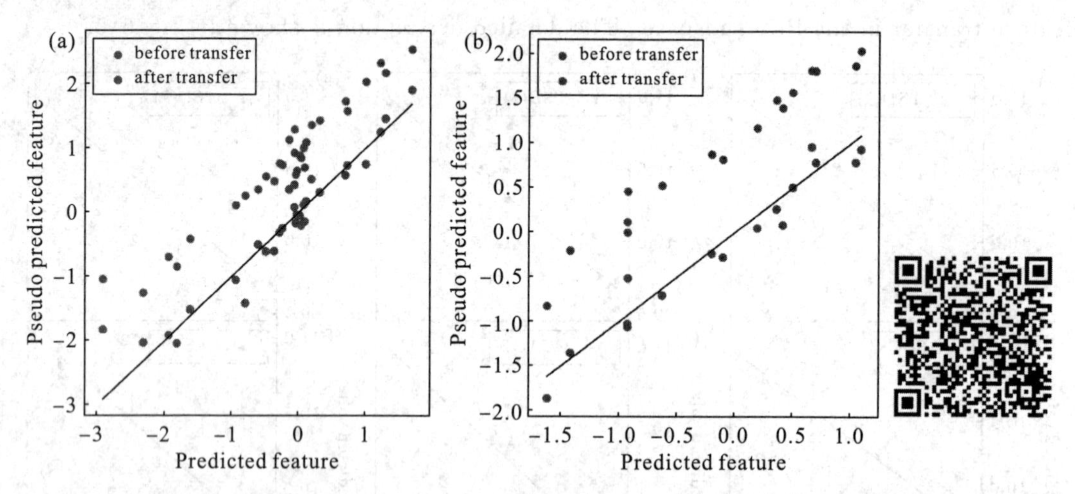

Fig. 6. 5　Plot for the differences between the feature before and after transfer in the partial least squares regression (PLS) subspace. (a) The differences of the first pseudo predicted feature of slave instrument standard set before and after transfer in PLS subspace, (b) The differences of the first pseudo predicted feature of slave instrument test set before and after transfer in PLS subspace

In order to evaluate the effect of the number of the samples in the standard set on different calibration methods, 5, 10, 15, 20, 25, and 30, standard samples were considered in the experiment. As can be seen from Table 6. A1 in the appendix, the RMSEP of MSC was relatively large, and the predictability of CCACT and GLSW were better than that of PDS, SBC and MSC. From 5 samples to 30 samples, the RMSEP of PLSCT was smaller than the RMSEP of PDS, SBC, CCACT, GLSW and MSC. Moreover, the RMSEP of PLSCT had been gradually stabilized when the number of the samples in the standard set from 20 to 30. So, we conclude that PLSCT had significantly better predictive performance than other models.

To further compare PLSCT with other models, the RMSEP improvement and p-value by Wilcoxon signed rank test are listed in Table 6. A2 in the appendix. The RMSEP improvement of PLSCT to PDS(3), PDS(5), PDS(7), SBC, CCACT, GLSW, MSC, Recalibration2 and Recalibration were as high as 35. 00575%, 34. 99841%, 34. 98937%, 41. 95097%, 37. 18537%, 30. 21822%, 85. 7502%, 8. 610493% and 2. 26298%, respectively. The Wilcoxon signed rank test shows statistically significant differences between PLSCT and other models (include Recalibration) at the 95% confidence level.

Appendix

Table 6. A1 RMSEP for Three Datasets Using Different Transfer Methods

	PDS			SBC	CCACT	GLSW	MSC	PLSCT	Recalibration 2[2]	Recalibration[3]
	$W^1=3$	$W^1=5$	$W^1=7$							
Corn dataset										
$N=5$	0.4142	0.4336	0.4354	0.5370	0.2411	0.4056	1.4302	0.1991	0.3538	0.2085
$N=10$	0.3753	0.3617	0.3729	0.4440	0.2545	0.3696	1.4302	0.1980	0.2237	
$N=15$	0.3507	0.3495	0.3357	0.4307	0.3663	0.3535	1.4302	0.2127	0.2425	
$N=20$	0.3440	0.3440	0.3440	0.3900	0.2841	0.3208	1.4302	0.2087	0.2379	
$N=25$	0.3373	0.3372	0.3366	0.3720	0.3528	0.3106	1.4302	0.2082	0.2314	
$N=30$	0.3136	0.3135	0.3135	0.3511	0.3245	0.2921	1.4302	0.2038	0.2230	
Wheat dataset										
$N=5$	8.2434	9.1587	8.4226	14.3731	1.6248	4.0835	1.5160	1.8478	2.7176	0.5308
$N=10$	8.5844	9.3534	10.8927	10.5310	1.2496	3.5824	1.5160	0.8588	2.2233	
$N=15$	2.1373	2.8513	3.2012	8.7159	1.5315	2.9205	1.5160	1.8280	1.3985	
$N=20$	1.9586	2.0927	2.2380	7.0482	0.9688	2.4743	1.5160	1.8263	0.4520	
$N=25$	1.5656	1.6480	1.7445	6.1945	1.0437	1.9804	1.5160	0.6850	2.3661	
$N=30$	1.3694	1.4468	1.5366	5.2635	0.7735	1.7085	1.5160	0.6604	2.2000	
Pharmaceutical tablet dataset										
$N=5$	4.7971	4.2899	4.4594	5.9983	4.1302	6.5988	4.2482	3.3202	5.8027	3.3160
$N=10$	4.1431	4.0098	4.0444	5.4720	4.1112	5.6721	4.2482	3.5821	5.5904	
$N=15$	3.9698	3.8314	3.8347	5.7069	3.9357	6.2284	4.2482	3.3834	5.8043	
$N=20$	3.9787	3.8789	3.9190	5.2838	3.8979	5.6511	4.2482	3.2794	5.0811	
$N=25$	3.9263	3.8416	3.7789	5.2514	4.0549	5.4809	4.2482	3.2765	4.9428	
$N=30$	3.8499	3.7931	3.7590	5.3699	3.8703	5.5354	4.2482	3.2195	4.2267	

[1] W stands for window size of PDS method;

[2] Recalibration 2: the calibration model of the standard set of the slave instrument;

[3] Recalibration: the calibration model of the calibration set of the slave instrument.

Table 6. A2　RMSEP Comparison of PLSCT and Other Methods, RMSEP Improvement and p-values by the Wilcoxon Signed Rank Test ($\alpha = 0.05$)
(The number of samples in the standard set is 30)

		PLSCT		
		Corn	Wheat	Pharmaceutical Tablet
PDS(3)[1]	h (%)[2]	35.00575	51.77389	16.3743
	p[3]	0.00717	3.17×10^{-9}	3.2×10^{-19}
PDS(5)[1]	h (%)	34.99841	54.35396	15.12146
	p	0.00717	2.23×10^{-9}	1.78×10^{-19}
PDS(7)[1]	h (%)	34.98937	57.02112	14.35178
	p	0.00717	2.23×10^{-9}	1.2×10^{-18}
SBC	h (%)	41.95097	87.45319	40.04516
	p	0.011286	7.56×10^{-10}	4.84×10^{-23}
CCACT	h (%)	37.18537	42.18862	16.81376
	p	0.004455	0.001161	4.37×10^{-21}
GLSW	h (%)	30.21822	61.34526	41.83697
	p	0.00717	8.53×10^{-10}	6.82×10^{-23}
MSC	h (%)	85.7502	56.43832	24.21448
	p	0.000531	1.57×10^{-6}	1.51×10^{-18}
Recalibration2	h (%)	8.610493	69.98222	23.82937
	p	0.017378	0.000231	3.05×10^{-23}
Recalibration	h (%)	2.26298	−24.4164	2.908651
	p	0.876722	9.06×10^{-5}	0.000198

[1] The number in brackets stands for window size of PDS method;
[2] h: the RMSEP improvement; [3] p: p-value by Wilcoxon signed rank test.

6.2.3.2　The Analysis of the Wheat Dataset

In Table 6.1, we can note that when no calibration transfer method was used, the difference between the RMSEP of directly using Calibration and the RMSEP of Recalibration was much smaller than the difference in corn dataset, in part because the difference between the two instruments in wheat dataset was relatively small.

Fig. 6.6 displays the comparison of the measured values and the predicted values from different models. From these plots, it is worth noting that the differences between measured values and predicted values in PLSCT were only slightly larger than Recalibration and smaller than any other methods.

Fig. 6.6

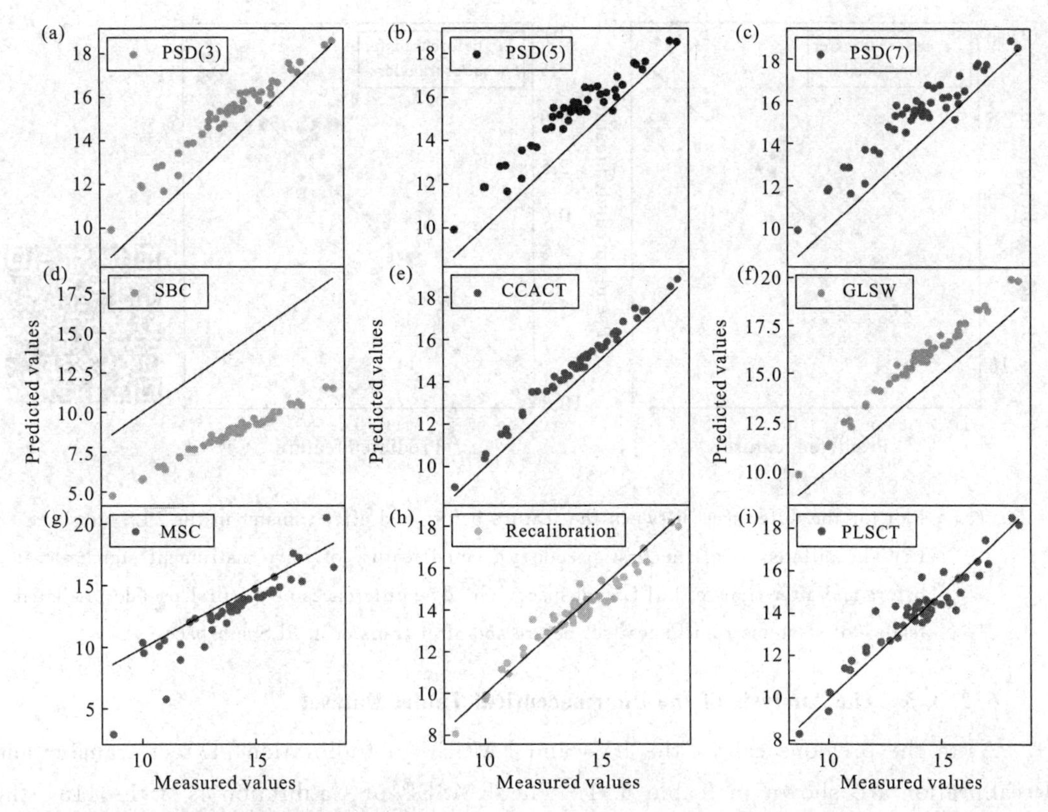

Fig. 6. 6 Measured values versus predicted values of protein content for wheat dataset as determined by (a) PDS(3), (b) PDS(5), (c) PDS(7), (d) SBC, (e) CCACT, (f) GLSW, (g) MSC, (h) Recalibration and (i) PLSCT

Since the spectra difference between the master instrument and the slave instrument was small in the wheat dataset, the effect of feature transfer was not obvious in the PLS subspace from Fig. 6. 7. However, the difference between the first pseudo predicted feature after transfer and the first predicted feature is still slightly smaller. The number of samples of the standard set in Fig. 6. 7(a) was 30.

The performances of the different methods on wheat samples are also shown in appendix Table 6. A1. The Table 6. A2 shows clearly that PLSCT has much lower prediction error than PDS, SBC, GLSW and MSC when the number of the samples in the standard set is 10, 25 and 30. When the number of the samples in the standard set was 30, the minimum RMSEP obtained by PLSCT was 0. 6604. The RMSEP of Recalibration2 fluctuated greatly, probably because there were outliers in the standard set of the slave instrument. These outliers also affect the performance of the SBC as shown in Fig. 6. 6(d).

The results by Wilcoxon signed rank test reveal that PLSCT is significantly different from PDS(3), PDS(5), PDS(7), SBC, CCACT, GLSW, MSC and Recalibration2 at 95% confidence level. The RMSEP improvement resulting from PLSCT compared with these models were 51. 77389%, 54. 35396%, 57. 02112%, 87. 45319%, 42. 18862%, 61. 34526%, 56. 43832% and 69. 98222%, respectively (shown in appendix Table 6. A2).

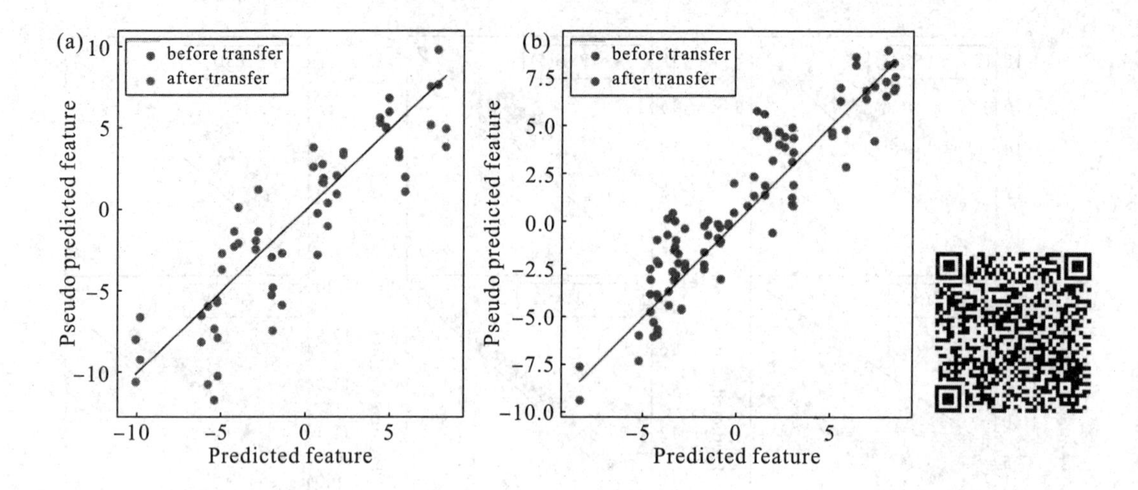

Fig. 6. 7 Plot for the differences between the feature before and after transfer in the PLS subspace. (a) The differences of the first pseudo predicted feature of slave instrument standard set before and after transfer in PLS subspace, (b) The differences of the first pseudo predicted feature of slave instrument test set before and after transfer in PLS subspace

6. 2. 3. 3 The Analysis of the Pharmaceutical Tablet Dataset

As in the previous cases, the LVs and RMSEP of Calibration, Direct transfer and Recalibration are shown in Table 6. 1. The RMSEP of Calibration is 3. 123115, the RMSEP of direct transfer is 4. 514284, the RMSEP of Recalibration was 3. 31598.

In the PLSCT model, the number of factors for constructing $\tilde{\boldsymbol{T}}_{\mathrm{std}}^{s\_m}$ and $\hat{\boldsymbol{T}}_{\mathrm{std}}^{m}$ was 4 when the number of the samples in the standard set was set to 25 and 30, as shown in Fig 6. 3(c), (d). When the number of the samples in the standard set was set to 30, the comparison between the predicted values and measured values is shown in Fig 6. 8. The results show that PLSCT has achieved the best performance.

Fig. 6. 8

Fig. 6. 9 displays the comparison of the first pseudo predicted feature of the slave instrument standard set and test set before and after transfer in the PLS subspace, where the number of samples of the standard set in Fig. 6. 9(a) was 30. From the two plots in Fig. 6. 9, the first pseudo predicted feature after transfer was significantly closer to the predicted feature of the master instrument, whether in the standard set or in the test set of the slave instrument.

From appendix Table 6. A1, as the number of the samples in the standard set increases, the performance of PLSCT gradually got better. The RMSEP of PLSCT gradually became stable when the number of samples in the standard set was 25 and 30, which were outperformed than PDS, SBC, CCACT, GLSW and MSC significantly. From the results in Table 6. A2, when the number of the samples in the standard set was greater than 20, the RMSEP of PLSCT was already less than that of Recalibration.

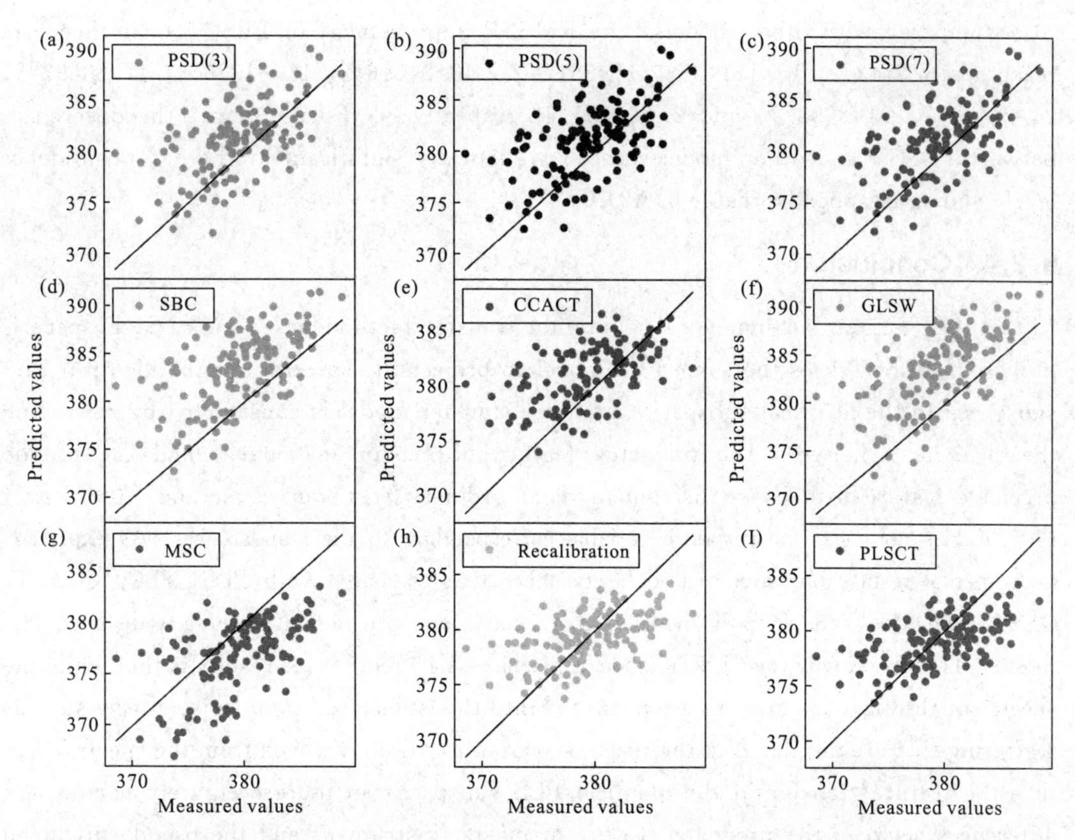

Fig. 6. 8 Measured values versus predicted values of pharmaceutical tablet dataset as determined by (a) PDS(3), (b) PDS(5), (c) PDS(7), (d) SBC, (e) CCACT, (f) GLSW, (g) MSC, (h) Recalibration and (i) PLSCT

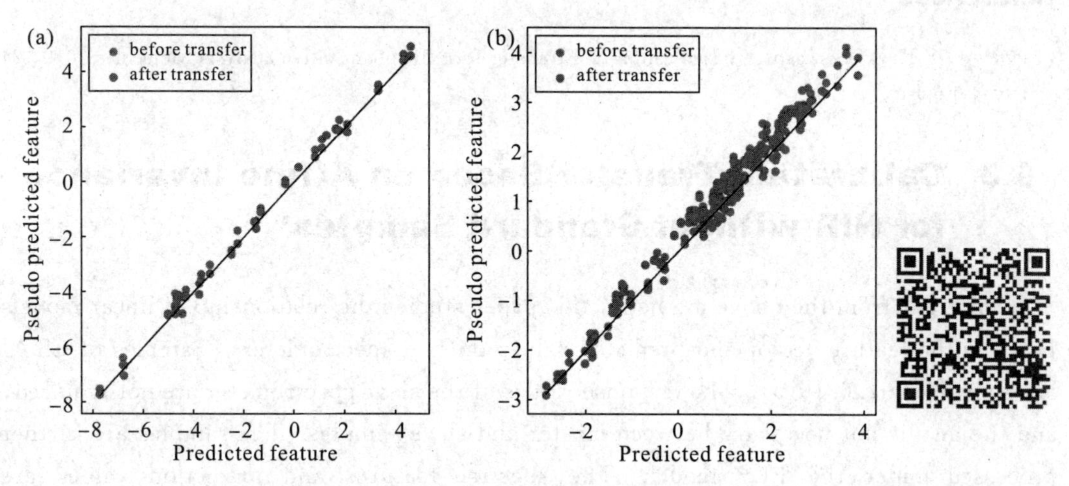

Fig. 6. 9 Plot for the differences between the feature before and after transfer in the PLS subspace. (a) The differences of the first pseudo predicted feature of slave instrument standard set before and after transfer in PLS subspace, (b) The differences of the first pseudo predicted feature of slave instrument test set before and after transfer in PLS subspace

Compared with other models, the RMSEP improvement of PLSCT over them can reach up 16.3743%, 15.12146%, 14.35178%, 40.04516%, 16.81376%, 41.83697%, 24.21448%, 23.82937% and 2.908651%, respectively. Furthermore, the differences between PLSCT and other models are all statistically significant at the 95% confidence level (shown in appendix Table 6.A2).

6.2.4 Conclusion

In this section, an ingenious calibration transfer method based on PLS subspace is proposed. PLSCT uses the same PLS model to project the spectra into the identical PLS subspace. In the identical subspace, a feature transfer model is constructed by narrowing the differences between the predicted feature of master instrument and the pseudo predicted feature of the slave instrument via an ordinary least squares method. Additional, PLSCT does not need the response variable corresponding to the standard set. As expected, experimental results on three real datasets show that compared with PDS, SBC, CCACT, GLSW, and MSC, the PLSCT model is more stable and can obtain more accurate prediction results. The reason why the PLSCT model can achieve such remarkable results is that while the spectra of the slave instrument are projected into this subspace, some noise effects such as scattering that are unrelated to the response variable will be removed from the spectra, and then the feature transfer in the identical PLS subspace can more accurately narrow the differences between the predicted feature of master instrument and the pseudo predicted feature of slave instrument.

References

[1] ZHAO Y H. PLS subspace based calibration transfer for NIR quantitative analysis Molecules 2019, 24 (7): 1289.

6.3 Calibration Transfer Based on Affine Invariance for NIR without Standard Samples

Different from the above methods, this paper studies the relationship of linear models between the feature vector and predicted values on two spectrometers. Samples of CTAI are shown in Fig. 6.10(a). The response values of the slave spectrometer are not required, and the map is not necessary between master and slave samples. The samples are further processed under the PLS model. The spectral features and prediction values are respectively obtained, and the processed samples are shown in Fig. 6.10(b). We obtain the linear models between the feature vector and the predicted values respectively. According to the linear models of two instruments, the relationship between the predicted values is further obtained. Firstly, the PLS model is built on the master instrument; Secondly, the feature matrices and predicted values are extracted according to the PLS model, respectively; Further, the angle and bias are calculated between two regression coefficients; Finally, the prediction values are corrected by affine transformation. The predictive performance of CTAI is verified by two NIR datasets.

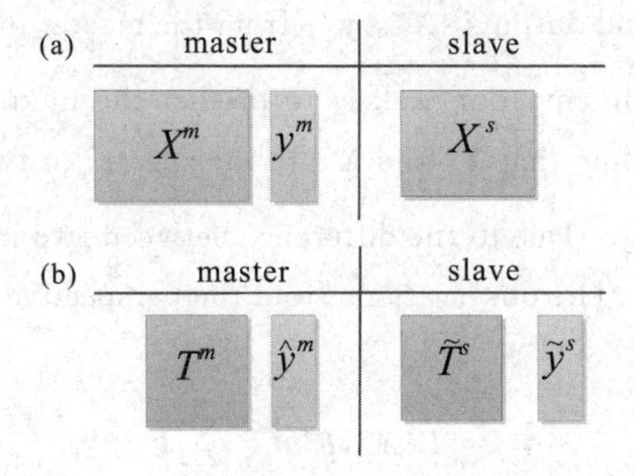

Fig. 6. 10 Data setting of CTAI. We assume the data to be available in Fig. 6. 10(a), and the data after being processed based on PLS model of the master instrument is shown in Fig. 6. 10. (b)

6.3.1 Theory

6. 3. 1. 1 Affine Transformation

This paper focuses on the rotation and translation properties of two-dimensional affine transformation. After transformation, the original line is still a straight line and the original parallel line is still parallel. Affine transformation is a transformation of coordinates. The derivation is written as Fig. 6. 11.

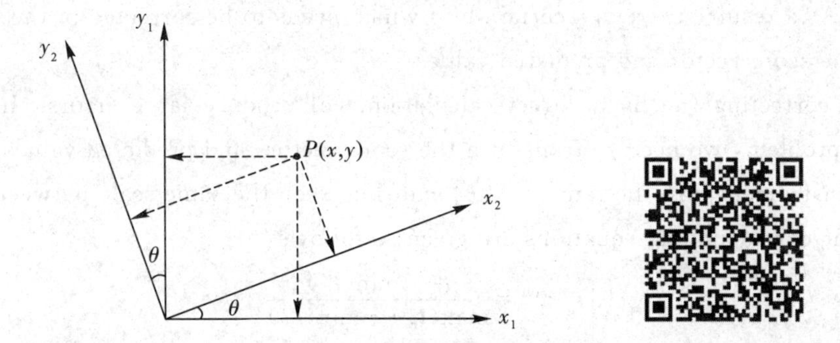

Fig. 6. 11 Derivation of affine transformation. In the coordinate system, the counterclockwise rotation of P is equivalent to the clockwise rotation of the coordinate system

Point P in the original coordinate system (black) is (x, y). A counterclockwise rotation of the point P is equivalent to clockwise rotation of the coordinate system. Thus, the point P in the black coordinate system is equivalent with the point P in the red coordinate system after the rotation. Based on this conclusion, we can determine the coordinates of the point P by simple stereo geometry, and then add the offset of the X axis and the Y axis based on this position, the formula is as follows:

$$\begin{cases} x' = x\cos\theta - y\sin\theta + \Delta x \\ y' = y\cos\theta + x\sin\theta + \Delta y \end{cases} \tag{6.3.1}$$

Where θ is the angle of rotation, Δx is the offset on the X axis, and Δy is the offset on the Y axis; x' and y' are coordinate in new coordinate system.

6.3.1.2 Calibration Transfer Method Based on Affine Transformation

Based on the inputs and outputs $\{\boldsymbol{X}^m, \boldsymbol{y}^m\}$ from the master instrument, and the inputs $\{\boldsymbol{X}^s\}$ from the slave instrument, our task is to predict the unknown outputs $\{\hat{\boldsymbol{y}}^s\}$ in the slave instrument. We assume that $\boldsymbol{X}^m$ and $\boldsymbol{X}^s$ are the spectra of two similar substances, $\boldsymbol{y}^m$ and $\hat{\boldsymbol{y}}^s$ are in the same range. Due to the difference between two instruments, the observed spectral data are different. The observations from the perspective of the master instrument model are as follows:

$$\begin{cases} \hat{\boldsymbol{y}}^m = \boldsymbol{F}(\boldsymbol{X}^m, \boldsymbol{\beta}^m) = \sum_{i=1}^{A} \boldsymbol{t}_i^m \boldsymbol{q}_i^m \\ \hat{\boldsymbol{y}}^s = \boldsymbol{F}(\boldsymbol{X}^s, \boldsymbol{\beta}^m) = \sum_{i=1}^{A} \tilde{\boldsymbol{t}}_i^s \boldsymbol{q}_i^m \end{cases} \tag{6.3.2}$$

Where $\boldsymbol{F}$ is the linear prediction function, which is obtained by partial least squares in this paper; $\boldsymbol{\beta}^m$ is the coefficient of the master model, $\hat{\boldsymbol{y}}^m$, $\boldsymbol{t}_i^m$ and $\boldsymbol{q}_i^m$ are the predicted values, the i-th column score vector and loading vector, respectively. Accordingly, $\hat{\boldsymbol{y}}^s$ and $\tilde{\boldsymbol{t}}_i^s$ are the biased predicted values and the i-th biased column score vector for the slave instrument, respectively.

Therefore, the score vectors and predicted values both of the two instruments are different. As a result, there is a certain bias which needs to be corrected in the coefficient between the score vector and predicted values.

When correcting the bias, direct calculation will produce large errors. In order to solve this problem, we need to transform the score vectors and predicted values of master and slave instrument into the range [0, 1] and thus keep the same scale between different values. The corresponding equations are given as follows:

$$\begin{cases} \boldsymbol{t}^{m\text{-norm}} = \dfrac{\boldsymbol{t}_i^m - \min(\boldsymbol{t}_i^m)}{\max(\boldsymbol{t}_i^m) - \min(\boldsymbol{t}_i^m)} \\ \hat{\boldsymbol{y}}^{m\text{-norm}} = \dfrac{\hat{\boldsymbol{y}}^m - \min(\hat{\boldsymbol{y}}^m)}{\max(\hat{\boldsymbol{y}}^m) - \min(\hat{\boldsymbol{y}}^m)} \\ \tilde{\boldsymbol{t}}^{s\text{-norm}} = \dfrac{\tilde{\boldsymbol{t}}^s - \min(\tilde{\boldsymbol{t}}_i^s)}{\max(\tilde{\boldsymbol{t}}_i^s) - \min(\tilde{\boldsymbol{t}}_i^s)} \\ \tilde{\boldsymbol{y}}^{s\text{-norm}} = \dfrac{\tilde{\boldsymbol{y}}^s - \min(\tilde{\boldsymbol{y}}^s)}{\max(\tilde{\boldsymbol{y}}^s) - \min(\tilde{\boldsymbol{y}}^s)} \end{cases} \tag{6.3.3}$$

Where $\boldsymbol{t}^{m\text{-norm}}$ and $\hat{\boldsymbol{y}}^{m\text{-norm}}$ are normalized score vector and predicted values of the master instrument respectively; $\tilde{\boldsymbol{t}}^{s\text{-norm}}$ and $\tilde{\boldsymbol{y}}^{s\text{-norm}}$ are normalized and biased score vector and predicted values.

Two linear regression equations between score vector and predicted values are as follows:

$$\begin{cases} \hat{\boldsymbol{y}}^{m\text{-norm}} = \boldsymbol{t}^{m\text{-norm}} \tan\theta_i^m + b_i^m \\ \tilde{\boldsymbol{y}}^{s\text{-norm}} = \tilde{\boldsymbol{t}}^{s\text{-norm}} \tan\tilde{\theta}_i^s + \tilde{b}_i^s \end{cases} \tag{6.3.4}$$

Where $\tan\theta_i^m$ and $\tan\tilde{\theta}_i^s$ are the regression coefficients (slopes) computed on the two instrument; b_i^m and $\tilde{b}_i^s$ are the intercepts, respectively.

In order to more intuitively reflect the difference between two instruments, it can be better understood from Fig. 6.12. The blue line is the regression coefficient between the score vector and predicted values. The black and red coordinate systems are the observations of the master and slave instrument, and there are differences from different observations.

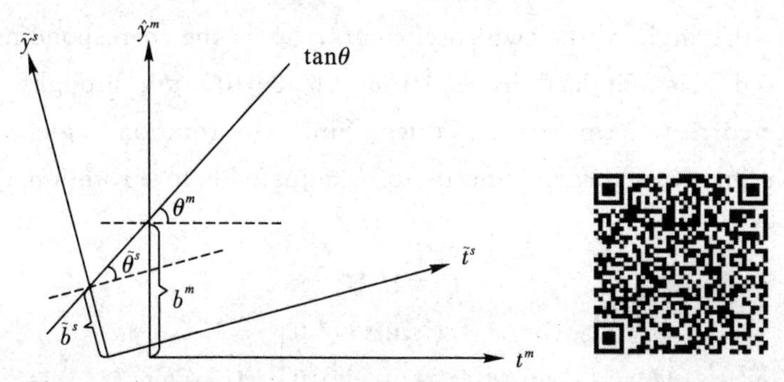

Fig. 6.12 Theory of CTAI. $\tan\theta$ is the coefficient between the feature vector and the predicted values. The angles and deviations observed under different instruments are different. We correct the predicted value of the slave instrument with the rotation and translation of affine transformation

The unknown slope and bias between two instruments are solved as follows:

Firstly, the regression coefficient $\boldsymbol{\beta}^m$, the weight $\boldsymbol{W}^m$ and loading $\boldsymbol{P}^m$ matrix of PLS are obtained.

Secondly, a linear regression both of master and slave instrument is performed respectively, and slopes and intercepts are determined.

On the grounds of the PLS model, the score matrices and predicted values are calculated as shown below:

$$\begin{cases} \boldsymbol{T}^m = \boldsymbol{X}^m \boldsymbol{W}^m (\boldsymbol{P}^m \boldsymbol{W}^m)^{-1}, \ \hat{\boldsymbol{y}}^m = \boldsymbol{X}^m \boldsymbol{\beta}^m \\ \tilde{\boldsymbol{T}}^s = \boldsymbol{X}^s \boldsymbol{W}^m (\boldsymbol{P}^m \boldsymbol{W}^m)^{-1}, \ \tilde{\boldsymbol{y}}^s = \boldsymbol{X}^s \boldsymbol{\beta}^m \end{cases} \tag{6.3.5}$$

Where $\boldsymbol{T}^m$ and $\tilde{\boldsymbol{T}}^s$ represent the score matrices of two instruments.

The score matrix $\boldsymbol{T}^m$, predicted values $\hat{\boldsymbol{y}}^m$, the score matrix $\tilde{\boldsymbol{T}}^s$ and predicted values $\hat{\boldsymbol{y}}^m$ are pre-processed using the equation (6.3.6).

According to score vector of each column and predicted values, the least square is used to compute the corresponding slopes and intercepts, respectively. The equations are as follows:

$$\begin{cases} \min\limits_{\theta_i^m, b_i^m} \left\| \hat{\boldsymbol{y}}^{m\text{-norm}} - \boldsymbol{T}_{\text{aug}}^m * \begin{bmatrix} \tan\boldsymbol{\theta}_i^m \\ b_i^m \end{bmatrix} \right\|^2 \\ \min\limits_{\tilde{\theta}_i^s, \tilde{b}_i^s} \left\| \tilde{\boldsymbol{y}}^{s\text{-norm}} - \boldsymbol{T}_{\text{aug}}^s * \begin{bmatrix} \tan\tilde{\boldsymbol{\theta}}_i^s \\ \tilde{b}_i^s \end{bmatrix} \right\|^2 \end{cases} \tag{6.3.6}$$

Where $\boldsymbol{T}_{\text{ang}}^m$ is an augmented matrix$[\boldsymbol{t}_i^{m\text{-norm}}, \boldsymbol{1}]$; $\boldsymbol{T}_{\text{aug}}^s$ is an augmented matrix$[\tilde{\boldsymbol{t}}_i^{s\text{-norm}}, \boldsymbol{1}]$; $\boldsymbol{1}$ is the column vector with all ones.

Finally, the angle and bias between two instruments are obtained. The equations for calculating the angle and bias are as follows:

$$\begin{cases} \Delta\theta_i = \theta_i^m - \tilde{\theta}_i^s \\ \Delta b_i = b_i^m - \tilde{b}_i^s \end{cases} \tag{6.3.7}$$

Where $\Delta\theta_i$ is the angle of the two coefficients; Δb_i is the corresponding bias.

The angle and bias obtained by equation (6.3.10) are brought into the affine transformation to correct the predicted values. Since the rotation angle is relative to the origin of the coordinate, each sample needs to be adjusted before rotation. The equation is shown as follows:

$$\hat{\boldsymbol{U}}_i = \tilde{\boldsymbol{U}}_i \boldsymbol{M}_i \tag{6.3.8}$$

Where the matrix $\boldsymbol{M}_i = \begin{bmatrix} \lambda_t \cos\Delta\theta_i & \lambda_t \sin\Delta\theta_i & 0 \\ \lambda_y \sin\Delta\theta_i & \lambda_y \cos\Delta\theta_i & 0 \\ 0 & b_i^m & 1 \end{bmatrix}$, $\tilde{\boldsymbol{U}}_i = [\tilde{\boldsymbol{t}}_i^{s\text{-test}}, \tilde{\boldsymbol{y}}^{s\text{-test}}, \boldsymbol{1}]$ and $\hat{\boldsymbol{U}}_i = [\hat{\boldsymbol{t}}_i^{s\text{-test}}, \hat{\boldsymbol{y}}_i^{s\text{-test}}, \boldsymbol{1}]$. In addition, $\lambda_t = \dfrac{(\tilde{\boldsymbol{t}}_i^{s\text{-test}} - \min(\tilde{\boldsymbol{t}}_i^s))}{(\max(\tilde{\boldsymbol{t}}_i^s) - \min(\tilde{\boldsymbol{t}}_i^s)) + \min(\tilde{\boldsymbol{t}}_i^s)} * (\max(\tilde{\boldsymbol{t}}_i^s) - \min(\tilde{\boldsymbol{t}}_i^s))$ and $\lambda_y = \dfrac{(\tilde{\boldsymbol{y}}_i^{s\text{-test}} - \min(\tilde{\boldsymbol{y}}_i^s))}{(\max(\tilde{\boldsymbol{y}}_i^s) - \min(\tilde{\boldsymbol{y}}_i^s)) + \min(\tilde{\boldsymbol{y}}_i^s)} * (\max(\tilde{\boldsymbol{y}}_i^s) - \min(\tilde{\boldsymbol{y}}_i^s))$ represent the corresponding scaling factors for feature vector and predicted values, respectively; $\tilde{\boldsymbol{t}}_i^{s\text{-test}}$ and $\tilde{\boldsymbol{y}}^{s\text{-test}}$ are biased score vector and predicted values of the test set; $\hat{\boldsymbol{y}}^{s\text{-test}}$ are corrected predicted values; $\hat{\boldsymbol{t}}_i^{s\text{-test}}$ is corrected score vector.

Each column score vector and predicted values are solved separately, and a prediction matrix is obtained. The mean of the prediction matrix is the final predicted values.

Therefore, according to the expansion of the predicted values, $\boldsymbol{\beta}^s$ is as follow:

$$\boldsymbol{\beta}^s = ((\boldsymbol{X}^s)^{\mathrm{T}} \boldsymbol{X}^s)(\boldsymbol{X}^s)^{\mathrm{T}} \left(\sum_i^A \frac{(\tilde{\boldsymbol{t}}_i^{s\text{-test}} * \lambda_i * \sin\Delta\theta_1 + (\hat{\boldsymbol{y}}^{s\text{-test}} - \tilde{b}_i^s * \boldsymbol{1}) * \lambda_y * \cos\Delta\theta_i + b_i^m)}{A} \right) \tag{6.3.9}$$

6.3.1.3 Summary of CTAI

Given calibration set of the master $(\boldsymbol{X}_{\text{cal}}^m, \boldsymbol{y}_{\text{cal}}^m)$, calibration set of the slave $\boldsymbol{X}_{\text{cal}}^s$ and test set $(\boldsymbol{X}_{\text{test}}^s, \boldsymbol{y}_{\text{test}}^s)$.

1. The PLS model is built on the calibration set ($\boldsymbol{X}_{\text{cal}}^{m}$, $\boldsymbol{y}_{\text{cal}}^{m}$) and the coefficient $\boldsymbol{\beta}^{m}$, the weight matrix $\boldsymbol{W}^{m}$ and the loading matrix $\boldsymbol{P}^{m}$ be obtained.

2. Modeling of affine transformation, and it consists of the two datasets ($\boldsymbol{X}_{\text{cal}}^{m}$, $\boldsymbol{y}_{\text{cal}}^{m}$) and $\boldsymbol{X}_{\text{cal}}^{s}$

a) Computing (T_{cal}^{m}, $\hat{\boldsymbol{y}}_{\text{cal}}^{m}$) and ($\tilde{T}_{\text{cal}}^{s}$, $\tilde{\boldsymbol{y}}_{\text{cal}}^{s}$) of master and slave instrument by equation (6.3.8).

b) (T_{cal}^{m}, $\hat{\boldsymbol{y}}_{\text{cal}}^{m}$) and ($\tilde{T}_{\text{cal}}^{s}$, $\tilde{\boldsymbol{y}}_{\text{cal}}^{s}$) are normalized separately by equation (6.3.6).

c) ($\tan\theta_i^{m}$, b_i^{m}) and ($\tan\tilde{\theta}_i^{s}$, $\tilde{\boldsymbol{b}}_i^{s}$) are calculated, respectively by equation (6.3.9).

d) Computing $\Delta\theta_i$ angle and Δb_i bias between master and slave instrument by equation (6.3.10).

3. Prediction

a) ($\tilde{T}_{\text{test}}^{s}$, $\tilde{\boldsymbol{y}}_{\text{test}}^{s}$) is obtained by equation (6.3.8).

b) The matrix $\boldsymbol{M}_i$ is introduced to correct predicted values by equation (6.3.11).

c) The corrected prediction values are accumulated. The mean values are the last result.

6.3.2 Experimental

6.3.2.1 Dataset Description

The corn and wheat datasets used in this section are described in detail in Section 3.6.3 and are therefore not described here.

6.3.2.2 Determination of the Optimal Parameters

Latent variables of PLS in CTAI are allowed to take values in the set from 1 to 15, and it is determined by the 10-fold cross-validation. The optimal number of latent variables is selected only when the lowest RMSECV.

Five methods were used for comparison, where the latent variable range and parameter optimization all of SBC, CCA, PDS and MSC in PLS are consistent with CTAI. In particular, the window size in PDS is searched from 3 to 16 in increments of 2, and is selected by 5-fold cross-validation. In addition, the dimensionality of the TCA space in TCR is estimated in the range from 1 to 24.

6.3.2.3 Model Performance Evaluation

In this experiment, root mean squared error RMSE is employed as indicators for parameter selection and model evaluation. Furthermore, RMSEC is the training error, RMSECV denotes the cross-validation error, and RMSEP indicates the prediction error of the test set. The RMSE calculation method is written as:

$$\text{RMSE}=\sqrt{\frac{(\boldsymbol{y}-\boldsymbol{y})^{\mathrm{T}}(\boldsymbol{y}-\hat{\boldsymbol{y}})}{n}} \tag{6.3.10}$$

Where $\hat{\boldsymbol{y}}$ is the predict value; $\boldsymbol{y}$ is the measured value; n represents the number of samples.

Bias and standard error (SE) are also utilized as reference indicators for model evaluation. The bias and SE are as follow:

$$\begin{cases} \text{bias} = \sum_{i}^{n} \frac{(\boldsymbol{y}_i - \hat{\boldsymbol{y}}_i)}{n} \\ \text{SE} = \sqrt{\frac{(y-\hat{y})^{\mathrm{T}}(\boldsymbol{y}-\hat{\boldsymbol{y}}) - \text{bias}}{n}} \end{cases} \tag{6.3.11}$$

Moreover, the Pearson correlation coefficient and corresponding testis used to determine if there is a linear relationship between the master instrument and the slave instrument. And One-Sample t-Test is utilized to determine whether a bias adjustment in predicted results should be implemented.

In order to compare CTAI and other methods further, another important parameter (h) is cited in order to compare the rate of improvement, defined as follows:

$$h = \left(1 - \frac{\text{RMSEP}}{\text{RMSEP}_{\text{other}}}\right) \times 100\% \tag{6.3.12}$$

Where RMSEP represents the prediction error of CTAI, and $\text{RMSEP}_{\text{other}}$ represents the others.

In addition, the Wilcoxon signed rank sum test at the 95% confidence level is used to determine whether there is a significant difference between CTAI and the others.

6.3.3 Results and Discussion

6.3.3.1 Analysis of the Corn Dataset

The training error, prediction error, and the correlation coefficients for the predicted vs. actual results about the PLS model of the corn dataset are shown in Table 6.2. We can see that RMSEPm of the PLS between the instrument m5spec are smaller than the RMSEPm of the instrument mp6spec. Thus, m5spec as the master instrument and mp6spec as the slave instrument is a more reasonable choice.

In order to more fully assess the predicted performance of CTAI, the methods MSC, TCR, CCA, SBC and PDS are tested. In this work, when PDS was performed, PLS was utilized to compute the transformation function. For the PLS model, the optimal number of latent variables is shown in Table 6.2. The optimal dimensionality of the subspace in TCR are 4, 6 10 and 10, respectively. In addition, optimal window sizes of PDS are all 3.

As shown in Table 6.3, we can see the correlation coefficients r_{pre} and corresponding p_{pre} value indicate the prediction values between the master instrument and the slave instrument are linearly correlated. And we can see that the t_{pre} is greater than the t critical value. We then know the bias adjustment in predicted results should be implemented. Furthermore, the RMSE of prediction without any correction for the slave instrument shows more error of prediction than the master instrument. The results of the CTAI corrected results in a significant reduction in RMSE of prediction. The same situation can be found between $\boldsymbol{y}^m$ and $\tilde{\boldsymbol{y}}^s$ in Table 6. 3. For the corn dataset, the effect of correction in CTAI is vividly described by Fig. 6. 13. It can be seen that the corrected predicted values of CTAI more close to the straight line, and RMSEP is greatly reduced.

Table. 6. 2 Summary of the PLS Models and Properties

Instrument	Reference Values	$RMSEC^m$	$RMSEP^m$	$RMSECV_{min}$ (LV)	$bias^m$	r^m	p^m
m5spec	moisture	0. 00599	0. 00764	0. 01066(14)	0. 00080	0. 99973	2. 6e−24
m5spec	oil	0. 02686	0. 05664	0. 05049(15)	−0. 01327	0. 93320	1. 3e−7
m5spec	protein	0. 05070	0. 10066	0. 11012(15)	0. 02814	0. 97632	1. 0e−10
m5spec	starch	0. 09539	0. 18993	0. 19227(15)	0. 01789	0. 97464	1. 6e−10
mp6spec	moisture	0. 09991	0. 15637	0. 14775(10)	−0. 02678	0. 92083	4. 2e−7
mp6spec	oil	0. 06052	0. 09098	0. 09872(12)	0. 01868	0. 87697	8. 2e−6
mp6spec	protein	0. 10101	0. 13338	0. 15043(12)	0. 02128	0. 96659	1. 1e−9
mp6spec	starch	0. 27636	0. 26723	0. 35978(9)	0. 02124	0. 93136	1. 6e−7
B1	protein	0. 32880	0. 33254	0. 50337(15)	0. 00906	0. 98508	2. 3e−38
B2	protein	0. 21636	0. 83755	0. 32441(15)	−0. 13124	0. 84850	7. 2e−15
B3	protein	0. 30288	0. 51567	0. 43896(15)	−0. 03400	0. 96009	3. 2e−28

RMSECm: Root Mean Square Error of calibration set

$RMSEP^m$: Root Mean Square Error of test set

$RMSECV_{min}$: Minimum Root Mean Square Error of Cross-Validation

LV: The optimal number of latent variables is selected only when the lowest RMSECV

r^m: Pearson correlation coefficient for predicted vs. actual values

p^m: p values corresponding to the Pearson correlation coefficient is obtained by test.

Moreover, the results listed in Table 6. 4 and Table 6. 5 show the difference between the 16 predictive corn samples by different methods, respectively. In general, the result of CTAI exhibits the best performance for prediction compared to other five methods. When moisture is used as the property, though there are no statistically significant differences, CTAI is greatly improved in predictive accuracy compared with CCA and TCR. There is a significant difference at the 95% confidence level between CTAI and MSC, SBC and PDS. When oil is used as the property, the results of the Wilcoxon signed rank sum test show significant difference compared with the results obtained from MSC and TCR. And the improvement rate of CTAI is improved compared with CCA, SBC and PDS. Other properties are similar with the property of oil, CTAI achieves better predictive performance.

Table. 6.3 Summary of the relevant results between uncorrected and CTAI corrected

Instrument		m5spec* -mp6spec				B1* -B2	B1* -B3	B3* -B2
Reference Values		moisture	oil	protein	starch	protein		
$\hat{y}^m$ vs $\hat{y}^s$	$RMSEP^u_{pre}$	1.60705	0.79890	2.06797	2.11743	0.69894	2.92541	1.23368
	$RMSEP_{pre}$	0.21255	0.06922	0.13195	0.33358	0.31537	0.62632	0.65398
	k_{pre}	0.64980	0.77129	0.94553	0.82527	0.88809	0.76290	0.86909
	r_{pre}	0.81644	0.89598	0.96286	0.92197	0.97594	0.87695	0.93715
	p_{pre}	1.1e−4	2.6e−6	2.3e−9	3.8e−7	2.0e−33	6.8e−17	1.3e−23
	t_{pre}	−15.429	19.335	−19.147	8.838	2.292	10.684	−3.826
y^m vs $\tilde{y}^s$	$RMSEP^u$	1.60762	0.81532	2.09665	2.10291	0.71977	2.90011	1.08008
	RMSEP	0.21095	0.08233	0.16614	0.34714	0.41419	0.68215	0.38446
	k	0.65191	0.53297	0.98736	0.79329	0.96898	0.85693	0.93896
	r	0.81922	0.78858	0.95844	0.91487	0.96770	0.89517	0.97796
	p	1.0e−4	2.8e−4	5.1e−9	6.9e−7	2.2e−30	1.8e−18	2.5e−34
	t	−15.437	19.657	−19.408	8.762	2.256	10.649	−3.701
$t_{critical\_value}$		2, 131	2, 131	2, 131	2, 131	2.010	2.010	2.010

*: the master instrument
$RMSEP^u_{pre}$: RMSEP of uncorrected slave instrument relative to primary instrument prediction
$RMSEP_{pre}$: RMSEP of CTAI corrected slave instrument relative to primary instrument prediction
k_{pre}: the slope between predicted values of uncorrected slave instrument and primary prediction
r_{pre}: correlation coefficient of uncorrected slave prediction relative to master prediction
p_{pre}: p values corresponding to the Pearson correlation coefficient is obtained by test.
t_{pre}: the result of One−Sample t−Test between uncorrected slave prediction and master prediction
$RMSEP^u$: RMSEP of uncorrected slave instrument relative to primary actual values
RMSEP: RMSEP of CTAI corrected slave instrument relative to primary actual values
k: the slope between predicted values of uncorrected slave instrument and primary actual values
r: Pearson correlation coefficient of uncorrected slave prediction relative to primary actual values
p: p values corresponding to the Pearson correlation coefficient is obtained by test.
t: the result of One-Sample t-Test between uncorrected slave prediction and master actual values
$t_{critical\_value}$: the t critical value for n-1 degrees of freedom at the significance level alpha = 0.05

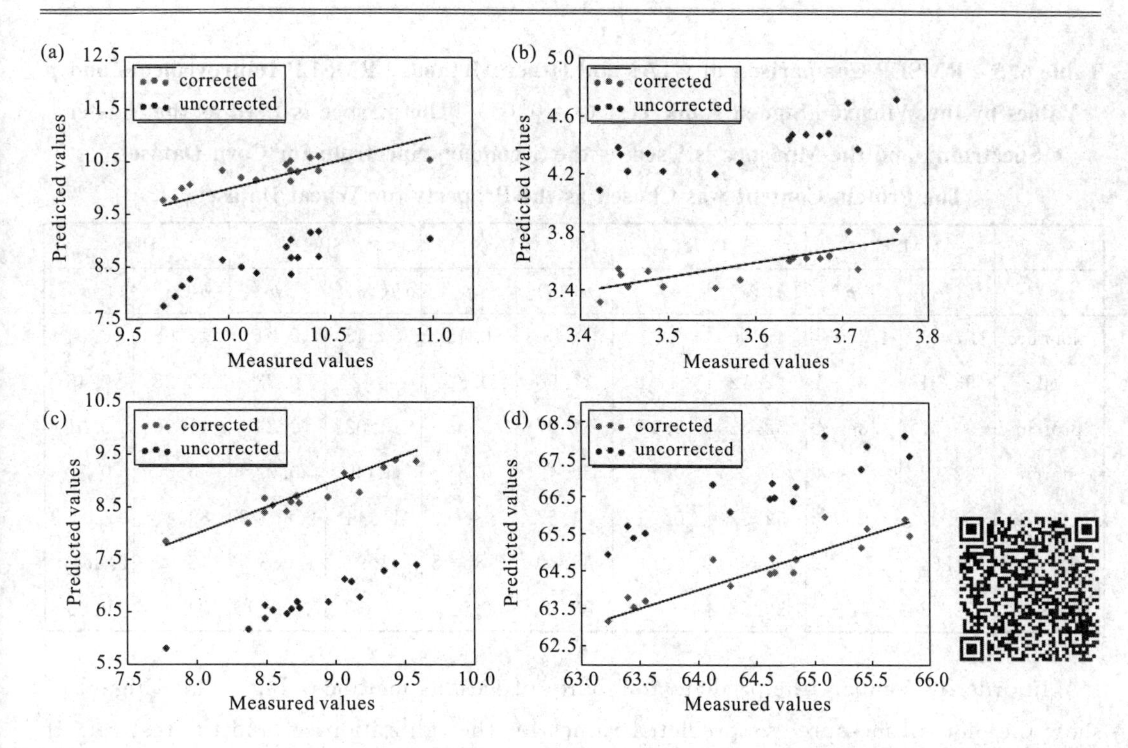

Fig. 6. 13 The relationship between the uncorrected and the corrected predict values for corn dataset by (a) moisture, (b) oil, (c) protein, (d) starch. The blue and red dots represent the uncorrected and the corrected predicted results for each sample, respectively

Table 6. 4 Summary of RMSEP and RMSEC of Different Methods. The m5spec is Used as the Master Spectrum, and the Mp6spec is Ued as the Secondary Spectrum for Corn Dataset. The Protein Content was Chosen as the Property for Wheat Dataset

method		CTAI	MSC	TCR	CCA	SBC	PDS
moisture	RMSEC	0. 22646	1. 92839	0. 61873	0. 15996(14[a])	0. 185067(5[a])	0. 14742(17[a])
	RMSEP	0. 21095	1. 66890	0. 39066	0. 23304(14[a])	0. 42574(5[a])	0. 24238(17[a])
oil	RMSEC	0. 08141	1. 21647	0. 14543	0. 15764(6[a])	0. 08423(23[a])	0. 10794(28[a])
	RMSEP	0. 08233	1. 23209	0. 14225	0. 11432(6[a])	0. 08361(23[a])	0. 09495(28[a])
protein	RMSEC	0. 17247	1. 77294	0. 28297	0. 27860(14[a])	0. 17422(6[a])	0. 24662(23[a])
	RMSEP	0. 16614	1. 80087	0. 35223	0. 39535(14[a])	0. 19101(6[a])	0. 28193(23[a])
starch	RMSEC	0. 39517	1. 89165	1. 21093	0. 33937(10[a])	0. 38426(23[a])	0. 62099(23[a])
	RMSEP	0. 34714	1. 93129	0. 79852	0. 85704(10[a])	0. 36969(23[a])	0. 78977(23[a])
B1*-B2	RMSEC	0. 55682	1. 31153	0. 99246	1. 11889(5[a])	0. 48509(6[a])	1. 36760(7[a])
	RMSEP	0. 41419	0. 92194	0. 86881	2. 68469(5[a])	0. 46770(6[a])	4. 09019(7[a])
B1*-B3	RMSEC	0. 81895	2. 91695	0. 84682	0. 68529(15[a])	1. 00007(8[a])	0. 57858(5[a])
	RMSEP	0. 68215	2. 40587	0. 72996	1. 10564(15[a])	0. 79294(8[a])	1. 33547(5[a])
B3*-B2	RMSEC	0. 54753	1. 25096	0. 76972	1. 57073(14[a])	0. 56236(5[a])	2. 10390(8[a])
	RMSEP	0. 38446	1. 38468	0. 63689	2. 29856(14[a])	0. 53534(5[a])	1. 83564(8[a])

[a]: number of standard samples

The number of samples for slave instrument with labels is 20 in TCR

Table 6.5　RMSEP Comparison of CTAI and Other Methods, RMSEP Improvements and *p* Values by the Wilcoxon Signed Rank Test ($\alpha=0.05$). The m5spec is Used as the Master Spectrum, and the Mp6spec is Used as the Secondary Spectrum for Corn Dataset. The Protein Content was Chosen as the Property for Wheat Dataset.

	MSC		TCR		CCA		SBC		PDS	
	h(%)	p	h(%)	p	h(%)	p	h(%)	p	h(%)	p
moisture	87.35	4.3e-4	46.00	0.53	9.48	0.43	50.45	0.01	12.96	0.04
oil	93.31	4.3e-4	42.12	0.01	27.98	0.32	1.52	0.23	13.28	0.46
protein	90.77	4.3e-4	52.83	0.09	57.97	0.03	13.02	0.23	41.06	0.01
starch	82.02	4.3e-4	56.52	0.23	59.49	0.83	6.09	0.02	56.04	0.75
B1*-B2	55.07	0.11	52.32	0.79	84.57	5.3e-9	11.44	2.6e-9	89.87	9.2e-3
B1*-B3	71.64	7.5e-10	6.55	0.11	38.30	1.8e-5	13.97	1.0e-5	48.92	9.8e-5
B3*-B2	72.23	3.1e-9	39.63	4.6e-3	83.27	0.02	28.18	7.5e-10	79.05	0.06

In order to compare the predictive stability of various methods, Fig. 6.14 - Fig. 6.17 show the plots of measured vs predicted values for the calibration set and the test set. If the model predicts better, the point will be more close to the straight line. As shown in Fig. 6.14 - Fig. 6.17, MSC deviates from the straight line and the prediction performance is the worst. When starch is used as the property, the predictive performance of CCA is also poor. Conversely, it was observed that CTAI is generally more close to the straight line than the other methods. In addition, the standard error has also achieved good results in CTAI compared with others. From the discussion above, one can easily conclude that CTAI can achieve the best RMSEPs.

Fig. 6.14

Fig. 6.15

Fig. 6.16

Fig. 6.17

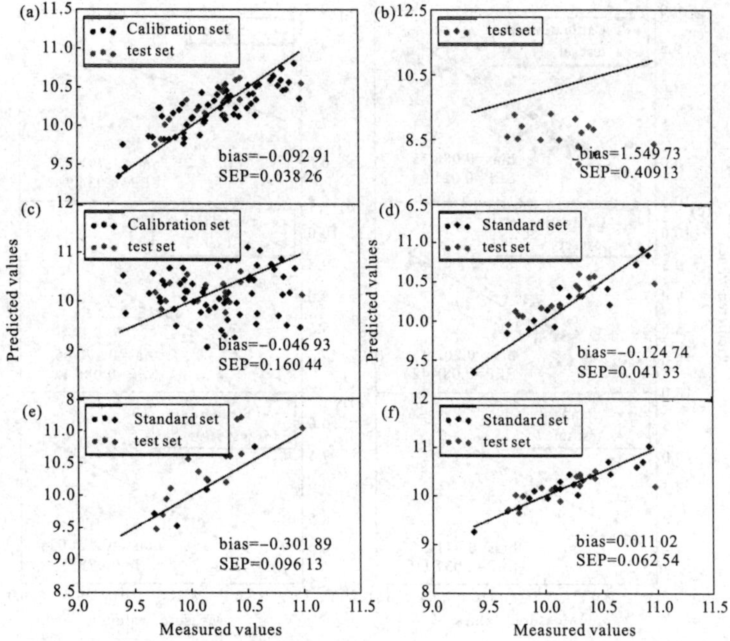

Fig. 6. 14 Moisture content predicted for Corn dataset as determined by (a) CTAI, (b) MSC, (c) TCR, (d) CCA, (e) SBC and (f) PDS. The blue and red dots represent the results for each sample in the train set and test set, respectively

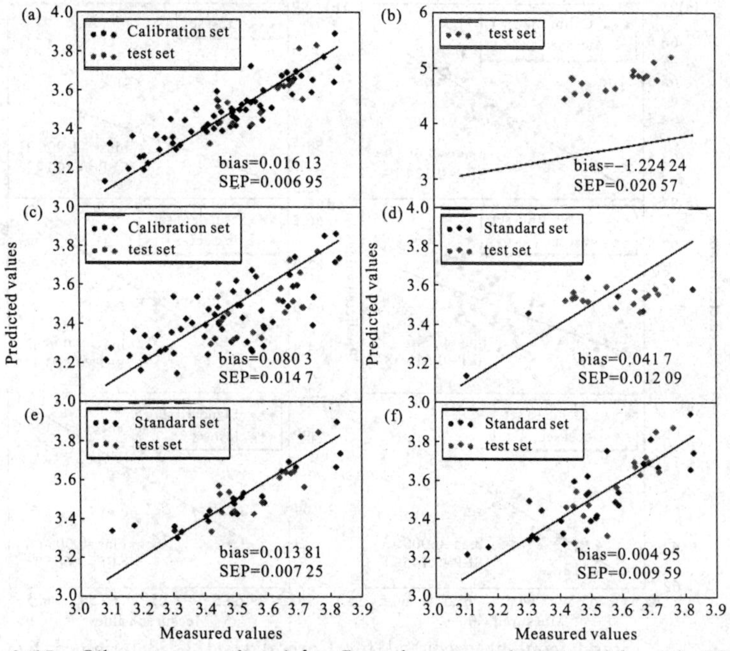

Fig. 6. 15 Oil content predicted for Corn dataset as determined by (a) CTAI, (b) MSC, (c) TCR, (d) CCA, (e) SBC and (f) PDS. The blue and red dots represent the results for each sample in the train set and test set, respectively

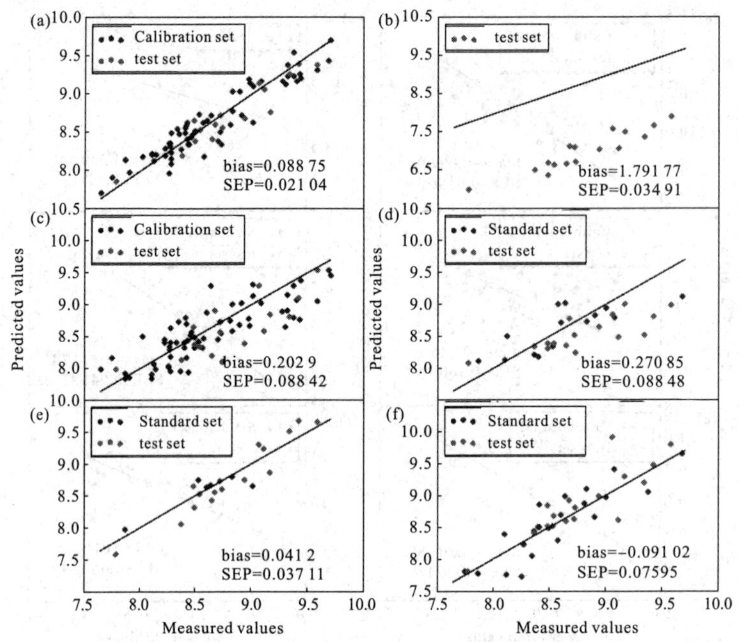

Fig. 6. 16　Protein content predicted for Corn dataset as determined by (a) CTAI, (b) MSC, (c) TCR, (d) CCA, (e) SBC and (f) PDS. The blue and red dots represent the results for each sample in the train set and test set, respectively

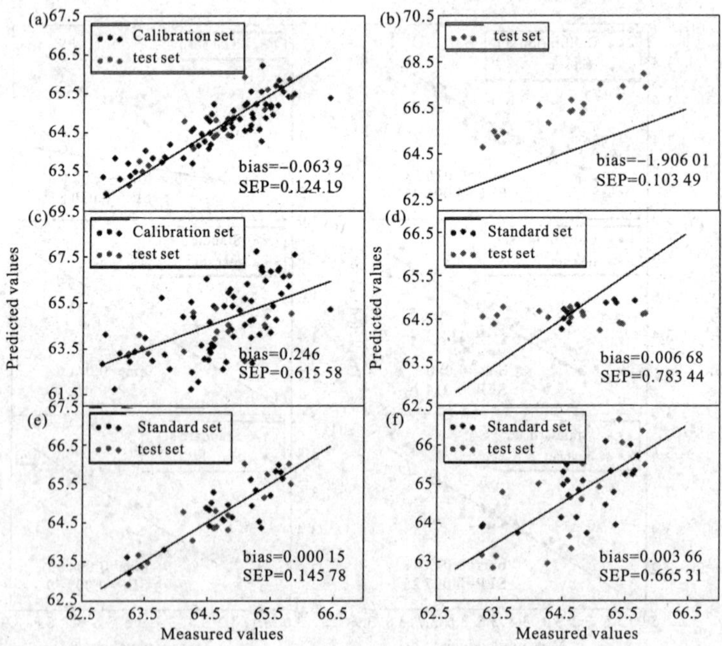

Fig. 6. 17　Starch content predicted for Corn dataset as determined by (a) CTAI, (b) MSC, (c) TCR, (d) CCA, (e) SBC and (f) PDS. The blue and red dots represent the results for each sample in the train set and test set, respectively

6.3.3.2 Analysis of the Wheat Dataset

The RMSEP of the PLS model is listed in Table 6.2. We can see that the predicted performance of the instrument B1 is better than B3, the instrument B3 is better than B2. Thus, three combinations (B1-B2; B1-B3; B3-B2) of the instruments B1, B2, B3 are used to analyze the wheat dataset. The first instrument of every combination stands for master instrument and the second instrument stands for slave instrument. For PLS model, the optimal number of latent variables are 14, 15 and 15, respectively; and the corresponding optimal dimensionality of the subspace in TCR are 17, 12 and 17. Besides, the optimal number of window sizes for B1-B2, B1-B3 and B3-B2 is 3, 9 and 13, respectively.

It can be seen from Table 6.4 that there is a linear relationship between the predicted values of the two instruments for wheat dataset. In addition, there is a significant bias between uncorrected predicted values of the slave instrument and predicted values of the master instrument. So, we can correct the predicted values of the slave instrument by affine transformation. The experimental results show that the prediction performance of CTAI corrected is significantly enhanced. We found the same phenomenon for the uncorrected prediction values of the slave instrument relative to the master instrument actual values. Furthermore, for the predicted performance of CTAI, Fig. 6.18 shows the difference between uncorrected and corrected prediction values for B1-B2, B1-B3 and B3-B2. It can be seen that CTAI plays an important role in the correction of predicted values.

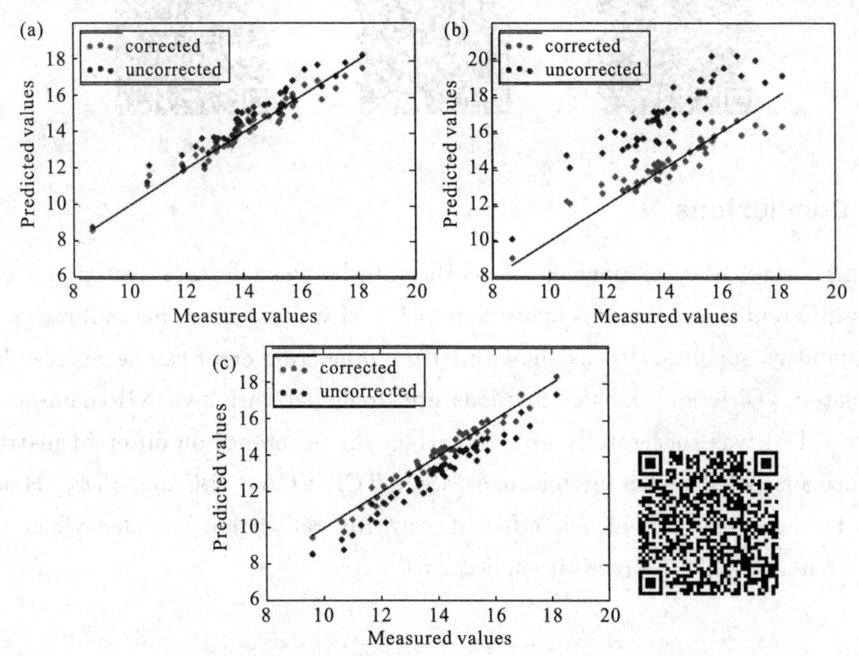

Fig. 6.18 The relationship between the uncorrected and the corrected predict values for wheat dataset by (a) B1-B2, (b) B1-B3, (c) B3-B2. The blue and red dots represent the uncorrected and the corrected predicted results for each sample, respectively

In addition, Table 6.4 lists the results of different methods for calibration set and test set. For the B1-B2, CTAI produces the lowest RMSEP. Further, a statistical testing is utilized to evaluate the RMSEP difference between the CTAI and other methods for the wheat dataset. The Wilcoxon signed rank sum test was performed and at the significance level $\alpha=0.05$. It can be seen that there is a statistically significant difference compared with CCA SBC and PDS. In addition, the improvements rate of prediction provided by CTAI for MSC and TCR are up to 55.07% and 52.32%, respectively. For the other two combinations, CTAI achieves the best predictive performance. Compared to the five methods, there are significant differences at the 95% confidence level or the RMSEP improvements are large.

To further display the prediction abilities of different models, the correlation between measured and predicted values obtained in Fig. 6.19 – Fig. 6.21. It can be seen that good correlations are found between expected and predicted concentrations, which confirm the good performance of CTAI. CTAI achieved the lowest standard error for three combinations. Moreover, the predictive abilities of PDS and CCA are poor for wheat dataset. For SBC, PDS and CCA, they require standard samples and TCR requires reference values of the slave instrument samples, both of which are expensive and difficult to obtain. Obviously, this means that CTAI shows much more outstanding performance.

Fig. 6.19

Fig. 6.20

Fig. 6.21

6.3.4 Conclusions

In this study, the relationship of coefficients between feature vector and predicted values on different instruments is investigated, CTAI was proposed for calibration transfer without standard samples. It was shown that the prediction error can be reduced by affine transformation. Different transfer methods are predicted with two NIR datasets, which prove that CTAI was successfully used to correct the difference on different instruments. The performance of CTAI is better than MSC, TCR CCA, SBC and PDS. Hence, the proposed method may provide an efficient way for calibration transfer when standard samples are unavailable in practical applications.

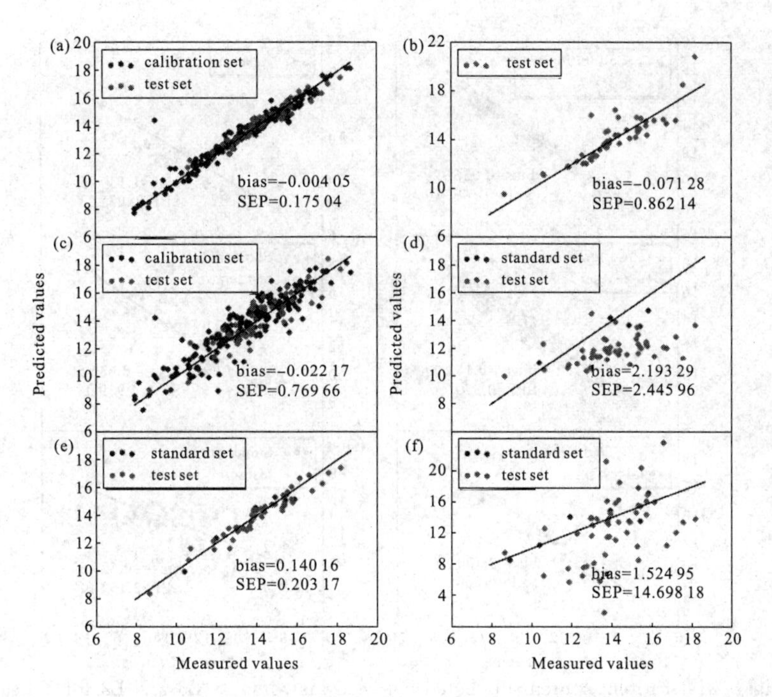

Fig. 6. 19　Protein content predicted between instruments B1 and B2 for Wheat dataset as determined by (a) CTAI, (b) MSC, (c) TCR, (d) CCA, (e) SBC and (f) PDS. The blue and red dots represent the results for each sample in the train set and test set, respectively

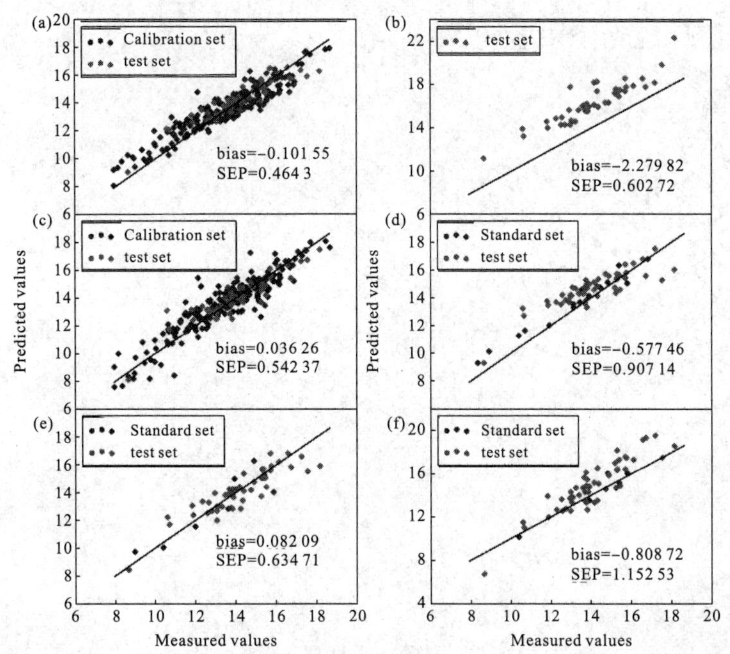

Fig. 6. 20　Protein content predicted between instruments B1 and B3 for Wheat dataset as determined by (a) CTAI, (b) MSC, (c) TCR, (d) CCA, (e) SBC and (f) PDS. The blue and red dots represent the results for each sample in the train set and test set, respectively

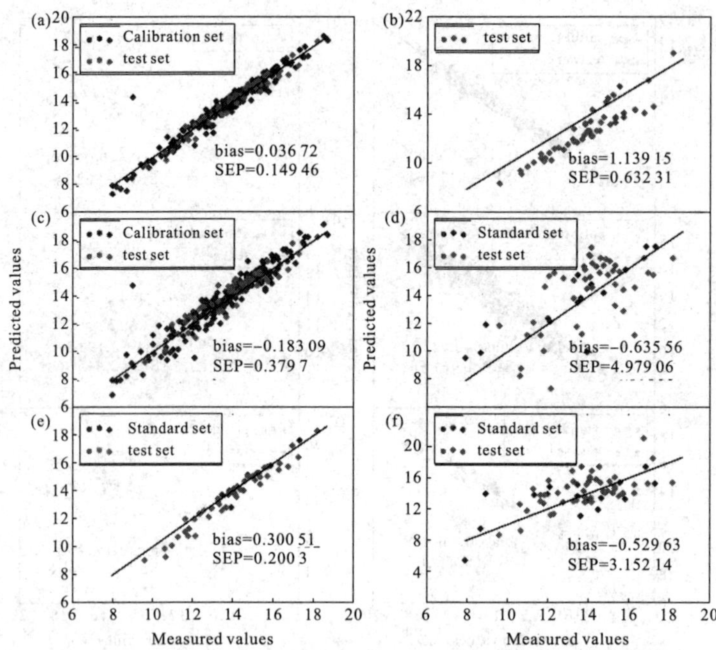

Fig. 6. 21　Protein content predicted between instruments B3 and B2 for Wheat dataset as determined by (a) CTAI, (b) MSC, (c) TCR, (d) CCA, (e) SBC and (f) PDS. The blue and red dots represent the results for each sample in the train set and test set, respectively